TECHNOLOGIES FOR SUSTAINABLE DEVELOPMENT

PROCEEDINGS OF THE 7TH NIRMA UNIVERSITY INTERNATIONAL CONFERENCE ON ENGINEERING (NUICONE 2019) NOVEMBER 21-22, 2019, AHMEDABAD, INDIA

Technologies for Sustainable Development

Edited by

Alka Mahajan, Parul Patel & Priyanka Sharma

CRC Press
Taylor & Francis Group
Boca Raton   London   New York

CRC Press is an imprint of the
Taylor & Francis Group, an **informa** business

A BALKEMA BOOK

CRC Press/Balkema is an imprint of the Taylor & Francis Group, an informa business

© 2020 Taylor & Francis Group, London, UK

Typeset by Integra Software Services Pvt. Ltd., Pondicherry, India

Library of Congress Cataloging-in-Publication Data

Applied for

Published by: CRC Press/Balkema
        Schipholweg 107C, 2316XC Leiden, The Netherlands
        e-mail: Pub.NL@taylorandfrancis.com
        www.routledge.com – www.taylorandfrancis.com

ISBN: 978-0-367-33737-7 (Hbk)
ISBN: 978-0-367-54585-7 (pbk)
ISBN: 978-0-429-32157-3 (eBook)
DOI: 10.1201/9780429321573
DOI: https://doi.org/10.1201/9780429321573

Technologies for Sustainable Development – Mahajan, Patel & Sharma (eds)
© 2020 Taylor & Francis Group, London, ISBN 978-0-367-33737-7

Table of contents

Electronics and communication engineering track

Instrumentation and control engineering track

Mechanical engineering track

Preface

This volume of proceedings from the conference provides an opportunity for readers to engage with a selection of refereed papers that were presented during the 7th International conference NUiCONE'19. Researchers from industry and academia were invited to present their research work in the areas as listed below. The research papers presented in these tracks have been published in this proceeding with the support of CRC Press, Taylor & Francis. This proceeding will provide a platform to proliferate new findings among the researchers.

- ❖ Advances in Transportation Engineering
- ❖ Emerging Trends in Water Resources and Environmental Engineering
- ❖ Construction Technology and Management
- ❖ Concrete and Structural Engineering
- ❖ Futuristic Power System
- ❖ Control of Power Electronics Converters, Drives and E-mobility
- ❖ Advanced Electrical Machines and Smart Apparatus
- ❖ Chemical process development and design
- ❖ Technologies and Green Environment
- ❖ Sustainable Manufacturing Processes
- ❖ Design and Analysis of Machine and Mechanism
- ❖ Energy Conservation and Management
- ❖ Advances in Networking Technologies
- ❖ Machine Intelligence / Computational Intelligence
- ❖ Autonomic Computing
- ❖ Control and Automation
- ❖ Electronic Communications
- ❖ Electronics Circuits and System Design
- ❖ Signal Processing

Introduction

Nirma University International Conference on Engineering, NUiCONE is a flagship event of the Institute of Technology, Nirma University, Ahmedabad, Gujarat, India. This conference follow the successful organization of four national conferences and six international conferences in the previous years.

We are delighted to present the knowledge sharing journey of the 7th international conference on Engineering, organised by the Institute of Technology, Nirma University during 21-22 November 2019. The main theme of the conference was 'Technologies for Sustainable Development'. The objective of the conference was to bring professional engineers working relentlessly for the sustainable development technologies from industry, academics, research organizations, and research scholars of matching interests on a common platform to share new ideas, experiences and knowledge in various fields of Engineering and Technology.

NUiCONE 2019 was a multidisciplinary event encompassing themes related to various disciplines of Engineering & Technology. There were a total of 20 sub-themes to cater the diversified needs of academics and industries. Around 180 research papers were received from various disciplines and only 67 papers were accepted for oral presentation after stringent quality checks. To ensure the quality, the papers were checked for plagiarism, language correction and were double blind review. These papers are going to be published by the reputed publisher CRC Press, Taylor and Francis.

More than 500 delegates from industry, experts, research organisations, postgraduate students and faculty members participated in the conference.

The expert speakers were invited from all across India and abroad covering researchers and eminent experts from academia, industry and research organizations such as NUVOCO vistas corp ltd, JKLakshmi cement, Zydus Cadila, Mitsubishi Electric India, AFCON, NVIDIA, Savvy Infrastructures Pvt. Ltd., EnerComp Solutions Pvt. Ltd, Ruminate, Bangalore, Robotics Automation and IoT, EroNkan Technologies Pvt. Ltd, Centre for Environmental Education, Ahmedabad, ISRO, IIT Madras, University of Victoria, Canada, Florida Atlantic University, USA etc.

Further, this year, the conference has been linked with many reputed associations of various engineering institutions like Institute of Engineers, Indian Institute of Chemical Engineers (IIChE), NVIDIA, Society of Power Engineers (SPE), International Society of Automation (ISA), Indian Green Building Council (IGBC), Green Business Certification Inc. (GBCI), American Society of Heating, Refrigerating and Air-Conditioning Engineers (ASHRAE), Synoptek, India Smart Grid Forum (ISGF), Institute of Electronics and Telecommunication Engineers, Society of Automotive Engineers INDIA (SAEINDIA) and Gujarat Council of Science and Technology (GUJCOST).

The conference was graced by the Chief Guest, Padma Shri Dr. Kiran Kumar. He is Vikram Sarabhai Professor at the Indian Space Research Organisation, Bangalore. He is a Member of the Space Commission, Government of India and Member, Governing Council, Indian Institute of Science, Bangalore. During the period from January 2015 to January 2018, he has served as Secretary, Department of Space, Chairman, Space Commission and Chairman, Indian Space Research Organisation.

Padma Shri Shri Amitava Roy was a guest of honour and keynote speaker. He is the Raja Ramanna Fellow, Department of Atomic Energy (Govt. of India) and Former distinguished scientist and Chief Executive, Nuclear Recycle Board, Bhabha Atomic Research Centre (Mumbai).

Padma Shri Dr. Karshanbhai Patel, Chairman Nirma Group of Industries and President Nirma University was the president of the function. Dr Anup K Singh, Director General, Nirma University motivated the gathering through his august presence.

List of reviewers (TRACK WISE)

Chemical Engineering Department

1. Dr. Jayesh Ruparelia, Head of Dept, IT, SoE, NU
2. Dr. Sanjay S. Patel, Professor, IT, SoE, NU
3. Dr. Milind Joshipura, Professor, IT, SoE, NU
4. Dr. Parin Shah, Associate Professor, Chemical Engg Dept, VGE College of Engg, Chandkheda
5. Dr. Nimish Shah, Associate Professor, IT, SoE, NU
6. Dr. R.K. Mewada, Head of Dept, Chemical Engg Dept, LE College of Engg, Morbi
7. Dr. Femina Patel, Head of Dept, Chemical Engg Dept, VGE College of Engg, Chandkheda
8. Dr. Nitin Bhate, Professor, Chemical Engg Dept, M.S.University, Vadodara
9. Dr. Sachin Parikh, Head of Dept., Chemical Engg Dept, LD College of Engg, Chandkheda
10. Dr. A.P.Vyas, Professor, Indrasheel University
11. Dr. M.S. Rao, Head of Dept, Chemical Engg Dept, DDU, Nadiad
12. Dr. Vimal Gandhi, Associate professor, Chemical Engg Dept, DDU, Nadiad

Civil Engineering Department

1. Dr. H. S. Patel, Principal, Government Engineering College (GEC), Patan
2. Dr. Paresh H. Shah, Dean, Faculty of Technology, CEPT University, Ahmedabad
3. Dr. Gargi Rajpara, Principal & Director, LDRP Institute of Technology & Research, Gandhinagar
4. Dr. Neeraj Shah, Principal, School of Engineering, P. P. Savani University, Surat
5. Dr. Vikram M. Patel, Professor & Principal, Department of Civil & Infrastructure Engineering, Adani Institute of Infrastructure Engineering, Ahmedabad
6. Dr. S. S. Kolte, Principal, Alpha College of Engineering & Technology, Ahmedabad
7. Dr. Shakil Malik, Principal, F. D. (Mubin) Institute of Engineering and Technology, Bahiyal
8. Dr. Yogesh D. Patil, Professor & Head, Applied Mechanics Department, Sardar Vallabhbhai National Institute of Technology (SVNIT), Surat
9. Dr. M. Mansoor Ahmed, Professor & Head, Department of Civil Engineering, Sardar Vallabhbhai National Institute of Technology (SVNIT), Surat
10. Dr. L. B. Zala, Professor & Head, Civil Engineering Department, Birla Vishvakarma Mahavidyalaya (BVM), Anand
11. Dr. Vijay R. Panchal, Professor & Head, M.S. Patel Department of Civil Engineering, Chandubhai S. Patel Institute of Technology, CHARUSAT, Changa
12. Dr. Urmil V. Dave, Professor & Head, Civil Engineering Department, School of Engineering, Institute of Technology, Nirma University, Ahmedabad
13. Dr. Tarak P. Vora, Associate Professor & Head, Faculty of Technology, Civil Engineering Department, Marwadi Education Foundation's Group of Institutions, Rajkot
14. Dr. B. J. Shah, Professor, Applied Mechanics Department, Government Engineering College (GEC), Modasa.
15. Dr. Gaurang J. Joshi, Professor, Department of Civil Engineering, SVNIT, Surat

16. Dr. Jignesh A. Amin, Professor, Department of Civil Engineering, Sardar Vallabhbhai Patel Institute of Technology (SVIT), Vasad
17. Dr. P. P. Lodha, Professor, Department of Civil Engineering, Vishwakarma Government Engineering College (VGEC), Chandkheda.
18. Dr. H. M. Patel, Professor, Civil Engineering Department, Faculty of Technology and Engineering, M. S. University, Baroda
19. Dr. J. N. Patel, Professor, Department of Civil Engineering, Sardar Vallabhbhai National Institute of Technology (SVNIT), Surat
20. Dr. Paresh V. Patel, Professor, Civil Engineering Department, School of Engineering, Institute of Technology, Nirma University, Ahmedabad
21. Dr. Parul R. Patel, Professor, Civil Engineering Department, School of Engineering, Institute of Technology, Nirma University, Ahmedabad
22. Dr. Sharadkumar P. Purohit, Professor, Civil Engineering Department, School of Engineering, Institute of Technology, Nirma University, Ahmedabad
23. Dr. Srinivas S. Arkatkar, Associate Professor, Department of Civil Engineering, Sardar Vallabhbhai National Institute of Technology (SVNIT), Surat
24. Dr. H. R. Dhananjaya, Associate Professor, Department of Civil Engineering, School of Technology, Pandit Deendayal Petroleum University (PDPU), Gandhinagar
25. Dr. Gaurang R. Vesmawala, Assistant Professor, Applied Mechanics Department, Sardar Vallabhbhai National Institute of Technology (SVNIT), Surat
26. Dr. Subhamoy Sen, Assistant Professor, School of Engineering, Indian Institute of Technology (IIT) Mandi, Kamad
27. Dr. Mahesh Reddy, Assistant Professor, Civil Engineering Department, Indian Institute of Technology (IIT) Mandi, Kamad
28. Dr. Rajesh Kumar, Assistant Professor, Civil Engineering Department, BITS Pilani, Pilani Campus, Rajasthan
29. Dr. Dilip A. Patel, Assistant Professor, Department of Civil Engineering, Sardar Vallabhbhai National Institute of Technology (SVNIT), Surat
30. Dr.Bindi Dave, Assistant Professor, Faculty of Technology, CEPT University, Ahmedabad
31. Dr. Trudeep Dave, Assistant Professor, Department of Civil Engineering, Institute of Infrastructure Technology Research and Management (IITRAM), Ahmedabad
32. Dr. Ujjaval Solanki, Assistant Professor, Department of Civil Engineering, Darshan Institute of Engineering & Technology, Rajkot
33. Dr. Kannan Iyer, Assistant Professor, Department of Civil Engineering, Institute of Infrastructure Technology Research and Management (IITRAM), Ahmedabad
34. Dr. Manas Bhoi, Assistant Professor, Department of Civil Engineering, School of Technology, Pandit Deendayal Petroleum University (PDPU), Gandhinagar
35. Dr. Kashyap Patel, Assistant Professor, Department of Civil Engineering, Institute of Infrastructure Technology Research and Management (IITRAM), Ahmedabad
36. Dr. Hasan M. Rangwala, Assistant Professor, Civil Engineering Department, School of Engineering, Institute of Technology, Nirma University, Ahmedabad
37. Dr. Digesh D. Joshi, Assistant Professor, Civil Engineering Department, School of Engineering, Institute of Technology, Nirma University, Ahmedabad

Computer Science and Engineering Department

1. Dr. Uttama Lahiri, Associate Professor, IIT Gandhinagar
2. Dr. Krishna Prasad Miyapuram, Assistant Professor, IIT Gandhinagar
3. Dr. Kamal Raj Pardasani, Professor, NIT Bhopal
4. Dr. B Kartikeyan, Scientist 'SF', SAC, ISRO, Ahmedabad
5. Dr. D K Swami, Professor, VNS Institute of Technology, Bhopal.
6. Dr. Girish Patnaik, Professor, SSBT College of Engineering and Technology, Jalgaon
7. Dr. Narayan Joshi, Proessor and Head, Dharmsinh Desai University, Nadiad
8. Dr. Narendra Patel, Associate Professor, BVM College of Engineering, V.V. Nagar
9. Dr. Manish Chaturvedi, Assistant Professor, Pandit Dindayal Petroleum University, Gandhinagar

10. Dr. Samir Patel, Assistant Professor, Pandit Dindayal Petroleum University, Gandhinagar
11. Dr. Amit Ganatra, Professor and Head, Charusat University, Changa
12. Dr. Nikhil Kothari, Professor, Dharmsinh Desai University, Nadiad
13. Dr. Sanjay Vij, Dean(Academics), ITM Universe, Vadodara
14. Prof. C.K. Bhensdadia, Professor, Dharmsinh Desai University, Nadiad
15. Dr. Savita Gandhi, Professor, Department of Computer Science Gujarat University
16. Dr. Paresh Virparia, Associate Prof., Sardar Patel University, V V Nagar
17. Dr. Vibha Patel, Professor, Vishwakarma Government Engineering College, Ahmedabad
18. Dr. Priti Srinivas Sajja, Professor, Sardar Patel University V V Nagar
19. Prof. Prashant Swadas, Associate Professor, BVM College of Engineering, V.V. Nagar
20. Dr. Mayur Vegad, Professor, BVM College of Engineering, V.V. Nagar
21. Dr. Harshad Prajapati, Associate Professor, Dharmsinh Desai University, Nadiad
22. Dr. Dhaval Patel, Assistant Professor, School of Engineering and Applied Science, Ahmedabad University
23. Dr. Hitesh Chhikaniwala, Associate Professor & Head, Adani Institute of Infrastructure Engg.
24. Dr. Sanjay Garg, Professor, Institute of Technology Nirma University
25. Dr. Priyanka Sharma, Professor, Institute of Technology Nirma University
26. Dr. Madhuri Bhavsar, Professor and Head, Institute of Technology Nirma University
27. Dr. K P Agrawal, Associate Professor, Institute of Technology Nirma University
28. Dr. Gaurang Raval, Associate Professor, Institute of Technology Nirma University
29. Dr. Sharada Valiveti, Associate Professor, Institute of Technology Nirma University
30. Dr. Vijay Ukani, Associate Professor, Institute of Technology Nirma University
31. Dr. Sudeep Tanwar, Associate Professor, Institute of Technology Nirma University
32. Dr. Priyank Thakkar, Associate Professor, Institute of Technology Nirma University
33. Dr. Ankit Thakkar, Associate Professor, Institute of Technology Nirma University
34. Dr. Swati Jain, Associate Professor, Institute of Technology Nirma University
35. Dr. Jaiprakash Verma, Associate Professor, Institute of Technology Nirma University
36. Dr. Saurin Parikh, Assistant Professor, Institute of Technology Nirma University

Electrical Engineering Department

1. Dr. Naran M. Pindoriya, Assistant Professor, Department of Electrical Engineering, Indian Institute of Technology, Gandhinagar
2. Dr. Priyesh Chauhan, Assistant Professor, Electrical Engineering, IITRAM, Ahmedabad
3. Dr. B. R. Parekh, Professor and Head, Department of Electrical Engineering, Birla Vishawakarma Mahavidyalaya Engineering College, Vallabh Vidhyanagar
4. Dr. Praghnesh Bhatt, Associate Professor, Electrical, School of Technology, Pandit Deendayal Petroleum University, Gandhinagar, Gujarat.
5. Dr. Kartik Pandya, Professor, Department of Electrical Engineering, C. S. Patel Institute of Technology, Charusat, Anand, Gujarat
6. Dr. Sandeep Chakravorty, Professor & Dean Academics, Indus Institute of Technology and Engineering, Indus University, Ahmedabad
7. Dr. R. B. Jadeja, Professor & Dean, Faculty of Engineering, Faculty of Technology, Marwadi Education Foundation Group of Institutions, Rajkot
8. Dr. Ketan Badgujar, Professor and Head, Electrical Engineering Department, Shantilal Shah Government Engineering College, Bhavnagar
9. Dr. N. G. Mishra, Professor, BVM Engineering College, V. V. Nagar
10. Dr. Jaydeep Chakravorty, Associate Professor and Head, Department of Electrical Engineering, Indus Institute of Technology and Engineering, Indus University, Ahmedabad.
11. Dr. Ritesh Patel, Professor, Department of Electrical Engineering, GCET, V.V. Nagar
12. Dr. C. V. Sheth, Associate Professor, Department of Electrical Engineering, GCET, V.V. Nagar
13. Dr. Siddharthsingh K. Chauhan, Associate Professor, Department of Electrical Engineering, Institute of Technology, Nirma University

14. Dr. Akhilesh A. Nimje, Associate Professor, Department of Electrical Engineering, Institute of Technology, Nirma University
15. Dr. Kuntal Bhattacharjee, Associate Professor, Department of Electrical Engineering, Institute of Technology, Nirma University
16. Dr. Pavan Khetrapal, Associate Professor, Department of Electrical Engineering, Institute of Technology, Nirma University
17. Prof. Amit Patel, Assistant Professor, Department of Electrical Engineering, Institute of Technology, Nirma University
18. Prof. Tejas Panchal, Assistant Professor, Department of Electrical Engineering, Institute of Technology, Nirma University
19. Prof. Manisha Shah, Assistant Professor, Department of Electrical Engineering, Institute of Technology, Nirma University
20. Prof. Sarika S. Kanojia, Assistant Professor, Department of Electrical Engineering, Institute of Technology, Nirma University
21. Prof. Chintan R. Mehta, Assistant Professor, Department of Electrical Engineering, Institute of Technology, Nirma University
22. Prof. Chanakya B. Bhatt, Assistant Professor, Department of Electrical Engineering, Institute of Technology, Nirma University
23. Prof. P. N. Kapil, Assistant Professor, Department of Electrical Engineering, Institute of Technology, Nirma University
24. Prof. Swapnil N. Jani, Assistant Professor, Department of Electrical Engineering, Institute of Technology, Nirma University
25. Prof. Hormaz Amrolia, Assistant Professor, Department of Electrical Engineering, Institute of Technology, Nirma University
26. Prof. Tarun Tailor, Assistant Professor, Department of Electrical Engineering, Institute of Technology, Nirma University
27. Prof. Chintan Patel, Assistant Professor, Department of Electrical Engineering, Institute of Technology, Nirma University
28. Prof. Satyanarayan Karri, Assistant Professor, Department of Electrical Engineering, Institute of Technology, Nirma University

Electronics and Communication Engineering Department

1. Mr. Durga Prasad Kethineedi, Sr. Engineer, Doordarshan Kendra, Ahmedabad.
2. Dr. Shruti Oza, Professor, College of Engineering, Bharati Vidyapeeth University (BDVU), Pune.
3. Dr. Rajesh A Thakkar, Professor, Vishwakarma Govt. Engineering College (VGEC), Chandkheda, Ahmedabad
4. Dr. Kiran Parmar, Professor, Former Professor, Gujarat Technological University (GTU).
5. Dr.Tanmay D Pawar, Professor, Birla Vishvakarma Mahavidyalaya Engineering College, Vallabh Vidhyanagar Nagar.
6. Mr. N N Maurya, Sr. Scientist, Doordarshan Kendra, New Delhi.
7. Dr. Shweta Shah, Professor, Sardar Vallabhbhai National Institute of Technology (SVNIT), Surat.
8. Dr. Upena Dalal, Professor, Sardar Vallabhbhai National Institute of Technology (SVNIT), Surat.
9. Dr. Sanjeev Gupta, Professor, Dhirubhai Ambani Institute of Information and Communication Technology (DAIICT), Gandhinagar.
10. Dr. Ashish Chittora, Assistant Professor, Birla Institute of Technology & Science (BITS), Goa.
11. Dr. Arun Nandurbarkar, Associate Professor, Lalbhai Dalpatbhai College of Engineering, Ahmedabad.
12. Dr. J M Rathod, Assistance Professor, Birla Vishvakarma Mahavidyalaya Engineering College, V.V. Nagar.

13. Dr. Hiren Mewada, Associate Professor, Charotar University of Science and Technology, Charusat.
14. Dr. Dilip K. Kothari, Professor, Electronics and Communication Department, Institute of Technology, Nirma University.
15. Dr. Dhaval A. Pujara, Professor, Electronics and Communication Department, Institute of Technology, Nirma University.
16. Dr. Usha Mehta, Professor, Electronics and Communication Department, Institute of Technology, Nirma University.
17. Dr. Yogesh N. Trivedi, Professor, Electronics and Communication Department, Institute of Technology, Nirma University.
18. Dr. Nagendra P. Gajjar, Professor, Electronics and Communication Department, Institute of Technology, Nirma University.
19. Dr. N. M. Devashrayee, Professor, Electronics and Communication Department, Institute of Technology, Nirma University.
20. Dr. Amisha P. Naik, Professor, Electronics and Communication Department, Institute of Technology, Nirma University.
21. Dr. Sachin Gajjar, Associate Professor, Electronics and Communication Department, Institute of Technology, Nirma University.
22. Dr. Mehul Naik, Associate Professor, Electronics and Communication Department, Institute of Technology, Nirma University.
23. Dr. Manisha A. Upadhyay, Senior Associate Professor, Electronics and Communication Department, Institute of Technology, Nirma University.
24. Dr. Mangal Singh, Associate Professor, Electronics and Communication Department, Institute of Technology, Nirma University.

Instrument and Control Engineering Department

1. Dr. Axay Mehta, Associate Professor, IITRAM, Ahmedabad
2. Dr. Vipul Shah, Professor, DDIT, Nadiad
3. Mr. R.B.Bhorania, Manager, Mitsubishi Electric India, Pune
4. Dr Ankit Shah, Assistant Professor, LDCE, Ahmedabad
5. Prof Kalpesh Pathak, Assistant Professor, GEC, Gandhinagar
6. Dr. Rajdeep Dasgupta, Assistant Professor, NIT Silchar
7. Dr Suryakant Gupta, Scientific Officer, Institute for Plasma Research, Gandhinagar
8. Mr Keyoor shah, Manager, Mamata Machinery Pvt Ltd, Changodar
9. Dr Dipak M Adhyaru, Professor, School of Technology, Nirma University
10. Dr H K Patel, Associate Professor, School of Technology, Nirma University
11. Dr J B Patel, Associate Professor School of Technology, Nirma University
12. Prof Sandip A Mehta, Assistant Professor, School of Technology, Nirma University
13. Dr Ankit Sharma, Assistant Professor, School of Technology, Nirma University

Mechanical Engineering Department

1. Dr Nilesh Ghetiya, Associate Professor, Mechanical Engineering Dept., Institute of Technology, Nirma University, Ahmedabad
2. Dr Amit Trivedi, HoD & Professor, Production Engineering Dept., Birla Vishwakarma Mahavidyalaya (Engineering College), Vallabh Vidyanagar, Anand
3. Dr Reena Trivedi, Associate Professor, Mechanical Engineering Dept., Institute of Technology, Nirma University, Ahmedabad
4. Dr Manoj Gour, Engineer-SE, Institute of Plasma Research (Cryogenics Division), Bhat, Gandhinagar.
5. Dr N K shah, Associate Professor, Mechanical Engineering Dept., Institute of Technology, Nirma University, Ahmedabad
6. Dr Mukul shrivastava, Assistant Professor, Vellore Institute of Technology, Vellore.
7. Dr Nirav Patel, Assistant Professor, Mechanical Engineering Dept., Pandit Deendayal Petroleum University, Raisan Village, Gandhinagar.

8. Dr B A Modi, Professor, Mechanical Engineering Dept., Institute of Technology, Nirma University, Ahmedabad

9. Dr. Mayur P. Sutaria, Professor, Mechanical Engineering Dept., Chandubhai S Patel Institute of Technology.

10. Dr Amit Sata, HoD & Professor, Mechanical Engineering Dept., Marwadi Institute of Technology, Rajkot

11. Dr B K Mavandiya, Associate Professor, Mechanical Engineering Dept., Institute of Technology, Nirma University, Ahmedabad

12. Dr. Piyush P. Gohil, Associate Professor, Mechanical Engineering Dept., MSU Baroda

13. Dr. Darshak Desai, HoD & Professor, Mechanical Dept., G H Patel College of Engineering & Technology, Bakrol Road, Vallabh Vidhyanagar

14. Dr S V Jain, Associate Professor, Mechanical Engineering Dept., Institute of Technology, Nirma University, Ahmedabad

15. Dr N M Bhatt, Director, Gandhinagar Institute of Technology, Vil. MotiBhoyan, Khatraj – Kalol Road,Tal. Kalol, Dist. Gandhinagar

16. Dr M B Panchal, Associate Professor, Mechanical Engg.Dept., Institute of Technology, Nirma University, Ahmedabad

17. Dr Siddhartha Tripathi, Assistant Professor, Mechanical Engg.Dept., BITS Goa.

18. Dr. S.J. Joshi, Associate Professor, Mechanical Engg.Dept., Institute of Technology, Nirma University, Ahmedabad

19. Dr Mitesh Mungla, Assistant Professor, Indus University, Ahmedabad

20. Dr Nirav P Patel, Assistant Professor, Mechanical Engg.Dept., Pandit Deendayal Petroleum University, Raisan Village, District Gandhinagar

21. Dr Jaykumar Vora, Assistant Professor, Mechanical Engg.Dept., Pandit Deendayal Petroleum University, Raisan Village, District Gandhinagar

22. Dr Vijay Duryodhan, Assistant Professor, IIT Bhillai

23. Dr J M Dave, Associate Professor, Mechanical Engineering Dept., Institute of Technology, Nirma University, Ahmedabad

24. Dr Vinod N Patel, Associate Professor, G H Patel College of Engineering& Technology, Bakrol Road, Vallabh Vidyanagar

25. Dr A M Achari, Associate Professor, Mechanical Engineering Dept., Institute of Technology, Nirma University, Ahmedabad

26. Amit Patel, Assistant Professor, Electrical engineering department, Institute of Technology, Nirma University, Ahmedabad

27. P. N. Kapil, Assistant Professor, Electrical engineering department, Institute of Technology, Nirma University, Ahmedabad

Technologies for Sustainable Development – Mahajan, Patel & Sharma (eds)
© 2020 Taylor & Francis Group, London, ISBN 978-0-367-33737-7

Organizing committee

Chief Patrons:

Shri K K Patel, Vice President, Nirma University
Dr Anup Singh, Director General, Nirma University

Patrons:

Dr Alka Mahajan, Director, IT, NU
Dr R N Patel, Additional Director, School of Engineering, IT,NU

Conference Chair:

Dr Parul R Patel, School of Engineering, IT, NU

Conference Co-Chair:

Dr Priyanka Sharma, School of Technology, IT, NU

Members:

Dr G R Nair (Executive Registrar, NU)
Dr S C Vora
Dr Dhaval Pujara
Dr D M Adhyaru
Dr Madhuri Bhavsar
Dr Urmil Dave
Dr V J Lakhera
Dr J P Ruparelisa
Dr Kunal Pathak
Dr S S Patel

Technical Committee

Dr. Usha Mehta (Chair)
Dr. N.D. Ghetiya (Co-Chair)
Prof. Digesh Joshi
Prof. Hemang Dalwadi
Dr. Absar Lakdawala
Dr. Madhusudan Achari
Dr. Nimish Shah
Prof. Nikita Choksi
Dr. Y N Trivedi
Dr. Richa Mishra

Prof. Bhavin Kakani
Prof. S.A. Mehta
Prof. Ankit Sharma
Dr. S. K. Chauhan
Dr. Akhilesh Nimje
Prof. Vijay Ukani
Prof. Sonia Mittal
Dr. Kunal Pathak

Registration and Logistic Committee

Dr. Sharad Purohit (Chair)
Dr. Manisha Upadhyay (Co-Chair)
Prof. Jhanvi Suthar
Dr. N.K. Shah
Dr. S G Pillai
Dr. Dhaval Shah
Prof. Alpesh Patel
Prof. A.N.Patel
Prof. Jitali Patel
Dr. Sandeep Malhotra

Catering Committee

Dr. Ankur Dwivedi (Chair)
Prof. Rushabh Shah (Co-Chair)
Prof. Vishal Vaidya

International Promotion Committee

Dr. Mehul Naik (Chair)
Dr. Saurin Parikh (Co-Chair)

Industry-Interaction, Finance and sponsorship Committee

Dr. K M Patel (Chair)
Dr. J. B. Patel (Co-Chair)
Prof. S. P. Thakkar
Prof. Ath Singhal
Dr.Amit Kumar
Prof. Jayesh Patel
Prof. Tejas Panchal
Prof. Priyank Thakkar
Dr. Bijal Yeolekar

Cyber Committee

Dr. Ankit Thakkar (Chair)
Prof. Sapan Mankad,
Prof. Pimal Khanpara

Publication Committee

Dr. Millind Joshipura (Chair)
Dr. Vaishali Dhare (Co-Chair)
Prof. Sunil D. Raiyani
Prof. Anand Bhatt
Prof. Leena Bora
Prof. Hardik Joshi
Prof. Harsh Kapadia
Dr. Kuntal Bhattacharjee
Prof. K P Agrawal
Dr. Ratna Rao

Anchoring Committee

Dr. Richa Mishra (Chair)
Prof. Ath Singhal

Media and Publicity Committee

Dr. N.P.Gajjar (Chair)
Prof. Hasan Rangwala
Prof. P.N.Kapil

International advisory committee

1. Dr.Banu Örmeci, Professor and Jarislowsky Chair in Water and Health Director, Global Water Institute, Canada Research Professor, Department of Civil and Environmental Engineering, Carleton University
2. Dr. Daniel P. Abrams, Professor Emeritus, Civil and Environmental Engineering, University of Illinois at Urbana-Champaign
3. Dr. Raman Patel, Ph. D., Adjunct Faculty, Civil and Urban Engineering, Tandon School of Eng., NYU, Six Metrotech, Brooklyn, NY
4. Dr.Rishi Gupta, FEC, P.Eng, Civil and Urban Engineering, Tandon School of Eng., NYU, Six Metrotech, Brooklyn, NY
5. Dr Alan Richardson, Associate Professor, Faculty of Engineering and Environment, Department of Mechanical and Construction Engineering, Northumbria University, Newcastle upon Tyne,
6. Dr. Yogesh M. Desai, Professor, Department of Civil Engineering, Indian Institute of Technology Bombay, Powai, Mumbai
7. Dr. Pradeep Kumar Ramancharla, Dr. Pradeep Kumar Ramancharla, International Institute of Information Technology (IIIT), Hyderabad
8. Dr. Amit Prashant, Professor, Civil Engineering, Indian Institute of Technology Palaj, Near GIFT City, Gandhinagar
9. Dr. Shiv Mohan, Scientist (RTD), Space Application Centre (SAC), ISRO, Ahmedabad
10. Prof I S Jawahir, Professor, Department of Mechanical Engineering, University of Kentucky, Lexington, KY, 40506, United States
11. Prof. A.K. Kulatunga, Professor, Department of Manufacturing & Industrial Engineering, Faculty of Engineering, University of Peradeniya, Peradeniya, Sri Lanka
12. Prof. Raghu Echempati, Professor, Mechanical engineering Department, Kettering University, Flint, MI USA
13. Shri Sanjay Desai, CEO, RBD Engineers, Ahmedabad, India
14. Shri Rajesh Sampat, Vice President, Inspiron Pvt.Ltd, Ahmedabad, India
15. Dr Suparna Mukharji, Professor, Chemical Engg, Dept, IIT-Bombay, Mumbai
16. Dr K K Pant, Professor, Chemical Engg Dept, IIT Delhi, Delhi, India
17. Dr Abinash Agrawal, Department of Earth & Environmental Sciences, and Environmental Science Program, Wright State University, Dayton, USA
18. Dr. Ram B Gupta, Associate Dean for Research, Professor, School of Engineering, Virginia Commonwealth University
19. K. Gopa kumar, Professor, Department of Electronic Systems Engineering Indian Institute of Science (IISc), Bangalore, India.
20. Prof. Ramesh Bansal, Professor & Group Head (Power), Department of Electrical, Electronic and Computer Engineering, University of Pretoria, Pretoria, South Arica
21. Prof. Akhtar Kalam, Professor & Discipline Leader, Engineering, Victoria University, Melbourne, Australia
22. Prof. Akshay Kumar Rathore, Associate Professor, Electrical and Computer Engineering, S-EV 5159 Engineering, Computer Science and Visual Arts Integrated Complex, 1515 ST. Catherine W.
23. Dr. I.N. Kar, Professor, IIT Delhi, Delhi
24. Dr. Sisil Kumarawadu, Professor, University of Moratuwa

25. Mr. Jayesh Gandhi, Managing Director, Harikrupa Automation Pvt Ltd
26. Dr. Rajkumar Nagpal, Senior Manager, Synopsys, Delhi
27. Prof. Abhijit Pandya, Professor, Department of computer and Electrical Engineering and Computer science, Florida Atlantic University
28. Dr. Manoj Gaur, Director, IIT Jammu
29. Prof BVR Chowdhary, Senior Executive Director/Professor, President's office, Nanyang Technological University
30. Dr Rajiv Ranjan, Professor, New Castle University, UK
31. Dr Shailendra Mishra, Professor, Majmaah University, Saudi Arabia
32. Dr M.M. Gore, Professor, Motilal Nehru National Institute of Technology, Allahabad
33. Dr. Ravi A V Kumar, Scientist SG & Head, Accelerator Division, Institute of Plasma Research, Ahmedabad
34. Dr. Sang Won Yoon, Associate Professor, University of Binghamton NY, USA
35. Dr. Devesh Jinwala, Professor, SVNIT, Surat (Currently on deputation at IIT, Jammu)
36. Dr Ashok Pandey, Associate Professor, Massachusetts Maritime Academy, Boston, USA
37. Dr K Kotecha, Professor, Dean, Symbiosis Institute of Technology, Pune
38. Prf. Nigam Dave, Director, SLS,PDPU
39. Dr. K. R. Pardasani, Professor, Mathematics Department, Maulana Azad National Institute of Technology, (MANIT) Bhopal (MP)
40. Dr Tajinder Pal Singh, Professor, Department of Mathematics, School of Technology, PDPU, Gujarat
41. Dr. Gaikwad S. N., Professor, Department of P G Studies and Research in Mathematics Gulbarga University, Gulbarga (Karnataka)
42. Dr Dhirendra Bahguna, Professor, Dept. of Mathematics, IIT Kanpu
43. Dr Geetanjali Panda, Associate Professor, Department of Mathematics Indian Institute of Technology, Kharagpur West Bengal India

Acknowledgments

NUiCONE'19 was a team effort. We are thankful to all authors for providing papers and presenting their research work. We are grateful to all program committee members and the reviewers, who worked very hard in reviewing papers and providing expert comments to authors. Due credit should be given to CRC Press, Taylor & Francis for publishing the proceedings of the research papers presented in the conference. Finally, we thank the hosting organization Nirma University management, our sponsors, International advisory committee members, organising committee members.

Chemical engineering track

Technologies for Sustainable Development – Mahajan, Patel & Sharma (eds)
© 2020 Taylor & Francis Group, London, ISBN 978-0-367-33737-7

Comparison of three solvents for extractive distillation of ethanol and water system

Tushar Perkar, Naitik Chokshi & Milind Joshipura*

Chemical Engineering Department, School of Engineering, Institute of Technology, Nirma University, Ahmedabad, Gujarat, India

ABSTRACT: In the present paper, a thorough analysis of simulation of extractive distillation to separate the Ethanol-Water azeotrope has been performed. An extensive literature survey has been performed. Aspen Plus has been used as the process simulator to simulate the process of extractive distillation (ED) for the Ethanol-Water system. Comparison of different entrainers: Glycerol, Ethylene Glycol and Dimethyl Sulfoxide (DMSO) for separation of the Ethanol-Water system has been done. Sensitivity analysis is used to optimize every flowsheet. The total cost measured by the Aspen Process Economic Analyser, along with the environmental impact of the solvents are used as the criteria to determine the best solvent. For the Ethanol Water system, Glycerol has been determined as the best solvent.

Keywords: Extractive Distillation (ED), entrainer, process simulators, Aspen Plus^{TM}, sensitivity analysis

1   INTRODUCTION

Distillation is one of the prominent techniques used in various chemical industries for the separation of two or more components. Distillation column contributes to more than 50% of the plant's energy consumption. Therefore, it is very important to appropriately design, fabricate and run the distillation column. Simple distillation column is able to separate all mixtures except azeotropic mixtures. In an azeotrope, the constitution of its liquid is the same as the vapor, and hence cannot be separated by simple distillation processes. Various methods have been devised to separate these constant boiling azeotropes, such as extractive distillation (ED), pressure- swing distillation, liquid-liquid extraction and so on. Amongst these, ED is most widely used in the industry. In ED, solvent is added to the mixture which has a higher affinity with one of the components in the azeotrope and thus facilitates the separation of the two compounds forming the azeotrope. Also, the solvent feed enters the distillation column at a stage different than that of the azeotropic feed. In ED, one of the components appears in the distillate and the other component along with the entrainer gets separated in the bottom product. The bottom product is fed into another column, known as the entrainer recovery column, where the entrainer and the component gets separated by simple distillation and thus the azeotrope is broken and the components get separated.

2   LITERATURE SURVEY

An intensive 360° survey was carried out, and was observed that most of the design and simulation work were performed by either Aspen Plus or Aspen Hysys simulator. It was also observed that few of the researchers have focused mainly on simulation of an azeotropic system (Az.S) using extractive distillation while very few enthusiasts had made an intensive research study of simulation for the separation of components in the Az.S along with incorporation of various aspects like flowsheet modification through heat integration, effects of different solvents for the separation of Az.S and process optimization. Taking motivation from these researchers, our study is based on the cost-effective comparison and process optimization of a close boiling point mixture.

3 ETHYL ALCOHOL AND WATER SYSTEM

Ethyl Alcohol is dominantly produced through fermentation of sugar and starch and is industrially also produced through the process of ethylene hydration. It is highly miscible in water. Ethanol has variety of uses in our everyday life such as it serves as raw materials, solvents, transportation fuels, preservatives, additives and also as an intermediate in various chemical industries. Ethyl alcohol and water azeotrope is often encountered during the ethyl alcohol production through fermentation process. Because of high utilization of ethyl alcohol around us, separation of ethyl alcohol water mixture proves to be mandatory. Ethyl alcohol forms a homogeneous azeotropic (positive or minimum boiling) mixture with water at 95.5 mol% at atmospheric pressure (shown in Figure 1) and therefore for further ethyl alcohol and water separation, simple distillation separation technique cannot be effectively used. In our research study, we concentrated on the use of extractive distillation technique for the effective dehydration of ethyl alcohol using different solvents as mentioned below:

    I. Glycerol (GLY)   II. Ethylene Glycol (EG)   III. Dimethyl Sulfoxide (DMSO)

4 STEADY STATE SIMULATION

Two distillation columns were considered in the flowsheet for ethyl alcohol-water ED system. The primary column called as Extractive Distillation Column (C1) is used for the effective separation of the components of the azeotropic mixture and the secondary column called as Entrainer Recovery Column (C2) is used for retaining the solvent used in the separation process. The azeotropic feed for all the solvents were assumed to be same in order to make an effective solvent comparison based on the results obtained through cost analysis and process optimization.

    The study uses below mentioned procedure for the simulations comprising ED, used for the separation of Az.S. Using these procedural guidelines, proper analyzation for the effects of different solvents on product purity and on the cost of process were observed.

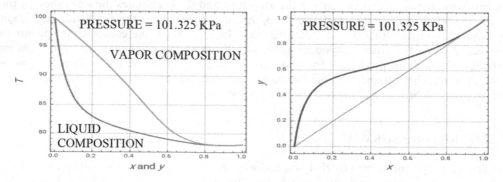

Figure 1.   VLE data for ethyl alcohol at atmospheric pressure.

Table 1.   Feed stream parameters.

Feed stream specifications

| Vapor Phase Fraction | Pressure (KPa) | Molar Flow Rate (Kmol/hr) | Mole Fractions | |
			Component 1 (Ethanol)	Component 2 (Water)
0 (Sat. liquid)	101.325	100	0.5	0.5

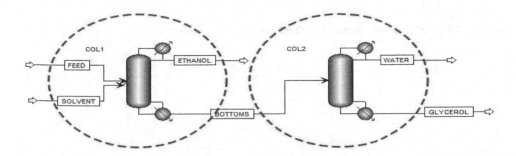

Figure 2..   Flowsheet for ethanol water system using glycerol as a solvent.

I.   Addition of components, thermodynamic model, Binary Interaction Parameters.
II. Defining the Feed and Solvent stream, followed by adding Radfrac Column (C1) and defining the parameters of the column. Run simulation till the column converges and required ethanol purity of 99% is achieved.
III. Adding the Entrainer Recovery column (C2) and adding the distillate of the C1 as feed to the C2. Run simulation till the column converges and 99% water purity is achieved.
IV. Perform the optimization of the flowsheet through Sensitivity Analysis and carryout Cost Estimation using Aspen Process Economic Analyser.

5   RESULTS AND SENSITIVITY ANALYSIS

All the sensitivity analysis mentioned in Table 2 were carried out for all the three solvents. Cost analysis for Ethyl alcohol and Water system using different solvents is highlighted in the Table 3. The results obtained for all the three solvents are encapsulated in the Table 4.

Table 2.   Details of sensitivity analysis.

Manipulated Variable	Observed Variable
Solvent Flowrate	Purity of Ethanol and Water in Distillates of two columns
Reflux Ratio	Purity of Ethanol in Distillate
Number of Stages	Purity of Ethanol in Distillate
Inlet stage of Feed	Purity of Ethanol in Distillate
Inlet stage of Solvent	Purity of Ethanol in Distillate
Inlet stage of Feed	Condenser Duty (Qc) and Reboiler Duty (Qd)
Inlet stage of Solvent	Condenser Duty (Qc) and Reboiler Duty (Qd)

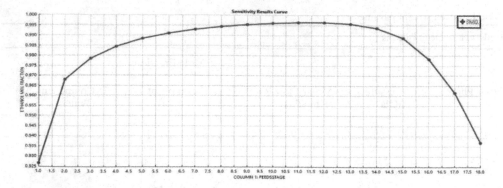

Figure 3.   Effect of feed stage on purity with ethylene glycol as a solvent.

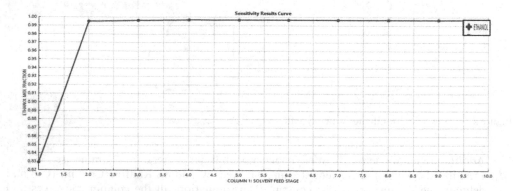

Figure 4.   Effect of solvent feed stage on purity with ethylene glycol as a solvent.

Table 3.   Cost analysis for ethyl alcohol and water system using different solvents.

	Glycerol	Ethylene Glycol	DMSO
Total Capital Cost [USD]	4,921,730	4,571,710	4,903,810
Total Operating Cost [USD/YEAR]	1,798,180	2,090,240	3,548,760
Total Cost [USD]	6,719,910	6,661,950	8,452,570
Total Utilities Cost [USD/YEAR]	449,683	747,064	2,067,630
Total Installed Cost [USD]	1,494,500	1,175,300	1,489,700

Table 4.   Summary table obtained after performing all the sensitivity analysis.

Ethyl alcohol - Water System

Solvent	Glycerol		Ethylene Glycol		DMSO	
Column	C1	C2	C1	C2	C1	C2
No of Stages	19	20	18	11	20	20
Distillate Rate (Kmol/Hr)	50	50	86.864	13.7	50	50
Reflux Ratio	0.5	3	1.225	0.25	3.3	3
Feed Stage	13	10	12	6	15	10
Solvent Stage	3	-	3	-	3	-
Pressure (KPa)	101.3	101..3	101.3	101.3	101.3	101.3
Solvent/Feed Ratio	0.55	-	0.324	-	1.25	-

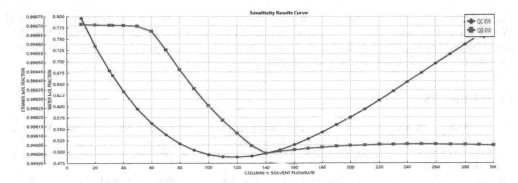

Figure 5.   Effect of solvent flowrate on purity with ethylene glycol as a solvent.

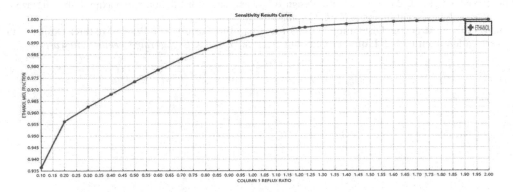

Figure 6.   Effect of reflux ratio on purity with ethylene glycol as a solvent.

6   CONCLUSIONS

An intensive analysis of a special type of distillation called ED was performed. A thorough literature review was carried out for the years 2016-2019 and the data was encapsulated. It helped us to narrow down our research work to focus upon the separation of ethyl alcohol and water as an Az.S through ED. This Az.S was simulated with three different solvents i.e. glycerol, ethylene glycol and DMSO. A comparative analysis was carried out using these three solvents on the basis of the results obtained through cost estimation and process optimization using sensitivity analysis. For ethanol and water system, glycerol proved to be the best solvent. Such an analysis eases the task of the chemical industries to opt for the best solvent for a particular Az.S. Such a solvent analysis for different Az.S can save a significant amount of money and can also contribute in profit maximization. On the basis of the results obtained through the sensitivity analysis for ethyl alcohol and water system, the solvents were ranked as follows:

I. Glycerol   II. Ethylene Glycol (EG)   III. Dimethyl Sulfoxide (DMSO)

The ranking of the solvents was done on the basis of the cost of the solvent, amount of the solvent used, environmental impact, cost of the overall process, and purity obtained. Total installed cost for the process is maximum for glycerol in comparison to other solvents used, but this cost will soon be nullified because the total operating cost and total utility cost (which are recurring costs for every year) of the process using Glycerol as a solvent are minimum, followed by EG and DMSO. The cost of the three solvents are Rs 70,71 and 80 per litre respectively, making Glycerol the cheapest. Also, glycerol is biodegradable and converts to various unreactive organic molecules in soil. This makes Glycerol the best solvent for Ethanol Water system, considering all the parameters used for ranking. In absence of Glycerol, EG could be used as the 2nd choice of the solvent. This is because, the total cost of the process and the cost of EG itself is quite less when compared to DMSO.

REFERENCES

Iván D.G & Botía D. 2009. Extractive Distillation of Acetone/Methanol Mixture Using Water as Solvent. *Industrial & Engineering Chemistry Research*.

Laroche, L. 1991. Homogeneous azeotropic distillation: Comparing entrainers. *The Canadian Journal of Chemical Engineering*.

Lladosa, E. 2011. Separation of di-n-propyl ether and n-propyl alcohol by extractive distillation and pres- sure-swing distillation: Computer simulation and economic optimization. *Chemical Engineering & Processing: Process Intensification*.

Luyben, W. 2013. Comparison of extractive distillation and pressure-swing distillation for ace- tone/ chloroform separation. *Computers & Chemical Engineering*.

Shen, Weifeng, Li, J. 2015. Systematic design of an extractive distillation for maximum-boiling azeo- tropes with heavy entrainers. *AIChE Journal*.

Yih, B. 2017. Extraction–Distillation Process for Separating Acetonitrile–Water Mixture: Design with verification from Liquid–Liquid Equilibrium Data. *Industrial & Engineering Chemistry Research*.

Technologies for Sustainable Development – Mahajan, Patel & Sharma (eds)
© 2020 Taylor & Francis Group, London, ISBN 978-0-367-33737-7

Photocatalytic degradation of Congo red dye using Visible Light Active (VLA) carbon doped titanium dioxide nanocomposite and its kinetics

Jammula Koteswararao* & Madhu Gattumane Motappa
Department of Chemical Engineering, MS Ramaiah Institute of Technology, Bangalore, Karnataka, India

Venkatesham Vuppala
Department of Chemical Engineering, Adhiyamaan College of Engineering, Hosur, Tamil Nadu, India

Nagaraju Kottam
Department of Chemistry, MS Ramaiah Institute of Technology, Bangalore, Karnataka, India

ABSTRACT: Degradation of Congo red dye was studied using synthesized nano TiO_2 and carbon under visible light. The nano materials were synthesized by sol-gel and co-precipitation doped TiO_2 methods. The synthesized nanoparticles were characterized by XRD and SEM. The band gap energy was found to be 2.88 eV. All the experiments conducted at 10ppm dye concentration studied for effect of catalyst dosage and pH on dye degradation. Maximum degradation was observed and low catalyst dosage of 75mg and pH 3. Carbon doped TiO_2 is more active when compared to un-doped TiO_2 for the degradation of Congo red. The kinetics of degradation of dye was found to be pseudo first order.

Keywords: C-doped TiO_2, band gap energy, photocatalytic degradation, Congo red dye

1 INTRODUCTION

In the recent days, most of the rural and urban community do not have access to safe water supply, as it gets contaminated due to the wastes coming from different origins. It is essential to look at water treatment with a greater degree of seriousness than it is viewed at present. It is observed that surface water requires more treatment which is simple, cheap and efficient. The textile industry is most polluting sector accountable to producing 1.3 million tons of dyes and pigments most which are made synthetically. Most of the dyes discharged into wastewater treatment processes and remain in the environment.

As safe water supply is not in abundance, treating and processing wastewater are among the top priorities in industry. Industrial pollutants generally have low solubility in water, which form separate layer on the surface of water affecting physical properties, like dyes are among the most harmful pollutants (Konstantinou et al., Molinari et al., 2004). Many dyes are resistant to conventional treatment process and remain unconverted for an extended period of time, because of their high thermal and photo stability and also resistance to biodegradation. This is highly detrimental to the environment and the degradation of these dyes is absolutely essential (Pandey et al., 2015; Bilgi and Demir, 2005; Vuppala et al., 2012; Venkatesham, 2014; Madhu et al., 2009, 2007a, 2006).

The use of nanomaterials and nanomaterial composites for water treatment is relatively new area compared to other processes. Nanoparticles are promising as they exhibit unique properties such as, large surface area and the ease with which they can be attached onto the solid matrices for enhanced treatment in aqueous and gaseous streams (Sowmay et al., 2018). Congo red is a secondary diazo dye, soluble in water yielding red colloidal solution. Congo red and other bright dyes are commonly used in the manufacture of silk clothing. It is highly toxic to many organisms and is a potential carcinogen and hence absolutely essential for treatment (Konstantinou et al., 2004, Yang et al., 2010; Chen and Wang, 2014; Thomas et al, and Nair et al., 2016).

The present work deals with the synthesis of TiO_2 and carbon doped TiO_2 nanocomposite by co-precipitation method and characterize the same using XRD and SEM and UV-Vis analysis. The synthesized materials are used to study the effect of variable on the visible light photocatalytic degradation of Congo red dye and to study the Kinetics of the reaction.

2 MATERIALS AND METHODS

2.1 Reagents

Titanium chloride (≥97.0%, 189.679 g/mol) and Titanium (IV) butoxide (97.0%, 340.321 g/mol) were purchased from Sigma Aldrich. Sulphuric Acid (98%, 98.08 g/mol) was procured from Chemlines. Ammonia LR grade (17.031g/mol) and Ammonium Hydroxide AR grade (35.05 g/mol) were purchased from Karnataka Fine Chemicals, Bangalore. Dextrose solution (180.156 g/mol), purchased from Johnson and Smith. All materials and reagents were used as received, unless mentioned.

2.2 Preparation of un-doped TiO_2

TiO_2 was synthesized by Sol-gel method using Titanium chloride as starting material. Calculated amount of $TiCl_4$ with 1ml of H_2SO_4 was taken in a beaker and diluted to 100 ml (Pandey et al., 2015; Bilgi and Demir, 2005; Pelaez et al., 2012). Ammonia was added to maintain the pH of the solution in the range of 7-8. The resulting gel precipitated was allowed to settle and the settled precipitate was washed with excesses water to free the solution of chloride and ammonium ions. The gelatinous white precipitate was filtered and dried to 100°C, finely ground and calcined at 550°C for 4.5 hours.

2.3 Preparation of C-doped TiO_2

Co-precipitation technique was employed to synthesize C-doped TiO_2. Calculated stoichiometric amounts of NH_4OH, Titanium (IV) butoxide and dextrose solution was added to double distilled water and stirred for some time. The resulting solution was slowly evaporated on a sand bath for about 40 minutes till all the water was evaporated. Further drying was done and this dried material was grounded and calcined at 400°C (Bilgi and Demir, 2005; Madhu et al., 2007b; Devi et al., 2009; Palanivelu et al., 2007).

2.4 Characterization

The synthesized material were characterized using XRD, SEM and UV-visible spectrometer to confirm nature, morphology and to estimate bandgap. UV-visible DRS technique is used find the visible light photocatalytic activity of the TiO_2 and carbon doped TiO_2 samples. A noticeable shift in the absorption edges was observed for carbon doped TiO_2 sample.

2.5 Diffuse Reflectance Spectroscopy (DRS)

Diffuse Reflectance Spectroscopy profiles shown in Figure 1 indicates the wave length as a function of absorbance of C-doped and un-doped TiO_2 nanomaterials. The band gap energy of synthesized nanomaterials was estimated using equation (1) (Tandon and Vats, 2016). The absorbance of C-doped TiO_2 value decreased with increase in wave length with reference to pure TiO_2.

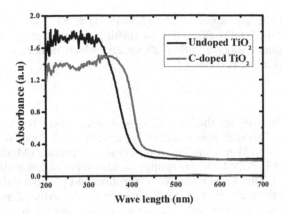

Figure 1. Diffuse Reflectance Spectroscopy NPs:(Red line) C-doped TiO_2 (Black line) un-doped TiO_2.

$$E_g = \frac{h \times C}{\lambda} \qquad (1)$$

In Equation (1), h is Plank's constant; h=6.626x10^{-34} J.s; c is the speed of light; C=3x10^8 m/s; λ is the cut off wavelength; conversion factor is 1.6x10^{-19} J.

Band gap energy (E_g) was calculated using equation (1) for pure TiO_2 at 400 nm wave length is 3.106 eV and similarly for C-doped TiO_2 at 430 nm is 2.88 eV. Doping introduces electronic states in the band gap, resulting in a diffused absorption spectrum. The doped samples have lower band gap compared to un-doped TiO_2 (3.2eV), these doped photo catalysts are likely to be photo catalytic active under visible light illumination.

2.6 *X-Ray diffraction*

The X-Ray diffraction of un-doped TiO_2 confirmed the anatase phase, which is in good agreement with JCPDS Card no. 73-1764 as shown in Fig. 2(a). X-Ray diffraction of C-doped TiO_2 as shown in Fig. 2(b) which is in good agreement with JCPDS card No.75-1621 (Rao et al., 2009), which yielded a higher relative amount of anatase phase as compared to un-doped TiO_2 (Rather et al., 2017).The crystal clear nature of nano TiO_2 was indicated by several sharp peaks shown in Fig.2(a). Ten distinct diffraction

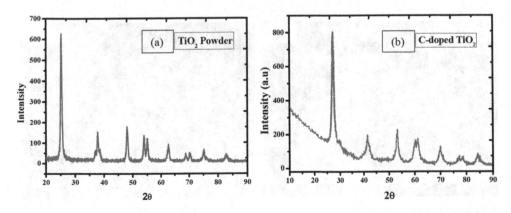

Figure 2. X-Ray diffraction spectra of synthesised TiO_2 NPs:(a) un-doped TiO_2 and (b) C-doped TiO_2

peaks can be seen at the 2θ values of 26, 37.5, 48, 31.8, 53, 54, 62.569, 71, 75, and 83°. XRD profiles of C-doped TiO_2 exhibited no shifting or broadening of peak at 2θ = 26°, but showed almost same peaks at other 2θ values, which noticeably indicated the presence of nano TiO_2 with the carbon.

2.7 *SEM analysis*

The surface morphology of synthesised TiO_2 was understood from the SEM images as shown in Figure 3 (a). This indicates nano TiO_2 particles which have agglomerated, spongy, structure, and size below 151 nm similar to reported in literature (Madhu et al., 2007). The surface morphology of C-doped TiO_2 is spongy in nature indicating a large number of active sites as in Figure 3 (b). This helps in better adsorption and photo catalytic activities. These pictures indicate that the scattering of dopant metals on the external surface of TiO_2 is not identical and doped carbon species have asymmetrical shaped nanoparticles which are lumps of small crystals.

2.8 *Experimental*

Batch experiments were conducted to find the effects of various parameters on visible light photo catalytic degradation of Congo red dye in aqueous solution. Initial concentration of the dye maintained at 10 mg/L for all the experimental trials and light intensity was maintained at 1700 lx. Initial dye concentration solution was prepared by the dilution technique using tap water. 50 ml of the dye solution was taken for each experimental run. To study the effect of catalyst dosage the dosage of the catalyst varied from 25mg to 125mg and was added to 50ml dye solution and exposed to visible light. The aliquots solution was withdrawn for every 5 minutes interval, centrifuged to measure the concentration. To study the effect of pH, initial concentration of the dye maintained at 10mg/L for the experimental trails and the pH was varied for 3 to 11pH.

3 RESULTS AND DISCUSSION

Batch experiments were conducted to study the visible light photo catalytic activity of synthesized un-doped and C-doped TiO_2 nanoparticles for the degradation of Cong red dye. Two important variables viz. catalyst dosage and pH was considered. Pseudo first order kinetics was considered to estimate the rate constant which is in good agreement with the experimental trials. No degradation of the dye was observed without photocatalyst.

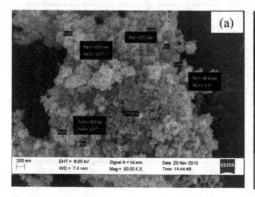

Figure 3. SEM image of synthesised TiO_2 NPs: (a) un-doped & (b) C-doped TiO_2 [100KX, 40.3 KX].

3.1 *Effect of catalyst dosage*

All the experimental trials were conducted at the initial Congo concentration of 10 mg/L maintained at pH 7. Catalyst dosage of un-doped catalyst and doped catalyst was varied from 25 – 125mg in 50 ml aqueous solution. The data for degradation (decrease in dye concentration) of dye with time for un-doped and doped catalyst are represented in Figure 4 (a) and (b) respectively. The data reveals that the degradation of dye increases with the increase in dosage for both catalysts. The increase in degradation may be attributed to the availability of large surface area of the adsorbent and equilibrium is attained at an adsorbent dosage of 75 mg. The trend shows that the absorbance of visible light declines gradually with increase in time. This is due to the decrease in concentration (ppm) of the dye as a result of photo catalytic degradation. The lowest absorbance value is seen at 30 minute for all dosages. Furthermore, the least concentration values of 2.46 (75.4% degradation) for 75mg dose of un-doped catalyst and 0.561ppm (94.39% degradation) at 50 mg dose of doped catalyst, suggesting lesser dosage of doped nanocomposite will be sufficient for effective degradation of dye by visible light than un-doped catalyst. The decrease in concentration of the dye is large initially and then steadily decreases due to lesser availability of active sites on the surface of the nanocomposite.

3.2 *Effect of pH*

Batch experiments were conducted with initial Congo red dye concentration of 10mg/L adsorbent dosage is 75mg for all the experimental runs. Solution pH varied form 3 – 11 covering acidic, neutral and basic medium. The experimental data obtained on decrease in concentration concentration (degradation) of Congo red dye at different pH for un-doped and doped catalyst are presented in Figure 5 (a) and (b) respectively.

The data shows that maximum degradation was possible at acidic pH (3) compared to and other pH for doped and un-doped TiO_2. Congo red is a anionic dye, the rate of degradation mainly depend on the electrostatic interaction between positively charged TiO_2 and Congo red anions results in high degradation at acidic medium. Under basic medium the decrease in the degradation of Congo red could be explained on the basis of amphoteric nature of TiO_2 catalyst. The negatively charged surface of TiO_2 catalyst and negatively charge Congo red dye anions result in electrostatic repulsion resulting in decreased degradation. The maximum degradation is seen at the 30 minute for all dosages. At pH 3, least dye concentration of 2.07 and 0.28 ppm for un-doped and doped catalyst respectively was observed, showing that degradation of dye is more effective at lower pH levels (acidic region) and on the contrary, degradation is least effective at higher pH levels (alkaline region).

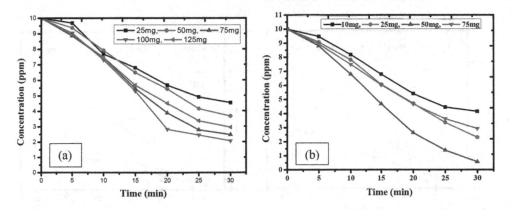

Figure 4. Effect of adsorbent dosage on initial concentration of dye (Co = 10ppm) with function of time (a) un-doped TiO_2 and (b) C-doped TiO_2.

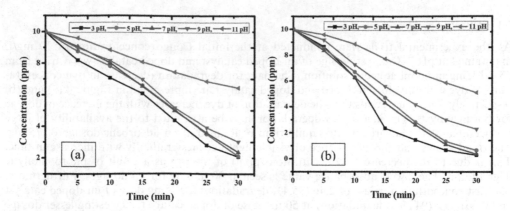

Figure 5. Effect of pH on initial concentration of dye (Dosage 75mg/50ml) with function of time (a) un-doped TiO_2 and (b) C-doped TiO_2.

3.3 *Kinetic studies*

ln $[C_o/C]$ Values were calculated and plotted against time as shown in Figure 6 (a) and (b). The linearity explains the data is in good agreement with pseudo first order kinetics, Equation (2) (Madhu et al., 2007a). The rate constants and regression values are listed in Tables 1 (a) (Effect of Catalyst dosage) and Table 1 (b) (effect of pH).

$$\ln[C_o/C] = kt \qquad (2)$$

C_O is initial concentration and C is concentration at time interval t

All regression values indicate the studies are above 0.95 representing best fit for pseudo first order kinetics. The value of rate constant gradually increases with increase in dosage suggesting higher amount of nanocomposite catalyst is more efficient in decomposition of the dye up to a certain dosage. The highest value of rate constant was observed at 100 mg dosage of TiO_2. The highest value of rate constant was observed at 75mg dosage of C-doped TiO_2.

From Fig.7 (a) and (b) the linearity of plots of ln $[C_o/C]$ versus time at various pH and catalyst dose of 75mg a pseudo first order kinetics for the degradation of dye can be suggested. The rate constants decrease with increase in pH of solution. Highest rate constant is observed at pH 3 (0.04723 min⁻¹) for un-doped catalyst. This result suggests that lower pH is more efficient in decomposition of the dye. In contrast highest rate constant is observed at pH 7 with a value of 0.10337 min⁻¹ for C-doped catalyst.

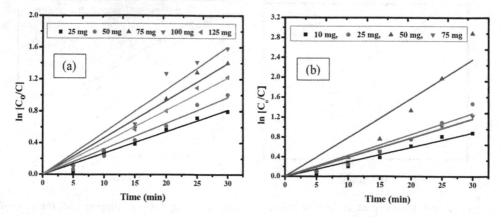

Figure 6. Effect of catalyst loading on First-Order Kinetics for Congo red dye solution (Co = 10ppm) (a) Un-doped (b) C-doped TiO_2.

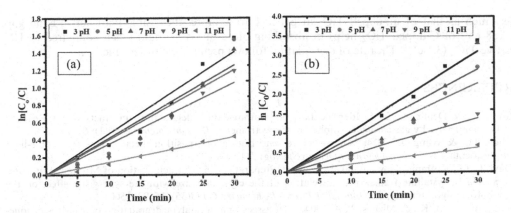

Figure 7. Effect of pH on First-Order Kinetics for degradation of Congo red dye solution (75mg/50ml; Co = 10ppm) using (a) un-doped TiO_2 (b) C-doped TiO_2.

Table 1. Pseudo first order rate constants and regression values.

Dosage (mg)	(a) Effect of catalyst dosage C_o = 10ppm, pH = 7				pH	(b) Effect of pH C_o = 10ppm, Dosage = 75mg/50ml			
	Un-doped TiO_2		C-doped TiO_2			Un-doped TiO_2		C-doped TiO_2	
	k (min^{-1})	R^2	k (min^{-1})	R^2		k (min^{-1})	R^2	k (min^{-1})	R^2
10	-	-	0.029	0.984	3	0.047	0.975	0.046	0.984
25	0.027	0.990	0.043	0.969	5	0.042	0.947	0.021	0.980
50	0.032	0.987	0.078	0.941	7	0.040	0.957	0.104	0.973
75	0.047	0.985	0.039	0.990	9	0.035	0.968	0.076	0.951
100	0.053	0.974	-	-	11	0.015	0.996	0.088	0.932
125	0.041	0.991	-	-					

3.4 Comparison of carbon doped TiO_2 with un-doped TiO_2

Dye degradation results using doped and un-doped TiO_2 were compared under identical conditions. Surface adsorption and photo catalytic degradation are the two removal mechanisms involved. It is found that photo catalytic degradation and surface adsorption are correlated. Since the reaction mixture is magnetically stirred, the dye molecule may get weakly adsorbed on the catalyst as the liquid film around the catalyst gets disturbed, thereby making surface reaction the rate controlling step. Since the rate is independent of concentration of OH^* radicals, the degradation follows pseudo first order. In case of doped TiO_2, it was found that adsorption is not significant when carried out in dark conditions. It was observed that 1 gm/L of carbon TiO_2 at pH 7 degraded 96.47% of Congo red after visible radiation of 30 minutes. The degradation is almost the same in acidic as well as neutral pH, whereas for un-doped TiO_2, there is a necessity to make the medium acidic for the un-doped catalyst to give a better performance.

4 CONCLUSION

Photocatalytic degradation of Congo red dye solution with carbon doped TiO_2 and un-doped TiO_2 was studied. The effects of amount of catalyst and pH on degradation of Congo red were examined. The effect of doping the catalyst with carbon on the degradation of Congo red dye under visible light was studied in detail. The effectiveness of C-TiO_2 was almost twice as much as that of un-doped TiO$_2$ at pH 3. Furthermore, the visible light active band gap (2.88 eV) of C-doped TiO$_2$ helps in absorbing more of visible light when compared to UV active TiO$_2$ (3.1 eV). The rate of degradation follows pseudo first order kinetics.

REFERENCES

Bilgi, S., & Demir, C. 2005. Identification of photooxidation degradation products of CI Reactive Orange 16 dye by gas chromatography–mass spectrometry. *Dyes and pigments 66*(1): 69–76.

Chen, H., & Wang, L. 2014. Nanostructure sensitization of transition metal oxides for visible-light photocatalysis. *Beilstein journal of nanotechnology 5*(1): 696–710.

Devi, L. G., Kottam, N., & Kumar, S. G. 2009. Preparation and characterization of Mn-doped titanates with a bicrystalline framework: correlation of the crystallite size with the synergistic effect on the photocatalytic activity. *The Journal of Physical Chemistry C 113*(35): 15593–15601.

Konstantinou, I. K., & Albanis, T. A. 2004. TiO$_2$-assisted photocatalytic degradation of azo dyes in aqueous solution: kinetic and mechanistic investigations: a review. *Applied Catalysis B: Environmental 49* (1): 1–14.

Madhu, G. M., Raj, M. A. L. A., Pai, K. V. K., & Rao, S. 2007. Photodegradation of methylene blue dye using UV/BaTiO$_3$, UV/H$_2$O$_2$ and UV/H$_2$O 2/BaTiO$_3$ oxidation processes. *Indian journal of chemical technology 14*(2): 139–144.

Madhu, G. M., Raj, M. L. A., & Pai, K. V. K. 2009. Titanium oxide (TiO). *Journal of environmental biology 30*(2): 259–264.

Madhu, G. M., Raj, M. L. A., Pai, K. V. K., & Rao, S. 2006. Photocatalytic degradation of Orange III. *Chem. Prod. Finder 25*: 19–24.

Madhu, G.M., Raj, M. L. A., & Pai, K. V. K. 2007. Titanium Oxide (TiO$_2$) assisted photocatalytic degradation of methylene blue.

Molinari, R., Pirillo, F., Falco, M., Loddo, V., & Palmisano, L. 2004. Photocatalytic degradation of dyes by using a membrane reactor. *Chemical Engineering and Processing: Process Intensification 43*(9): 1103–1114.

Nair, A. K., Kumar, B. V., & Jagadeeshbabu, P. E. 2016. Photocatalytic Degradation of Congo Red Dye Using Silver Doped TiO$_2$ Nanosheets. *In Recent Advances in Chemical Engineering* (pp. 211–217). Springer, Singapore.

Palanivelu, K., Im, J. S., & Lee, Y. S. 2007. Carbon doping of TiO$_2$ for visible light photo catalysis-a review. *Carbon letters 8*(3): 214–224.

Pandey, A., Kalal, S., Ameta, C., Ameta, R., Kumar, S., & Punjabi, P. B. 2015. Synthesis, characterization and application of naïve and nano-sized titanium dioxide as a photocatalyst for degradation of methylene blue. *Journal of Saudi Chemical Society 19*(5): 528–536.

Pelaez, M., Nolan, N. T., Pillai, S. C., Seery, M. K., Falaras, P., Kontos, A. G., & Entezari, M. H. (2012). A review on the visible light active titanium dioxide photocatalysts for environmental applications. *Applied Catalysis B: Environmental 125*: 331–349.

Rao, C. N. R., Biswas, K., Subrahmanyam, K. S., & Govindaraj, A. 2009. Graphene, the new nanocarbon. *Journal of Materials Chemistry 19*(17): 2457–2469.

Rather, R. A., Singh, S., & Pal, B. 2017. A C3N4 surface passivated highly photoactive Au-TiO$_2$ tubular nanostructure for the efficient H$_2$ production from water under sunlight irradiation. *Applied Catalysis B: Environmental 213*: 9–17.

Sowmya, S. R., Madhu, G. M., & Hashir, M. 2018. Studies on Nano-Engineered TiO2 Photo Catalyst for Effective Degradation of Dye. In *IOP Conference Series: Materials Science and Engineering* (Vol. 310, No. 1, p. 012026). IOP Publishing.

Tandon, S., & Vats, S. 2016. Microbial Biosynthesis of Cadmium Sulfide (CdS) Nanoparticles and their Characterization. *European journal of pharmaceutical and medical research 3*(9): 545–550.

Thomas, M., Naikoo, G. A., Sheikh, M. U. D., Bano, M., & Khan, F. 2016. Effective photocatalytic degradation of Congo red dye using alginate/carboxymethyl cellulose/TiO$_2$ nanocomposite hydrogel under direct sunlight irradiation. *Journal of Photochemistry and Photobiology A: Chemistry 327*: 33–43.

Venkatesham, V. 2014. Development of nano metal oxides and coupled metal oxides for heavy metal adsorption and photocatalytic degradation of azo dyes.

Vuppala, V., Motappa, M. G., Venkata, S. S., & Sadashivaiah, P. H. 2012. Photocatalytic degradation of methylene blue using a zinc oxide-cerium oxide catalyst. *European Journal of Chemistry 3*(2): 191–195.

Yang, G., Jiang, Z., Shi, H., Xiao, T., & Yan, Z. 2010. Preparation of highly visible-light active N-doped TiO$_2$ photocatalyst. *Journal of Materials Chemistry 20*(25): 5301–5309.

Technologies for Sustainable Development – Mahajan, Patel & Sharma (eds)
© 2020 Taylor & Francis Group, London, ISBN 978-0-367-33737-7

Treatment of congo red dye wastewater using spinel catalysts

Kishankumar Patel, Avinash Vagjiani & Sanjay Patel
Department of Chemical Engineering, School of Engineering, Institute of Technology, Nirma University, Ahmedabad, India

ABSTRACT: The release of untreated dyes into water bodies leads to serious environmental complications because of high toxicity and high Chemical Oxygen Demand (COD). There are many techniques available to degrade dye color out of which the adsorption method, owing to its economic feasibility and simplicity in design makes it more preferable. In this contribution, the effectiveness of adsorption capacity of $CuFe_2O_4$ synthesized by sol-gel and co-precipitation method for congo red (CR) dye was investigated. The degradation of CR was investigated by UV-Visible Spectrophotometer and COD analyzer. The effects of factors like pH of the solution, temperature and the initial concentration of CR dye solution was studied. Langmuir and Freundlich Isotherms were also modeled. The present work discusses the batch operation for color degradation. The optimum color removal for 100 ppm of CR dye solution was obtained at neutral pH 7 having 97.41% removal and at 50°C having 99.45% removal. Using Langmuir isothermal model, the maximum adsorption capacity of $CuFe_2O_4$ for CR obtained is 333.33 mg g^{-1}.

Keywords: Congo Red (CR), Chemical Oxygen Demand (COD), Langmuir and Freundlich Isotherm

1 INTRODUCTION

The ever increasing population growth, urbanization and industrialization across the globe have a lot of risk to the ecosystem. Currently, there are more than $1.0*10^5$ dyes available commercially whose production was greater than $7.0*10^5$ tons per annum. Confederation of Indian textile predicts that textile mills discharge $1.2*10^3$ million liters per day (MLD) of dye waste water without proper treatment into the water bodies. There exist 3441 textile mills in India. Textile and dye industries generate large amount of colored water which has high COD and BOD contents and other nutrients which need to be treated before disposing it off into the water bodies. The COD of Textile waste water range in between 3300-3500 mg/l and the BOD in the range of 2100-2300 mg/l. Dyes remain in the ecosystem for long duration due to high thermal and photo stability to resist biodegradation.

Spinel ferrites have general formula AB_2O_4. A and B are metallic cations located at two different crystallographic sites, A site is tetrahedral and B site is octahedral and consists mainly of Fe(III) as the main constituent. Very little research is done on spinel catalyst for treating waste water. The separation of catalyst from large mass of water is expensive and time consuming thus had limited application in industrial field but knowing that spinel ferrites have super magnetic behavior, the catalyst can be separated easily and the physical and chemical properties can be enhanced by the varied combinations that are possible. Thus, a new field of research has emerged.

2 MATERIALS AND METHOD

Commercially available congo red having 99% purity was obtained from HPLC. Crystal form of copper nitrate, ferric nitrate and citric acid were obtained from CDH Pvt. Ltd. Sodium hydroxide pellets obtained from SRL and dilute hydrochloric acid was used to adjust the pH.

2.1 Adsorbate

Congo Red (CR) is a benzidine-based anionic diazo dye toxic to many organisms and is suspected carcinogen and mutagen. CR is difficult to degrade due to its structural stability. It has a molecular weight of 696.66 g/mol and has a maximum absorbance wavelength at 497 nm. In India, the removal of congo red is suitable due to simple process control steps.

2.2 Synthesis of $CuFe_2O_4$

$CuFe_2O_4$ was synthesized by two methods i.e. sol-gel method and co-precipitation method. In the co-precipitation method, the metal nitrates were dissolved in water and later mixed together. The pH of the solution was increased to 10 and the solution was stirred properly. To enable the precipitation the solution was washed with water in a filter cloth setup until the pH of the resulting filtrate was not the same as that of the washing medium (water). The resulting precipitates were pre-heated in hot air oven at 120°C. The pre-treated precipitate was calcined in muffle furnace at 750°C for 5 h to obtain $CuFe_2O_4$.

In the sol-gel method, the metal nitrates were dissolved in water and citric acid was added to it in 1:1 ratio of citric acid and the metal cations. The contents were properly stirred. Ethylene glycol (4:1 molar ratio of citric acid) was added which acts as a gelating agent. The contents were then placed in a hot water bath at 100°C until the gel was formed. The gel was then heated in hot air oven at 120°C and calcined in muffle furnace at 750°C for 5 h to obtain $CuFe_2O_4$.

2.3 Batch adsorption experiments

The batch experiments of 500 ml were carried out in batches on a magnetic stirrer. Specific amount of the catalyst ($CuFe_2O_4$) was added and constant stirring conditions were maintained. The samples were injected out at particular intervals, filtered in Axiva filter paper and analyzed.

Adsorption of congo red on spinel catalyst is effective. Hydrogen bond interaction between the dye and the hydroxyl groups present in the spinel catalyst causes the adsorption of cationic dye on to the catalyst and thus reducing the color from wastewater. Adsorption technique is the most promising technologies owing to its simplicity of design, ease of operation, low cost, potential for regeneration, sludge free operation and high retention efficacy when applied with the proper adsorbent and the only disadvantage of adsorption process being that the catalyst needs to be regenerated after use.

3 RESULTS AND DISCUSSION

3.1 Effect of pH

pH as a parameter plays a crucial role in governing the adsorption-desorption and photo degradation of dye on the catalyst surface. To determine the optimum pH for color removal of congo red dye, various experiments were carried out at different pH values, both in acidic and basic medium with a experimental run time of 2 h. The concentration of dye was 100ppm and the catalyst $CuFe_2O_4$ dosage was 0.5g for all the batches that were carried out and the volume of dye sample was 500ml. The pH was maintained by adding 0.1N NaOH and 0.1N H_2SO_4. Samples were taken after every 30 min interval and filtered using Axiva filter paper. As shown in Figure 2, the pH 7 showed the maximum percentage color degradation of 97.41%. Similar results were obtained by Harun N.H. et al. and Manohar R. Patil et al. at pH 7.

3.2 Effect of temperature

Temperature plays an important role in adsorption. A batch process about the color removal of congo red dye was evaluated at different temperatures of 25, 35 and 50°C and continuous stirring was provided for 2 h. The capacity of the adsorbent to adsorb dye molecules increases with increase in the operating temperature. As shown in Figure 2, the maximum color removal of 99.45% was observed at 50°C.

3.3 Effect of initial concentration

The initial dye concentration is another important parameter that affects adsorption. Experiments at six different concentrations 100, 125, 150, 175, 200 and 1000 ppm respectively of 500 ml dye solution were carried out. The amount of catalyst ($CuFe_2O_4$ co-precipitation) dosage in each reactor was kept constant that is 0.5 g and 1 g for 1000 ppm and it was continuously stirred for 2 h. After every 30 minutes the sample were ejected out. As the initial concentration increases the adsorption capacity decreases. The dye removal decreased slightly from 97.41% at 100ppm to 94.14% at 200 ppm and the dye removal at 1000 ppm was 94.8%.

3.4 Chemical oxygen demand

COD removal is one of the important aspects related to the wastewater. The COD of 175 ppm, 200 ppm and 1000 ppm samples were carried out in HACH DRB 200 for digestion, followed by titration and result analysis. The percentage COD removal of 175 ppm, 200 ppm and 1000 ppm were 82.14%, 85% and 93.4% respectively. The effective COD removal of 1000 ppm sample can be attributed to the increased catalyst dosage.

3.5 Adsorption isotherm modeling

Adsorption isotherm provides adsorptive characteristics of the given adsorbent which helps in proper design of Adsorption process. Two commonly isotherms are commonly used, Langmuir and Freundlich. The Freundlich isotherm is used to delineate heterogeneous system and reversible adsorption. Langmuir isotherm assumes that the structure of absorbent is homogenous and all sorption sites are identical.

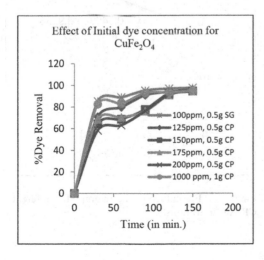

Figure 1.   Effect of initial dye concentration.

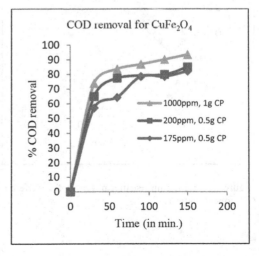

Figure 2.   %COD removal.

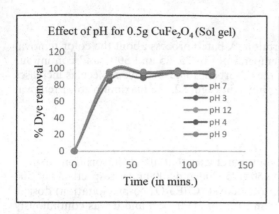

Figure 3.   Effect of different pH.

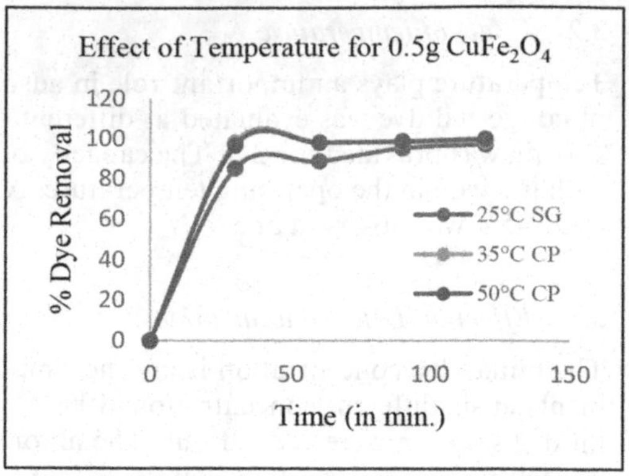

Figure 4.   Effect of different temperatures.

Langmuir isotherm equation

$$\frac{1}{q_e} = \frac{1}{q_{max}} + \frac{1}{K_l * q_{max}} * \frac{1}{C_e}$$

Freundlich isotherm equation

$$log q_e = log K_f + \frac{1}{n} log C_e$$

Q_e - equilibrium adsorption capacity per gram dry weight of the adsorbent, mg g^{-1}, C_0- initial concentration of solution (mg L^{-1}), C_e- equilibrium concentration (mg L^{-1}), V is volume of solution(L), W is weight of adsorbent(g). Q_{max}- maximum amount of adsorption on the adsorbent surface (mg g^{-1}),K_l-langmuir constant related to the energy of adsorption(Lmg^{-1}).

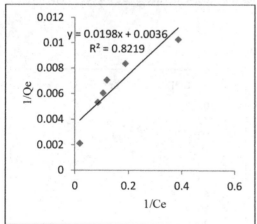

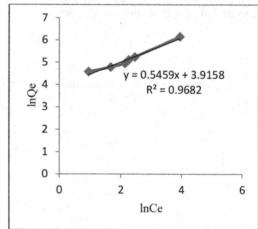

Figure 5.   Langmuir isotherm graph for different CR concentrations.

Figure 6.   Freundlich isotherm graph for different CR concentrations.

Table 1. Adsorption isotherms.

Langmuir isotherm			Freundlich isotherm		
Q_{max}	K_l	R^2	N	K_f	R^2
333.33	1.579	0.821	1.835	50.15	0.968

4 CONCLUSION

$CuFe_2O_4$ can be used effectively for the reduction of congo red dye concentration in wastewater. The maximum color removal was observed at neutral pH. The optimum temperature is 50°C for the reduction of color which was confirmed by varying temperatures from 25-50°C. Higher the dye concentration, higher the COD and less effective color removal for the same quantity of catalyst used.

REFERENCES

Chequer, Farah Maria Drumond, Gisele Augusto Rodrigues de Oliveira, Elisa Raquel Anastácio Ferraz, Juliano Carvalho Cardoso, Maria Valnice Boldrin Zanoni and Danielle Palma de Oliveiraet al. "Textile dyes: dyeing process and environmental impact." *Eco-friendly textile dyeing and finishing*. IntechOpen, 2013.

Divyapriya, Govindaraj, Indumathi M. Nambi and Jaganathan Senthilnathan. "Nanocatalysts in Fenton based advanced oxidation process for water and wastewater treatment." *Journal of Bionanoscience* 10 (2016): 356–368.

Aswin, K.R., Sidhaarth, J. Jeyanthi, S. Baskar and M. Vinod Kumar. Adsorption of Congo Red Dye Using Cobalt Ferrite Nanoparticles. *International Journal of Civil Engineering and Technology*, 9 (2018): 1335–1347.

Edokpayi, Joshua N., John O. Odiyo and Olatunde S. Durowoju. "Impact of wastewater on surface water quality in developing countries: a case study of South Africa." *Water quality*. Intech Open, 2017.

Gómez-Pastora, Jenifer, Eugenio Bringas and Inmaculada Ortiz. "Recent progress and future challenges on the use of high performance magnetic nano-adsorbents in environmental applications." *Chemical Engineering Journal* 256 (2014): 187–204.

Gonawala, Kartik H. and Mehali J. Mehta. "Removal of color from different dye wastewater by using ferric oxide as an adsorbent." *Int J Eng Res Appl* 4.5 (2014): 102–109.

Harun, N. H., *M.N.A. Rahman, W.F.W. Kamarudin, Z Irwan, A Muhammud, N.E.F.M. Akhir, M. R. Yaafar*. "Photocatalytic degradation of Congo red dye based on titanium dioxide using solar and UV lamp." *Journal of Fundamental and Applied Sciences* 10.1S (2018): 832–846.

Hatwar C.R., Deshpande P.S., Dafare S., Talwatkar C.B., Bhavsar R.S and Hatwar L.R. "Photocatalytic Degradation Of Congo Red Dye On Combustion Synthesized $CaZrO_3$ Catalyst Under Solar Light Irradiation." *International Journal of Current Advanced Research* 5 (2016): 1170–1174.

Katal, Reza, H. Zare, S. O. Rastegar and P. Mavaddat. "Removal of dye and chemical oxygen demand (COD) reduction from textile industrial wastewater using hybrid bioreactors." *Environmental Engineering & Management Journal (EEMJ)* 13 (2014): 43–50.

Kefeni, Kebede K., Bhekie B. Mamba, and Titus AM Msagati. "Application of spinel ferrite nanoparticles in water and wastewater treatment: a review." *Separation and Purification Technology* 188 (2017): 399–422.

Manohar R. Patil and Annapurna Jha. "Synthesis of Magnetic Nano Sized Cobalt Ferrite Thin Film by Chemical Bath Deposition Method and their Photocatalytic Application for Removal of Congo red Dye." *Journal of Applicable Chemistry* 7 (2018): 779–784.

Markandeya, S., P. Shukla and D. Mohan. "Toxicity of Disperse Dyes and its Removal from Wastewater Using Various Adsorbents: A Review." *Research Journal of Environmental Toxicology* 11 (2017): 72–89.

Saini, Rummi Devi. "Textile organic dyes: polluting effects and elimination methods from textile waste water." *Int J Chem Eng Res* 9 (2017): 121–136.

Valenzuela, Raúl. "Novel applications of ferrites." *Physics Research International* (2012) 1–9.

Wang, Lixia, Jian-Chen Li, Yingqi Wang, Lijun Zhao, Qing Jiang. "Adsorption capability for Congo red on nanocrystalline MFe_2O_4 (M= Mn, Fe, Co, Ni) spinel ferrites." *Chemical Engineering Journal* 181 (2012): 72–79.

Analysis of swelling and de-swelling kinetics and water absorbency in ionic medium for (Acrylamide-co-Acrylic Acid) hydrogel

Nimish Shah & Ankur Dwivedi

Department of Chemical Engineering, School of Engineering, Institute of Technology, Nirma University, Ahmedabad, India

ABSTRACT: Hydrogels are three dimensional cross linked polymer networks, linked through chemical or physical interactions, and has the capacity to absorb large amounts of water. Hydrogels that are biocompatible have medicinal applications. The water absorbency of gels may depend on pH, temperature, ionic strength. Acrylamide (AAm) and Acrylic acid (AAc) were polymerised by solution polymerisation to synthesise Poly (AAm-co-AAc). The equilibrium swelling was compared by varying the initial monomer proportion. The equilibrium swelling was compared by varying ionic strength of water. And, the polymers were characterised by FTIR spectroscopy.

Keywords: Hydrogels, cross-linking, kinetics, equilibrium swelling, polymer gels

1 INTRODUCTION

Hydrogels are insoluble, soft, three dimensional cross linked polymer networks that has the ability to absorb great amount of water and swell [1]. The cross linking is possible due to physical or chemical interaction [2]. Physical interactions include hydrogen bonds, crystallisation, and ionic interactions whereas chemical interactions includes linking with functional groups, high energy radiation and free radical polymerisation [3]. Newtonian behaviour is observed in case of low entanglement, and as cross links increases, hydrogels show viscoelastic or elastic behaviour [4]. The degree of cross linking defines the mechanical strength of the gel, strength increases upon increasing networks but results in low absorption [5]. Hydrogels with high strength can be synthesised when both chemical and physical networks are incorporated together. Various cross linking agents that can be used are N, N'-methylene bisacrylamide (N, N'-MBA), 3-oxapentamethylene dimethacrylate, ethylene glycol dimethacrylate (EGDMA) [6]. Because of their porosity, softness, biocompatibility [7], it resembles a natural tissue and hence has wide biomedical applications like as in lenses, drug delivery, hygiene products [8]. Hydrogels can be classified based on type of cross linking, appearance, charge and polymeric composition [9]. The swelling and de-swelling can be controlled using external environmental stimuli such as pH, temperature, ionic strength, electric field [10, 11]. The hydrophilic nature of gels is due to the presence of functional groups like – OH, –CONH, –COOH, –SO$_3$H, and –NH$_2$ [12]. Structure of hydrogels can be linear homopolymer, linear copolymer or interpenetrated polymer network [13]. Hydrogels are found in nature and can be synthetic or semi-synthetic [14]. Hydrogels can be synthesised by bulk polymerisation, solution polymerisation, suspension polymerisation, and polymerisation by irradiation [15]. Some of the desired technical features of hydrogels are high absorption capacity, rate of absorption, high durability and stability when swelled, biodegradability [16]. In this paper, hydrogels were synthesised by solution polymerisation of Acrylamide (AAm) and Acrylic acid (AAc) using N, N'-methylene bisacrylamide (N, N'-MBA) as crosslinking agent and Potassium per sulphate (KPS) and N, N, N', N'-tetramethylenediamine (TEMED) as the redox initiator [17]. Equilibrium swelling was measured for different monomer ratios. The dependence of equilibrium swelling was observed by varying the ionic strength of water. The samples were characterised by FTIR spectrums.

2 EXPERIMENTAL

2.1 Materials

Acrylamide (AAm) and Acrylic acid (AAc) as monomers, *N, N'*-methylene bisacrylamide (*N, N'*-MBA) as cross-linker, Potassium per sulphate (KPS) as initiator, N, N, N', N'-tetramethylenediamine (TEMED) as an activator, methanol and distilled water. Distilled water was used for synthesis, solution preparation and swelling measurements.

2.2 Method of synthesis

Poly (AAm-co-AAc) hydrogels were synthesised by solution polymerisation and were chemically cross-linked. Different monomer ratios AAm/AAc were taken for synthesis (80/20, 70/30, 60/40 and 50/50).

Acrylamide was dissolved in 50 mL distilled water and 1 mL TEMED was added to it. The solution was stirred and heated up to 60 °C throughout the reaction. After 20 minutes, Acrylic acid and 0.05 g MBA was added. After 10 minutes, KPS solution (0.1 g in 10 mL distilled water) was added drop wise. After 5-10 minutes gel like structure was formed. The gel was cut into small pieces and then washed with excess methanol to dehydrate until the polymer becomes hard and white. The polymer was then sun dried for 24 hours and dried product was crushed with mortar and pestle to form powder. Various experiments carried out with different compositions as mentioned in Table 1.

2.3 Swelling performance

From each sample, different samples of 0.1 g each were immersed in 500 mL distilled water at room temperature, and swelled weight was measured at different time intervals until equilibrium was achieved. Water absorbency at equilibrium can be measured as equation 1:

$$Water\,Absorbency = \frac{Ws - W}{W} \tag{1}$$

Where Ws and W are swelled and dry weights of hydrogel respectively. Absorbency is grams of water absorbed by the gel per gram of dried gel.

2.4 Effect of ionic strength on equilibrium swelling

Each sample (0.1 g) was immersed in 500 mL NaCl solution with varying concentrations. The concentration varied from 0.1 M to 0.5 M. The swelled gel was weighed after equilibrium was attained.

Table 1. Feed composition of poly (AAm-co-AAc) hydrogels.

Sr. No.	Monomer Ratio (AAm/AAc)	Weight of AAm (g)	Weight of AAc (g)	Weight of MBS (g)	Weight of KPS (g)
1.	80/20	5.68	1.44	0.05	0.1
2.	70/30	4.97	2.16	0.05	0.1
3.	60/40	4.26	2.88	0.05	0.1
4.	50/50	3.55	3.60	0.05	0.1

3 RESULTS AND DISCUSSION

3.1 *Swelling performance*

Rate of swelling curve was measured with data points as shown in Table 2 from initial weight to the time it takes to reach equilibrium weight. Also it was observed that with increase in AAm concentration, absorption capacity decreases from 50% to 70% AAm then increases for 80% AAm. This decrease may be attributed to the reduced number of COO⁻ groups which has high affinity towards holding water [17]. The trend also depends upon the number of hydrophilic PAAm chains [18], cross linking density, average molecular weight between cross-links [19].

3.1.1 *Swelling kinetics*

Rate of the swelling curve (Figure 1) was quantified with data points; as shown in Table 2 from initial weight to the time it takes to reach equilibrium weight. From the curves, it is observed that the rates are virtually kindred, but the time taken to reach equilibrium weight is different. The difference may be due to varying AAm concentration and the space between the cross-links. The first tribulation was carried out to engender a model with Microsoft Excel, utilizing trend lines feature, excel generate equations, and R^2 values to check model fitting. Users of MS Excel must be aware that, R^2 values should be proximately 1; then one can verbalize that the model fits correctly.

Exponential, linear, polynomial, and power equation models were generated. The kinetics Microsoft excel suggested sixth order polynomial equations for swelling of all polymers under study. The multiplication coefficients of higher order terms are quite more minuscule. If the minute coefficient terms are neglected, all these models become the second/third order

Table 2.   Equilibrium water absorbency.

Monomer Ratio (AAm/AAc)	Equilibrium swelled weight (g)	Water absorbency (g water/g sample)
80/20	27.618	275.18
70/30	12.906	128.06
60/40	17.415	173.15
50/50	18.316	182.16

Table 3.2.   Swelling rate data.

Time (minutes)	Initial weight W (g)	Total Swelled weight Ws (g) Polymer type (AAm/AAc Ratio)			
		A	B	C	D
0	0.1	0.1	0.1	0.1	0.1
1	0.1	8.29	3.36	10.24	5.46
2	0.1	10.68	4.79	12.1	6.24
4	0.1	13.9	6.44	15.43	8.1
8	0.1	17.67	9.91	15.79	10.43
12	0.1	21.9	11.9	16.53	11.3
15	0.1	23.74	12.45	17.27	11.84
30	0.1	26.88	12.88	17.15	13.83
45	0.1	27.16	12.84	16.87	15.75
60	0.1	28.03	13.48	17.48	18.15
120	0.1	27.84	12.92	18.11	18.38
180	0.1	28.18	12.87	17.61	18.42

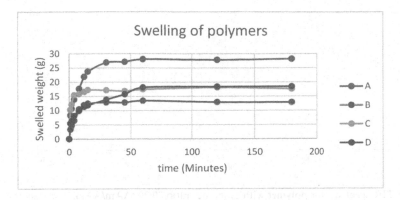

Figure 1. The swelling rate for different polymers (different ratios monomers).

equation only. According to MS excel generated R^2 values, all these models are loosely fitting with experimental data. So another models generated.

The study of water uptake graph carried out, it is exhibiting characteristics of the replication of response of the exponential first and second order systems to step input. The generalized model was prepared for such systems and optimized it by minimizing error with different mathematical methods.

$$y(t) = A\left(1 - \left(1 + \frac{t}{\tau}\right)e^{-t/\tau}\right). \tag{2}$$

Values of time constant τ have been optimized to minimize error. Error has been calculated and minimized with both arithmetic minimization and least square method. for polymer A the model and other parameter values are as below:

With the least square method, the values are optimized as below:

The model is:

$$y(t) = 27.618\left(1 - \left(1 + \frac{t}{1.628122}\right)e^{-t/1.628122}\right)... \tag{3}$$

Amplitude 27.618 and Time constant τ = 1.628122 minutes.

4 CHARACTERISATION

4.1 *FTIR spectroscopy*

For analysis of polymer compounds one of the most widely employed techniques is Fourier Transform Infrared Spectroscopy(FTIR). The spectra of polymeric flocculants obtained in the range of the 4000-400 cm^{-1} are depicted in Figure 2. We have opted to represent FTIR of these samples because it exhibits the best results for all four characteristics: conversion coefficient, residual monomer content, intrinsic viscosity and linearity constant. The major characteristic FTIR frequencies for the samples are given here.

The FTIR spectra attest the formation of copolymer of acrylamide and acrylic acid as is evidents from the characteristic bands appeared in the range of 3700–3500 cm^{-1} (O–H stretching from carboxylic acid group) [20]. At 1670-1690 cm^{-1}, a stretching of the C=O group from the AMD and AA unit appears in all spectra. At this frequency there is additionally the OH angular deformation from dihydrogen monoxide, which expounds the esse of this band in the polyacrylate spectrum [21]. The characteristic peak which appeared in the

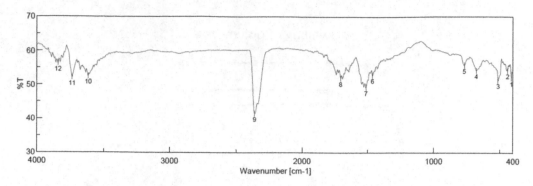

Figure 2. FTIR spectrum for polymer with monomer ratio 80/20 (AAm/AAc).

range of 1650-1580 cm^{-1} is due to the NH$_2$ bending from AMD. The absorption band in the region 1475-1445 cm^{-1}, is due to the CH$_2$ bending vibrations. The band at 2350 cm^{-1} appeared in all spectra is due to atmospheric carbon dioxide which should be neglected.

5 CONCLUSION

Poly (Acrylamide-co-Acrylic acid) hydrogels were synthesised with varying monomer ratios using solution polymerisation. The equilibrium swelled weight was measured and compared, and it was inferred that hydrogel with 80/20 (AAm/AAc) monomer ratio had the highest absorbency of 275.18 g H$_2$O per g sample. The swelling kinetics were plotted to compare the rates of swelling and model fitting was done, and all the rates differed but not significantly. De-swelling kinetics were also plotted to observe the nature of de-swelling. Upon increasing the ionic strength, the equilibrium swelled weight decreased. From the FTIR spectrums, it can be inferred that samples were synthesised appropriately and the peaks observed relate to the peaks of poly (acrylamide-co-acrylic acid).

REFERENCES

1. Gong, Jian Ping. "Friction and lubrication of hydrogels—its richness and complexity." *Soft matter* 2.7 (2006): 544–552.
2. Ma, Shuanhong, et al. "Structural hydrogels." *Polymer* 98 (2016): 516–535.
3. Hennink, Wim E., and Cornelus F. van Nostrum. "Novel crosslinking methods to design hydrogels." *Advanced drug delivery reviews* 64 (2012): 223–236.
4. Akhtar, Muhammad Faheem, Muhammad Hanif, and Nazar Muhammad Ranjha. "Methods of synthesis of hydrogels... A review." *Saudi Pharmaceutical Journal* 24.5 (2016): 554–559.
5. Haque, Md Anamul, Takayuki Kurokawa, and Jian Ping Gong. "Super tough double network hydrogels and their application as biomaterials." *Polymer* 53.9 (2012): 1805–1822.
6. Begam, Tamanna, A. K. Nagpal, and Reena Singhal. "A comparative study of swelling properties of hydrogels based on poly (acrylamide-co-methyl methacrylate) containing physical and chemical crosslinks." *Journal of applied polymer science* 89.3 (2003): 779–786.
7. Hoffman, Allan S. "Hydrogels for biomedical applications." *Advanced drug delivery reviews* 64 (2012): 18–23.
8. Caló, Enrica, and Vitaliy V. Khutoryanskiy. "Biomedical applications of hydrogels: A review of patents and commercial products." *European Polymer Journal* 65 (2015): 252–267.
9. Gyarmati, Benjámin, Árpád Némethy, and András Szilágyi. "Reversible disulphide formation in polymer networks: A versatile functional group from synthesis to applications." *European Polymer Journal* 49.6 (2013): 1268–1286.
10. Liu, Fang, and Marek W. Urban. "Recent advances and challenges in designing stimuli-responsive polymers." *Progress in polymer science* 35.1-2 (2010): 3–23.

11. Tomatsu, Itsuro, Ke Peng, and Alexander Kros. "Photoresponsive hydrogels for biomedical applications." *Advanced drug delivery reviews* 63.14-15 (2011): 1257–1266.
12. Okay, O. "General properties of hydrogels." *Hydrogel sensors and actuators*. Springer Berlin Heidelberg, 2009. 1–14.
13. Saini, Komal. "Preparation method, Properties and Crosslinking of hydrogel: a review." *PharmaTutor* 5.1 (2017): 27–36.
14. Samchenko, Yu, Z. Ulberg, and O. Korotych. "Multipurpose smart hydrogel systems." *Advances in colloid and interface science* 168.1-2 (2011): 247–262.
15. Ahmed, Enas M. "Hydrogel: Preparation, characterization, and applications: A review." *Journal of advanced research* 6.2 (2015): 105–121.
16. Zohuriaan-Mehr, Mohammad J., and Kourosh Kabiri. "Superabsorbent polymer materials: a review." *Iranian polymer journal* 17.6 (2008): 451.
17. Mutar, Mohammed A., and Rafid K. Kmal. "Preparation of copolymer of acrylamide and acrylic acid and its application for slow release sodium nitrate fertilizer." *Al-Qadisiyah Journal Of Pure Science* 17.4 (2017): 71–83.
18. Tomar, Rajive Singh, et al. "Synthesis of poly (acrylamide-co-acrylic acid) based superabsorbent hydrogels: Study of network parameters and swelling behaviour." *Polymer-Plastics Technology and Engineering* 46.5 (2007): 481–488.
19. Flory, Paul J., and John Rehner Jr. "Statistical mechanics of cross-linked polymer networks I. Rubberlike elasticity." *The Journal of Chemical Physics* 11.11 (1943): 512–520.
20. Nesrinne S, Djamel A. Synthesis, characterization and rheological behavior of pH sensitive poly (acrylamide-co-acrylic acid) hydrogels. Arabian Journal of Chemistry. 2013.
21. Magalhaes ASG, Almeida MPA No, Bezerra MN, Ricardo NMPS and Feitosa JPA. Application of ftir in the determination of acrylate content in poly(sodium acrylate-co-acrylamide) superabsorbent hydrogels. Quimica Nova. 2012; 35(7):1464–1467.

Technologies for Sustainable Development – Mahajan, Patel & Sharma (eds)
© 2020 Taylor & Francis Group, London, ISBN 978-0-367-33737-7

Synthesis of silver nanoparticles by using soluble starch and its application in detection of Hg^{2+} ions from wastewater

Rashi Sultania
Chemical Engineering Department, Pandit Deendayal Petroleum University, Gandhinagar, India

Parwathi Pillai & Swapnil Dharaskar
Chemical Engineering Department, Institute of Technology, Nirma University, Ahmedabad, India

Jayesh Ruparelia*
Chemical Engineering Department, Pandit Deendayal Petroleum University, Gandhinagar, India

ABSTRACT: Silver nanoparticles (Ag NPs)are in great trend these days because of their miraculous and unique properties which is quite helpful in the medical field. Various methods over the years have been devised to synthesize silver nanoparticles. In this research one such green method is presented i.e. synthesis of Ag NPs using soluble starch. The synthesis was carried out by using different amount of starch-10/1/0.5/0.1/0.05g. For each sample UV-Vis spectroscopy was performed and the results were carried out, which confirmed the formation of Ag NP's which was used for detection of Mercury ions. Here starch is used as a reducing agent as well as a stabilizing agent. This method is not only fast but it is also easy to perform without any toxic by-products. The peak for all the samples ranged between 415-420 nm. The peak shifts to 296-284 nm on addition of Hg^{2+} ions.

Keywords: silver nanoparticles, green synthesis, starch, mercury

1 INTRODUCTION

Silver nanoparticles have been researched extensively over a lot of years due to its specialized properties owing to its quantum size and large surface energy [1]. Therefore it can be used in sensing other metal ions, pharmaceuticals, optoelectronics, medicine and antibacterial properties.The concept of nanotechnology and nanoparticles came into picture a long time back in the history by physicist Richard Feynman who initiated a talk with the topic "There's Plenty of Room at the Bottom" at an American Physical Society meeting at the California Institute of Technology (CalTech) on December 29,1959 [2].There are various ways to produce nanoparticles such as Extrusion, Gas and Liquid atomisation, Arc Discharge Plasma Method, High flux solar furnace, Lithography, Laser Ablation, Sonochemical processing, Chemical Reduction and Biological Reduction [4-13], but the study is still going on to find more efficient ways to synthesize them. Among them one of the best methods is Chemical reduction due to their easy formation and comparatively less time used for their formation, which is the method used in this research [14-16]. For chemical reduction, two essential components used: (a) Reducing agent (b) Stabilising Agent. Reducing agent reduces the particle size to its nanorange and Stabilising Agent forms a protective layer around NPs and prevents agglomeration of the reduced particles.

Various reducing agents are used for formation of AgNPs such as sodium citrate, ascorbates, sodium borohydride, elemental hydrogen, polyol process, Tollen's reagent,

---

* Corresponding author: jr@nirmauni.ac.in

N-N dimethylformamide and Starch [17]. In this research, starch is used as reducing agent for synthesis because starch has free aldehyde and ketone groups, which enables them to be oxidised to gluconate and thus reducing them to Nanoparticle range, and thus acts as a reducing agent. Starch possesses a large number of OH groups which act as active sites for metal ions, and thus, has a good control over its size, shape and dispersion of AgNPs and thus are used as Stabilizing agent.The Aldehyde group present in soluble starch is responsible for the reduction of silver nitrate in its nano form while Starch self- stabilises AgNPs [18].

Heavy metal ions are extremely toxic for human health and hence its detection and removal from drinking water and wastewater is very important. Mercury is one of the very highly toxic metals, accumulating in the food chain and the concentration goes on increasing rapidly in the loop. According to World Health Organisation, maximum contaminant level of mercury in Drinking water is $2*10^{-6}$ g/L .Mercury causes disruption of any tissue that comes in contact with it. Its short term effect includes renal and neurological disturbances. Ingestion of 500 mg of mercury (II) chloride causes severe poisoning and sometimes even death in humans [19].

There is a limitation to detect mercury up to concentration of $0.6*10^{-6}$ g/L by ICP and $5*10^{-6}$ g/L by Flame AAS method. Further, detection method of Mercury by the synthesised silver Nanoparticles also discussed. Silver nanoparticles change their golden-yellow colour to colourless on detection of Mercury. The same pattern is observed in the solid matrix too [19].

Inexpensive, non-toxic, renewable and natural polymeric reducing agent has been used in this one- step green synthesis.

2 MATERIALS AND METHODS

2.1 *Materials*

Soluble starch, Silver Nitrate, Glutaraldehyde, Poly Vinyl Alcohol, Mercuric Chloride, Copper Sulphate were all procured from Central Drug House (P) ltd., New Delhi. All chemicals and solvents used in this work were of analytical grade and were used without further purification. REMI 5 MLH magnetic hotplate with stirrer was used for the entire process.

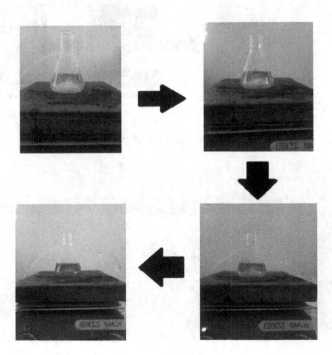

Figure 1.   Change in color observed after every 5 minutes for 0.05g starch sample.

2.2 Silver nanoparticle solution preparation

Different amount of starch i.e. 10/1/0.5/0.1/0.05/0.01 g was taken into 6 different Erlenmeyer flasks. 100 ml of distilled water was then added to the flasks. 3 ml of 50 mM $AgNO_3$ solution was added into each of the flasks and was kept on continuous heating on a heating mantle with magnetic stirring. The formation of nanoparticles was observed by a change in color as shown in Figure 1. Each sample had a slight different variation in their color. Each solution took different time to reduce to the nano range. The temperature of the entire process was kept to 80-85 C. Here the starch acts as a reducing as well as a stabilizing agent.

2.2.1 Conversion to solid matrix

A 10 % (w/v) PVA (Poly Vinyl Alcohol) solution was prepared by adding 10g of PVA in 100 ml of distilled water. The solution was continuously stirred for about 3-4 hours. After that 6ml of the produced AgNP solution to it and 2ml Glutaraldehyde was added which acts as a cross-linker. It was stirred again for around 5 minutes and the complete solution was casted on a dish and was subjected to natural drying for 4-5 days.This method is shown by Figure 2.

2.2.2 Application in sensing

The produced AgNP solution was first used for the colorimetric detection of Mercury (Hg^{2+}). Three different $HgCl_2$ solution were taken for detection i.e. 5 g/ L 0.5 g/L, 0.05 g/L.Upon addition of Hg^{2+}ions to AgNPs colloids, a complete color change from yellow to colorless was observed as shown in Figure 3. These solutions were analyzed by U.V.

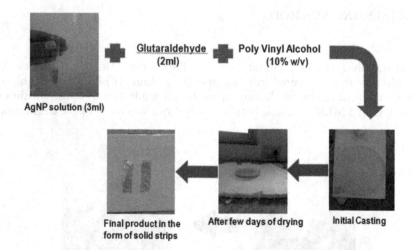

Figure 2.   Graphical flow chart for formation of nano solid strips.

Figure 3.   Change in color on addition of Mercury ions.

3 RESULTS AND DISCUSSION

3.1 *Detection of Hg (II) ions[20]*

The peak for all the samples ranged between 415-420 nm. The peak shifts to 296-284 nm on addition of Hg2+ ions. These values are shown by Figure 4.The change in color observed when the Mercury ions were added was due to a chemical reaction between Mercury ions and the formed AgNPs. This reaction forced the nanoparticles to come together and form a cluster and form bigger particles. Analysis by Horiba Scientific Nano Partial, Nano particle analyze, particle size observed to be 138.3 nm, and after addition of Mercury ions to be of 226 nm. The analysis confirms this. The pattern was also observed in the UV-Visible spectroscopy study. The peak shifted towards the left i.e. Hypsocromic shift. Also as the concentration of Hg^{2+} ions decreased, the absorption decreases as shown in Figure 4.

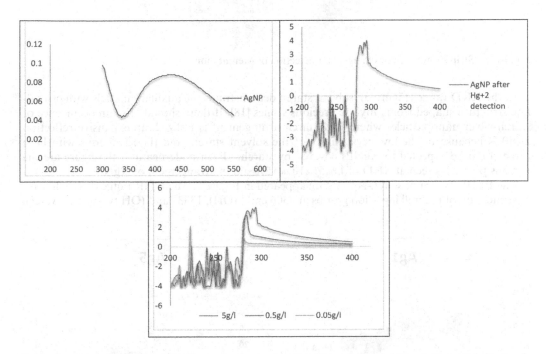

Figure 4. (From clockwise: UV-Vis spectroscopy showing the peak for AgNP solution - 0.5g/L starch sample, UV-Vis peak after addition of Hg2+ ions to AgNP solution, Compasrision of peak for three different HgCl2 samples).

Table 1. Summarisation of different observations and results observed for different samples.

Starch Sample (g/L)	50mM AgNO$_3$ solution (ml)	Time required (mins)	Final color	λ max (nm)	Stability
100	3	25	Red yellow	-	Agglomerated
10	3	35	Dark yellow	418	Stable
5	3	45	Light yellow	417	Stable
1	3	30	Golden yelow	420	Stable
0.5	3	20	Bright yellow	418	Stable

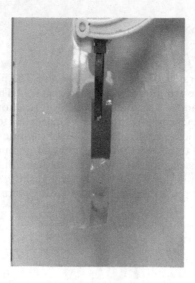

Figure 5.   Strip showing color change after detection of Mercury ions.

In the XRD range (Figure 6), solvent starch demonstrates the auxiliary models with crystalline structure shaped from amylose precious stones [18]. If there should arise an occurrence of starch-silver nanoparticles where 15% starch arrangement is utilized, an expansive reflection at 40 is because of the low crystallinity of the solvent starch, and the silver tops with little force could be expected to AgNPs covered by starch. Given underneath in the figure are the various pinnacles seen at XRD of 1%,5%,10%,15% starch arrangement of AgNPs.

The FTIR spectra of starch-AgNPs are appeared in Figure 7. The FTIR range of starch demonstrates run of the mill ingestion groups at 3306 cm^{-1} (OH), 1333 cm^{-1} (OH twisting of water),

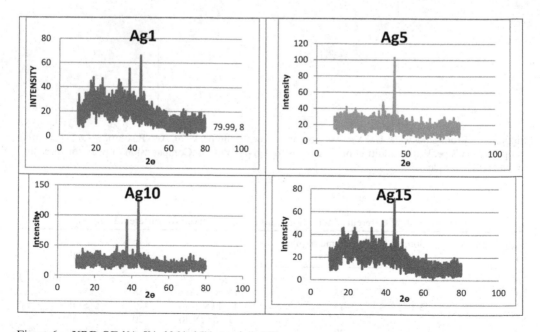

Figure 6.   XRD OF 1%, 5%, 10 %, 15% starch AgNP.

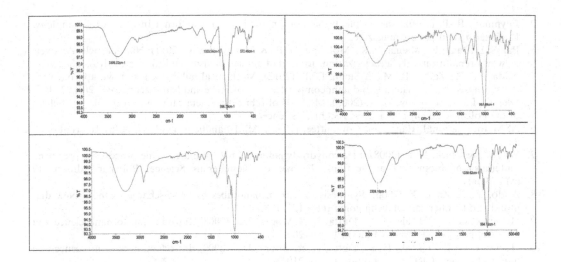

Figure 7.   From clockwise, FTIR of 1%, 5%,10%, 15% Starch AgNP respectively.

996 cm^{-1} (COH bending) [18]. The FTIR range of 15% starch demonstrates a comparable example to that of unadulterated starch A move in frequencies is watched for the pinnacles associated with the – OH useful gathering, at3296 cm−1 (OH extending) and 996 cm^{-1} and 930 cm^{-1} (COH twisting), clearly showing the connection ofOH⁻ bunches with AgNPs .

3.2   Selectivity test

After this experiment selectivity tests were carried out with other metal ions: Ba^{2+},Mg^{2+},K^{+}, Pb^{2+}, Co^{3+}, Ni^{2+}, Ca^{2+}. AgNPs were tested with the metal ions and all retained their color. Hence these AgNPs only detects Hg(II) ions

4   CONCLUSION

Over the years, silver nanoparticles have been synthesized using a variety of different methods but this method is a green method involving just a single step. Here starch is used as a reducing agent as well as a stabilizing agent. This method is not only fast but it is also easy to perform. It does not produce any toxic by-products. Starch here acts both as a reducing and stabilizing agent, and hence, reduces the cost of production by reducing the raw materials used. Hence this method is cost effective too. In this research, under specified reaction conditions, 0.5g/L was the last sample for which reduction was observed. The peak for all the samples ranged between 415-420 nm [21]. The peak shifts to 296-284 nm on addition of Hg^{2+} ions. Initially when the solution was casted it was light yellowish in color but it gradually acquired brownish color as the days passed. Solid strips were cut down from the produced film. The same test for Mercury was again performed with the solid strips and the same pattern for color change was observed but it took more time to get decolorized. Thus, this research proves that immobilized nano particles in polymeric matrix can detect of Hg^{2+} ions in solution.

REFERENCES

1. Vigneshwaran, N., Nachane, R. P., Balasubramanya, R. H., & Varadarajan, P. V. (2006). A novel one-pot 'green'synthesis of stable silver nanoparticles using soluble starch. Carbohydrate research, 341(12),2012–2018.

2. Feynman, R. P. (1960). There's plenty of room at the bottom. California Institute of Technology, Engineering and Science magazine.
3. Pal, S. L., Jana, U., Manna, P. K., Mohanta, G. P., & Manavalan, R. (2011). Nanoparticle: An overview of preparation and characterization. Journal of applied pharmaceutical science, 1(6),228–234.
4. Yadav, T. P., Yadav, R. M., & Singh, D. P. (2012). Mechanical milling: a top down approach for the synthesis of nanomaterials and nanocomposites. Nanoscience and Nanotechnology, 2(3),22–48.
5. Olejnik, L., & Rosochowski, A. (2005). Methods of fabricating metals for nano-technology. Bulletin of the Polish Academy of Sciences: Technical Sciences.
6. ASM International. Handbook Committee. (1998). ASM handbook (Vol. 7). ASM International, 25–50.
7. Wu, Y., & Clark, R. L. (2008). Electrohydrodynamic atomization: a versatile process for preparing materials for biomedical applications. Journal of Biomaterials Science, Polymer Edition, 19 (5),573–601.
8. Ando, Y., & Zhao, X. (2006). Synthesis of carbon nanotubes by arc-discharge method. New diamond and frontier carbon technology, 16(3),123–138.
9. Hornyak, G. L., Tibbals, H. F., Dutta, J., & Moore, J. J. (2008). Introduction to nanoscience and nanotechnology. CRC press, Edition 1, 197–200.
10. Hornyak, G. L., Tibbals, H. F., Dutta, J., & Moore, J. J. (2008). Introduction to nanoscience and nanotechnology. CRC press, Edition 1, 203–210.
11. Johnson, R. W., Hultqvist, A., & Bent, S. F. (2014). A brief review of atomic layer deposition: from fundamentals to applications. Materials today, 17(5),236–246.
12. Xu, H., Zeiger, B. W., & Suslick, K. S. (2013). Sonochemical synthesis of nanomaterials. Chemical Society Reviews, 42(7),2555–2567.
13. Ahmad, A., Mukherjee, P., Senapati, S., Mandal, D., Khan, M. I., Kumar, R., & Sastry, M. (2003). Extracellular biosynthesis of silver nanoparticles using the fungus Fusarium oxysporum. *Colloids and surfaces B: Biointerfaces, 28*(4), 313–318.
14. Bin Ahmad, M., Lim, J. J., Shameli, K., Ibrahim, N. A., & Tay, M. Y. (2011). Synthesis of silver nanoparticles in chitosan, gelatin and chitosan/gelatin bionanocomposites by a chemical reducing agent and their characterization. *Molecules, 16*(9), 7237–7248.
15. Van Dong, P., Ha, C. H., & Kasbohm, J. (2012). Chemical synthesis and antibacterial activity of novel-shaped silver nanoparticles. International Nano Letters, 2(1), 9.
16. Guzmán, M. G., Dille, J., & Godet, S. (2009). Synthesis of silver nanoparticles by chemical reduction method and their antibacterial activity. *Int J Chem Biomol Eng, 2*(3), 104–111.
17. Zielińska, A., Skwarek, E., Zaleska, A., Gazda, M., & Hupka, J. (2009). Preparation of silver nanoparticles with controlled particle size. *Procedia Chemistry, 1*(2), 1560–1566.
18. Raghavendra, G. M., Jung, J., & Seo, J. (2016). Step-reduced synthesis of starch-silver nanoparticles. International journal of biological macromolecules, 86, 126–128.
19. World Health Organization – WHO, (1997). Surveillance and control of community supplies. In Surveillance and control of community supplies, *Guidelines for drinking-water quality*, 2nd ed. *Vol. 3.*
20. Jeevika, A., & Shankaran, D. R. (2016). Functionalized silver nanoparticles probe for visual colorimetric sensing of mercury. Materials Research Bulletin, 83, 48–55.
21. Ramesh, G. V., & Radhakrishnan, T. P. (2011). A universal sensor for mercury (Hg, HgI, HgII) based on silver nanoparticle-embedded polymer thin film. ACS applied materials & interfaces, 3 (4),988–994.

$Co_3O_4/\gamma\text{-}Al_2O_3$ as a heterogeneous catalyst for ozonation of dye wastewater

Nikita P. Chokshi, Meet Kundariya, Jay Patel & J.P. Ruparelia*
Chemical Engineering Department, Institute of Technology, Nirma University, Ahmedabad, India

ABSTRACT: Water is a precious resource but due to the increased population and industrialization, the amount of water used for various purposes like drinking, sanitization and industrial use has also increased. The industrial wastewater is many a times coloured due to the type of impurities present in it. The wastewater from the dye industries is generally coloured and before discharging it to the water bodies, it must be treated. Ozonation – an Advanced Oxidation Process (AOP) – is employed for the colour removal treatment of the wastewater. To improve the colour removal rate from the wastewater, catalysts are used during ozonation. The paper discusses about the preparation of cobalt oxide catalyst supported on $\gamma\text{-}Al_2O_3$ by the impregnation method for catalyzing the ozonation reaction. The catalyst was characterized by BET, XRD, SEM and EDX techniques. Also the parameters like effect of pH, initial dye concentration, ozone flow rate and the amount of catalyst were discussed. It was found that the synthesized catalyst work satisfactorily for color removal of RB5.

Keywords: dye wastewater, advanced oxidation process, catalytic ozonation, cobalt supported catalyst

1 INTRODUCTION

Fresh water is a very important but a scarce resource. The growing population has increased the demand of water. Thus there is a need to purify the water, especially to remove the colour from it. The colour is imparted to the water by discharging dye effluents from industries like textile, leather, paper and pulp, dye manufacturing, etc. The methods of colour removal like biological oxidation, thermal destruction, physical treatment (adsorption, floatation, coagulation, etc.) and chemical oxidation have been explored but they cannot be employed at a large scale owing to their respective disadvantages (M. Stoyanova, 2014), (Barbara Kasprzyk-Hordern, (2003)). As ozone has a very high oxidation and disinfection potential it has gained a lot of attention in wastewater treatment. Ozone treated water has improved colour, taste, odor and reduced amount of organic and inorganic compounds. However, its drawbacks are relatively low solubility and low stability in water, higher production cost and partial or incomplete oxidation of the organic compounds and slow and selective reactions has limited its use in water treatment technology (M. Stoyanova, 2014), (B. Langlais), (A.R. Tehrani-Bagha, 2010). To improvise ozonation efficiency and economic feasibility, Advanced Oxidation Processes (AOPs) have been studied. AOPs are among the most effective chemical oxidation techniques used to completely destroy the colour causing impurities and have been studied extensively (K. Rajeshwar, 2008), (S. Esplugas, 2002), (Sanja Papić, 2006), (BolarinwaAyodele, 2013). These methods rely on the production of hydroxyl free radicals (OH•) for water treatment. These are one of the most reactive radicals and strong oxidants and their reactions are non-selective. Catalytic ozonation is an AOP. Catalytic ozonation utilizes catalysts which allows

---

* Corresponding author: jr@nirmauni.ac.in

a controlled decomposition of ozone and hydroxyl free radicals. Reactions of hydroxyl radicals are fast and non-selective compared to that of molecular ozonation reactions. Thus catalytic ozonation enables fast degradation of organic pollutants and also the mineralization of micro-pollutants as well as organic matter (natural) more effectively (Jacek Nawrocki, 2010). Catalytic ozonation can be classified into two classes: (i) homogeneous catalytic ozonation, where ozone activation is achieved by ions in the solution and (ii) heterogeneous catalytic ozonation, where ozone activation is achieved by metal oxides or metals/metal oxides on supports. The reported mechanism for heterogeneous catalytic ozonation can be given as follows: (i) Chemical adsorption of ozone on the catalyst surface leading to the generation of active species that react with non- chemisorbed organic molecule or the color causing impurities, (ii) Chemical adsorption of organic molecules (associative or dissociative) on the surface of the catalyst and its further reaction with gaseous or aqueous ozone, and (iii) Chemical adsorption of both ozone and organic molecules on the catalyst surface which leads to the interaction of the chemisorbed species. A lot of research has been done on increasing the efficiency of ozonation by the use of different catalysts. Jun Chen et al. (Jun Chen, 2015) prepared MgO (I) and MgO (II) through vacuum calcination and homogeneous precipitation respectively for the ozonation of 4-chlorophenol. Xiaosu Dong et al. (Xiaosu Dong, 2016) prepared $Cu/ZnO/ZrO_2$ by co-precipitation method for the synthesis of methanol by CO_2 hydrogenation method. P.C.C. Faria et al. (P.C.C.Faria, 2009) prepared Mn_3O_4, Co_3O_4, CeO_2 and CeO_2-Co_3O_4 by co- precipitation method for catalytic ozonation of aniline, sulfanilic acid and an azo dye (C.I. Acid Blue 113). Seema Singh et al. (Seema Singh, 2017) prepared Mn_3O_4, Co_3O_4 and NiO by hydrothermal approach and precipitation for catalytic per-oxidation of pyridine in aqueous solution. Babak Roshani et al. (Babak Roshani, 2014) prepared Mn (10 wt %)/γ-Al_2O_3, Cu (10 wt%)/γ-Al_2O_3 and Mn-Cu (5-5wt%)/γ-Al_2O_3 using dry impregnation and co-impregnation methods for mineralization of benzotriazole (BTZ) by heterogeneous catalytic ozonation. Ye Wang et al. (Ye Wang, 2016) prepared Mn/γ-Al_2O_3 using impregnation method for degradation of phenol in aqueous solution by catalytic ozonation. R.M. Navarro et al. (R.M. Navarro, 2005) prepared Ce/γ-Al_2O_3 using impregnation method for production of hydrogen by oxidative reforming of ethanol. Da Wang et al. (Da Wang, 2016) prepared Co_3O_4/γ-Al_2O_3 using impregnation method for Fischer-Tropsch synthesis. Enling Hu et al. (Enling Hu, 2016) prepared Co_3O_4/MCA using impregnation method for catalytic ozonation of a dyeing effluent containing C.I. Reactive Blue 19. The paper focuses on the use of heterogeneous supported catalyst (Co_3O_4/γ-Al_2O_3) for decolourization of dye wastewater using ozonation.

2 MATERIALS AND METHODS

2.1 Materials

Cobalt nitrate [Co $(NO_3)_2$ •$6H_2O$], sodium hydroxide (NaOH), hydrochloric acid (HCl) and potassium iodide (KI) were procured locally. Gamma alumina (γ-Al_2O_3) balls, which were procured from Sorbead India, were crushed down to get fine powdered particles. Reactive Black-5 (RB5) dye was used as obtained without any further processing.

2.2 Catalyst preparation and characterization

10% Co_3O_4 supported on γ-Al_2O_3 was prepared by impregnation method. 3.62 grams of [Co $(NO_3)_2$ • $6H_2O$] was mixed with 20 ml of distilled water. The solution was added to the powdered γ-Al_2O_3 slowly. The contents then were mixed in a mechanical shaker for 3 hours. The thoroughly mixed contents were dried in an oven at 110 °C overnight. . Catalyst was obtained by calcining dried material from the oven in a muffle furnace at 400 °C for 6 hours (Da Wang, 2016). XRD analysis of catalyst was carried out using XRD with an X-pert MPD system (Philips) Cu-Kα radiation (λ= 1.5406 A). The catalyst BET surface area and pore volume were determined using nitrogen (-195.126 °C) with Micromeritics 3Flex version 4.01. The morphological characteristics and semi-quantitative elemental analysis of catalyst were determined by SEM and energy dispersive of X-Ray (EDX).

2.3 Experimental procedure

The experimental set up for the ozonation experiment is shown in Figure 1. The prepared catalyst was used to treat the RB5 (Reactive Black 5) dye wastewater by ozonation method. A sample of 1 liter of 100 ppm dye concentration was introduced in the glass column reactor, with a magnetic stirrer to keep the catalyst in suspension form. A known amount of catalyst was added each time and the samples were collected at different time intervals. Initially in each experiment the adsorption capacity of the catalyst was checked in absence of ozone for about 20 minutes. After that the oxygen flow was started and sent to the ozonator to produce ozone and mixture of ozone and oxygen is bubbled through the dye wastewater through the sparger. The exit gases coming from the glass column passed through KI bottles to trap the residual ozone gas. The same experiment was done without catalyst also by keeping all other parameters same. The collected samples were filtered using Whatman filter paper and then analyzed by using UV spectrophotometer at a wavelength of 599 nm. All the experiments were performed at normal room temperature and pressure.

3 RESULTS AND DISCUSSION

3.1 Effect of initial dye concentration

Figure 2 shows the performance of catalyst at varying initial dye concentration of 100 ppm, 300 ppm, 500 ppm and 1000 ppm. Other parameters were kept as- 0.5 gram catalyst, 1 liter RB5 dye wastewater, 30 LPH ozone flowrate, pH=7. Observations show that an increased initial dye concentration leads to delayed colour removal. This study suggests that as the initial dye concentration increase from 100ppm to 1000ppm, more by-products will form in the solution and ozone is consumed for degradation of original dye as well as for these by-products. It shows that more amount of ozone is require to oxidize the intermediates formed.

3.2 Effect of ozone flowrate

Figure 3 shows the performance of the catalyst at varying flowrates- 30 LPH, 40 LPH and 60 LPH. The other parameters were kept as: 0.5 gram catalyst, 100 ppm solution of dye wastewater (1 liter), pH=7. Observations show that an increased ozone flowrate leads to a faster colour removal. It is because of more amount of ozone or hydroxyl radicals are available due to increase in ozone flow rate.

3.3 Effect of pH

pH is a crucial parameter in catalytic ozonation. Generally at lower pH ozone helps in degradation of pollutant where in basic medium hydroxyl radical is responsible for degradation of dye. To analyze the effect of pH the experiments were done at different initial pH- acidic (pH=3), neutral (pH=7) and basic (pH=12). The other parameters were kept as: 0.5 gram

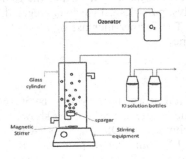

Figure 1.   Experimental setup for ozonation.

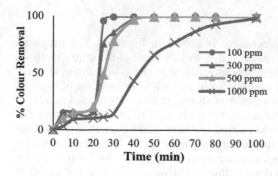

Figure 2. Effect of initial dye concentration.

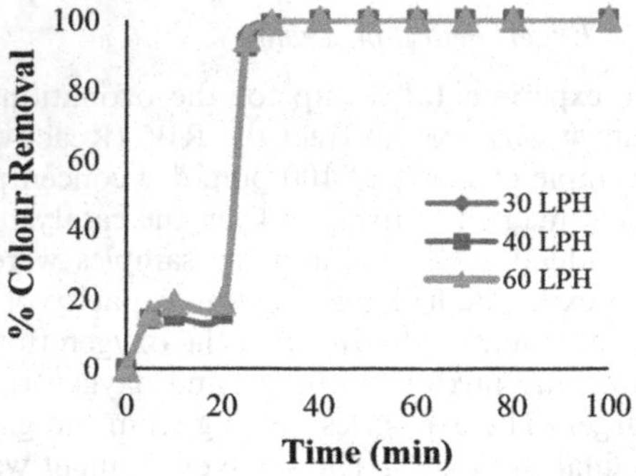

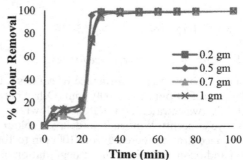

Figure 3. Effect of ozone flowrate.

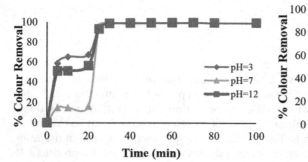

Figure 4. Effect of pH.

Figure 5. Effect of amount of catalyst.

catalyst, 100 ppm dye wastewater solution (1 liter), 30 LPH ozone flowrate. Observations show that the colour removal is faster in acidic medium than in basic medium. For metal oxide type catalysts, the neutral or negatively charged sites are proposed as the active sites to carry out the decomposition of ozone into hydroxyl radicals. Thus the catalyst showed better catalytic activity at a lower pH (Minghao Sui, 2012).

3.4 *Effect of amount of catalyst*

Figure 5 shows the performance of catalyst with different amount of catalyst was used- 0.2 gram, 0.5 gram, 0.7 gram and 1 gram. The other parameters were kept as: 100 ppm solution of dye wastewater (1 liter), 30 LPH ozone flowrate, pH=7. The initial 20 min time was for adsorption only and after 20 min of adsorption the supply of ozone starts. As seen in Figure 5 the color removal efficiency increases with increase in catalyst dosage as more active sites are available. But if we increase the catalyst weight further the removal efficiency decreased due to the fact that the excess catalyst particles get easily aggregated, thus reducing the number of active sites (Ye Wang, 2016). The further increase in catalyst quantity have no remarkable change in color removal efficiency.

3.5 *Characterization of catalyst*

The BET surface area of the catalyst was found out to be 56.1662 m^2/g. XRD pattern of the prepared catalyst is shown in the Figure 6. The analysis showed the presence of only two compounds in the catalyst i.e. Boehmite (JCPDS no.: 98-003-6340) and Co_3O_4 (JCPDS no.: 01-078-1970). Figure 7 shows the EDX analysis of prepared catalyst. As shown in Figure 7 the weight percentage of Al and Co are 38.95% and 7.41% respectively. Figure 8 shows the SEM

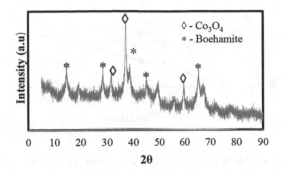

Figure 6.   XRD and EDX analysis of 10% Co₃O₄ supported on γ-Al₂O₃.

Element	Weight %	Atomic %
O K	53.64	68.12
Al K	38.95	29.33
Co K	7.41	2.55

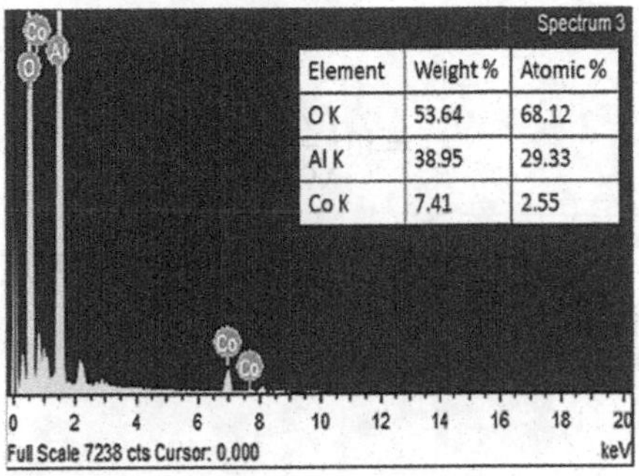

Figure 7.   EDX analysis of prepared catalyst.

images of the prepared catalyst at different magnification. As seen, the surface is rough so it can be helpful in adsorption of dye molecules.

3.6   *Kinetic study*

The prepared catalyst was used in ozonation of aqueous RB5. The kinetic data of the catalytic ozonation of RB5 in the presence of catalyst were shown in Figure 9.

   The overall color removal can be expressed by pseudo-first order reaction.

$$-\frac{dc_{RB5}}{dt} = k_1 c_{O3} c_{RB5} + k_2 c_{OH.} c_{RB5} \tag{1}$$

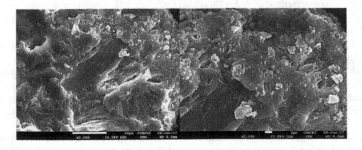

Figure 8.   SEM images of 10% Co₃O₄ supported on γ-Al₂O₃.

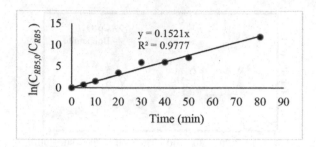

Figure 9. Pseudo first order plot of RB5 by catalytic ozonation (Dye Concentration 500mg/L, gas flow rate 30 LPH, catalyst dosage 0.5 g/L).

Where k_1 and k_2 were the reaction rate constant of the reaction of the RB5 with molecular ozone and hydroxyl radical respectively.
Then equation (2) can be expressed as:

$$kt = \ln\left(\frac{c_{RB5}}{c_{RB50}}\right) \tag{2}$$

Where c_{RB50} denotes the initial concentration of RB5 and k represented as $k_1 c_{O3} + k_2 c_{OH}$. The plot of ln ($C_{RB5,0}/C_{RB5}$) with time gives a straight line fits (R^2=0.977). The value of the rate constant is 0.1521 min^{-1}.

4 CONCLUSION

The effect of 10% Co_3O_4 supported on γ-Al_2O_3 on the ozonation of dye wastewater containing RB5 was investigated through various experiments. It was concluded that an acidic medium is more preferable over basic medium for the given catalyst. Also an increased ozone flowrate gives rapidly color removal of RB5. As the concentration of synthetic dye wastewater increases more time is required for decolourization as compared to the one with lower concentration. The kinetics of the color removal of RB5 was conforming to pseudo-first order rate equation.

ACKNOWLEDGMENT

The authors would like to thank Nirma University for providing all financial help to carry out experimental work.

REFERENCES

A.R. Tehrani-Bagha, N. M. (2010). Degradation of a persistent organic dye from colored textile waste-water by ozonation. *Desalination 260*, 34–38.
B. Langlais, D. R. (n.d.). Ozone in Water Treatment: Application and Engineering,. *Lewis Publishers, Chelsea, Michigan, USA, 1991.*
Babak Roshani, I. M. (2014). Catalytic ozonation of benzotriazole over alumina supported transition metal oxide catalysts in water. *Separation and Purification Technologies, 135*, 158–164.
Barbara Kasprzyk-Hordern, M. Z. ((2003)). Catalytic ozonation and methods of enhancing molecular ozone reactions in water treatment. *Applied Catalysis B: Environmental 46*, 639–669.
BolarinwaAyodele, O. (2013). Effect of phosphoric acid treatment on kaolinite supported ferrioxalate catalyst for the degradation of amoxicillin in batch photo-Fenton process. *Applied Clay Science, 72*, 74–83.
Da Wang, C. C. (2016). High thermal conductive core-shell structured Al2O3@Al composite supported cobalt catalyst for Fischer-Tropsch synthesis. *Applied Catalysis A: General, 527*, 60–71.

Enling Hu, S. S.-m.-l. (2016). Regeneration and reuse of highly polluting textile dyeing effluents through catalytic ozonation with carbon aerogel catalysts. *Journal of Cleaner Production, 137*, 1055–1065.

Jacek Nawrocki, B.-H. (2010). The efficiency and mechanisms of catalytic ozonation. *Applied Catalysis B: Environmental, 99*(1-2), 27–42.

Jun Chen, S. T. (2015). Catalytic performance of MgO with different exposed crystal facets towards the ozonation of 4-ChlorophenolCatalytic performance of MgO with different exposed crystal facets towards the ozonation of 4-Chlorophenol. *Applied Catalysis A General*, 118–125.

K. Rajeshwar, M. O.-A. (2008). *Journal of Photochemistry and Photobiology C: Photochemistry Reviews, 9* 171–192.

M. Stoyanova, I. S. (2014). Catalytic performance of supported nanosized cobalt and iron–cobalt mixed oxides on MgO in oxidative degradation of Acid Orange 7 azo dye with peroxymonosulfate. *Applied Catalysis A: General 476*, 121–132.

Minghao Sui, S. X. (2012). Heterogeneous catalytic ozonation of ciprofloxacin in water with carbon nanotube supported manganese oxides as catalyst. *Journal of Hazardous Materials, 227– 228*, 227–236.

P.C.C.Faria, D. J. (2009). Cerium, manganese and cobalt oxides as catalysts for the ozonation of selected organic compounds. *Chemosphere, 74*(6), 818–824.

R.M. Navarro, M. A.-G.-S. (2005). Production of hydrogen by oxidative reforming of ethanol over Pt catalysts supported on Al2O3 modified with Ce and La. *Applied Catalysis B Environmental, 55*(4), 229–241.

S. Esplugas, J. G. (2002). Comparison of different advanced oxidation processes for phenol degradation. *Water Res. 36*, 1034–1042.

Sanja Papić, N. K. (2006). Advanced oxidation processes in azo dye wastewater treatment. *Water environment research, 78*(6), 572–579.

Seema Singh, S.-L. L. (2017). Catalytic performance of hierarchical metal oxides for per-oxidative degradation of pyridine in aqueous solution. *Chemical Engineering Journal, 309*, 753–765.

Xiaosu Dong, F. L. (2016). CO2 hydrogenation to methanol over Cu/ZnO/ZrO2 catalysts prepared by precipitation-reduction method. *Applied Catalysis B: Environmental*, 8–17.

Ye Wang, W. Y. (2016). The role of Mn-doping for catalytic ozonation of phenol using Mn/γ-Al2O3 nanocatalyst: Performance and mechanism. *Journal of Environmental Chemical Engineering, 4*(3), 3415–3425.

A sustainable bioprocess for lipase production using seawater and the byproduct obtained from coconut oil industries

R. Raval & A. Verma
Department of Biotechnology, Manipal Institute of Technology, MAHE – deemed to be university, Manipal, India

K. Raval
Department of Chemical Engineering, National institute of technology Karnataka, Surathkal, Mangalore, India

ABSTRACT: Globally lipases are the most attractive source of research, as it has numerous applications in various industries like food industry, paper and pulp industry, preparation of beverages etc. A lipase producing bacterium, *Pseudomonas stutzeri*, was isolated from sea water. The bacterial culture was introduced to the physical and chemical mutagens and then allowed to grow on the solid media. A number of mutated clones were produced which were further followed by examining their lipase activity. There was a significant increase in the extracellular lipase activity i.e. 13, 56 and 14 folds increase in the case of UV mutation, sodium azide, and NTG respectively. Further, the mutants were subcultured and stability was observed in NTG mutants. The lipase production from the NTG mutants was optimized using Response Surface Methodology (RSM). The maximum lipase activity of 1132.6 U/ml was obtained which was about 7 folds higher than the parent strain using the process which utilized the residual coconut cake, a byprtoduct of coconut oil industries and the sea water which makes the process sustainable.

1 INTRODUCTION

Lipase (triacylglycerol acylhydrolase, EC 3.1.1.3) helps as a catalyst for the hydrolysis of the carboxyl ester bonds in triacylglycerols to create diacylglycerols, monoacylglycerols, unsaturated fats, and glycerol. The lipase enzyme has numerous applications in the field of food, dairy, pulp & paper, leather, detergents, organic chemical synthesis and biofuel industries (Hasan, Shah and Hameed, 2006). Numerous sources are identified for lipase production. The enzyme activity in promising microbial strains was enhanced by recombinant DNA technology, enzyme engineering and media optimization. However, many bioprocesses used costly oils such as olive oil to produce lipases (Treichel *et al.*, 2010). Almost all the submerged fermentation based bioprocesses for lipase production use potable water for production, which makes the bioprocess unsustainable. This prompted researchers to use agro waste residues to produce lipases (Treichel *et al.*, 2010). Enzymes obtained from marine microorganisms have advantage of robustness and long term stability in harsh environments. Therefore, this study is based on production of lipase from marine microorganism, especially, *Pseudomonas stutzeri*. Moreover, the bioprocess uses locally available raw materials i.e., coconut cake obtained from the coconut oil industries.

2 MATERIALS AND METHODS

2.1 *Microorganism subculturing and mutation*

The marine *Pseudomonas stutzeri* strain was procured from [MTCC 11712] Chandigarh, Luria Bertaini Miller broth, Glucose, $CuSO_4$, $MgSO_4$, $MgCl_2$, $CaCl_2$, Triton-X, NaK, Na_2CO_3, DNS reagent, and Agar were from Himedia. and Folin-Ciocalteu's phenol reagent from Merck, Para nitrophenyl laurate along with all the mutagens (sodium azide & NTG) were from the sigma Aldrich. For media optimization, coconut meal and sea water were used. UV-VIS spectrophotometer (Genesys 10S Thermo scientific) was used for measurement of optical density. To enhance the activity of lipase enzyme from *Pseudomonas stutzeri* UV irradiation, sodium azide mutation and methylnitronitrosoguanidine were used.

2.2 Lipase activity

To 0.8 ml of 0.05 M phosphate buffer (pH 8.0), 0.1 ml enzyme and 0.1 ml of 0.01 M p-nitrophenyllaurate was added. The reaction was carried out at 37°C for 10 min, after which 0.25 ml 0.1 M $Na_2CO_3$was added. The mixture was centrifuged and the activity was determined by measuring absorbance at 420 nm. One unit of enzyme activity is defined as the amount of enzyme, which liberates 1 mole of p-nitrophenol from pNP-laurate as substrate per min under standard assay conditions (Kaur, 2017).

2.3 Optimization of biological and nutritional parameters on lipase production

2.3.1 Effect of the different growth media

The effect of growth media on lipase production was studied by inoculating the culture into Luria-Bertini (LB) broth and the another media comprise coconut meal, glucose, MgSO4 CuSO4 and triton-X.

2.3.2 Effect of inoculum percentage

After the mutant stability studies the NTG mutants showed highest stability and it has two mutants i.e pigmented and non-pigmented, the effect of inoculum age and percent were conducted on both types.

The effect of inoculum percentage on lipase production was studied by changing the inoculum concentration of 5%, 10%, 15% and 20%. Shake flasks of 100 ml volume with a filling volume of 5 ml were used at 200 rpm and 37°C in an incubator shaker.

2.3.3 Effect of inoculum age

An inoculum of age 12,18,24,30,36,48 and 60 hour was used to study its effect on lipase production. Shake flasks of 100 ml volume with a filling volume of 5 ml were used at 200 rpm and 37° C in an incubator shaker.

2.3.4 Effect of trace elements

To study the effect of various mineral salts on the lipase activity, mineral salts were added to the medium. Trace minerals such as MgSo4, CuSo4, MgCl2 and CaCl2 were used. The concentration of trace element was kept 0.2g/l.

2.4 Statistical optimization of lipase production

After selection of important media constituents for lipase biosynthesis using one factor at a time approach, optimization of the significant variables (Glucose, Coconut meal cake, $MgSO_4$ and $CuSO_4$) were carried out by the central composite design (CCD) for enhancing the lipase activity. Four independent variables were studied at three different levels (-1,0,+1) as given in the Table 1 in the set of 30 experiments. the experiments were conducted in triplicates and the mean value of lipase activity (U/ml) was reported as the response (Y). Following generalized polynomial model

Table 1. Coded values and independent variables used for optimization.

Independent variables	Symbols	Coded levels		
		-1	0	1
Coconut meal(g/L)	A	10	20	30
Glucose(g/L)	B	10	20	30
MgSo4(g/L)	C	0.01	0.5	1
CuSo4(g/L)	D	0.01	0.5	1

was used for regression analysis of the data and linear, quadratic and polynomial terms were selected for optimization. StateEase 7 was used for statistical optimization.

$$Y = \beta_{\circ} + \sum_{i=1}^{4} \beta_i x_i + \sum_{i=1}^{4} \beta_{ii} x_i^2 + \sum_{i<j}^{4} \beta_{ij} x_i x_j \qquad (1)$$

Where, Y represents the response variables, $\beta_{\circ}$ is a constant, β_i, β_{ii} and β_{ij} are the linear, quadratic and cross-product coefficients, respectively. x_i and x_j are the levels of the independent variables.

Four independent variables were studied at three different levels (-1,0,+1) in set of 30 experiments as mentioned in Table 1.

3 RESULTS AND DISCUSSION

Figure 1 shows the results of the mutation study. The highest activity of lipase was observed in mutation by sodium azide treatment. However, this mutant was not stable. The activity of the mutant dropped drastically after two subculturings.

The methynitronitrosoguanidine treated mutant showed maximum stability after three subculturing. A stable lipase activity of about 600 U/mL was observed after several subculturing. Age of microorganisms also called as the age of inoculum play crucial role in the health of any bioprocess. If microorganisms age is too young or too old, then the biosynthesis of any compound will not occur at an optimum level. Usually, optimum inoculum age is different for different bio products synthesis and therefore needs to be optimized separately for each bio product (Shuler and Kargi 1992). After investigating

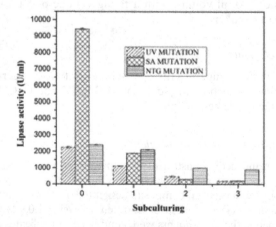

Figure 1. Stability test for mutants obtained from UV irradiation, Sodium azide and methylnitronitrosoguandiine treatment.

different inoculum age for their lipase production, the 24 hour old inoculum gave the highest lipase production (889 U/ml). Microbial concentration plays an essential role in many microbial systems. A low inoculum size may increase lag phase or non- productive phase of a bioprocess. The large inoculum size may reduce the productive phase of a bioprocess due to relatively fast depletion of nutrients (Kumdam, Murthy and Gummadi, 2013) . Higher inoculum concentrations may exhaust the medium nutrients at a much faster rate without increasing productivity and thus optimum inoculum percentage is essential for a successful bioprocess. In this study, the highest lipase production of 957 U/ml was achieved when 15 % inoculum was used. After optimizing the process parameters, the lipase activity was increased from 600 U/ml to 957 U/ml. Optimization of operating conditions reduced bioprocess time from about 72 hours to 24 hours. Table highlights the optimized operational, nutritional and biological parameters. 15 % inoculum of 24 hour old culture at 37°C temperature, pH 7 at 200 rpm produced maximum lipase activity.

4 STATISTICAL OPTIMIZATION

CCD was performed using above mentioned four independent variables at different levels (-1, 0, +1) in set of 30 experiments for optimization of lipase activity. The experiments were conducted in triplicates and the mean value of lipase activity (U/ml) was recorded as the Y-response. The data were fitted to various models.

The reduced cubic model gave the best fit out of all the models chosen for optimization. Therefore, the reduced cubic model was simplified by excluding insignificant model terms. Following were the model terms for the prediction of lipase activity.

Lipase activity =
$493.95+103.8*A-78.9*B+539.17*C+127.11*D+15.06*AB-3.98*AC-15.51*AD+1.75*BC-101.05*BD+77.56*CD-64.58*A^2-76.96*B^2+144.74*C^2-57.043*D^2-91.75*ABC-48.31*ABD-29.26*ACD+144.05*A^2B-385.94*A^2C-215.76*A^2D-145.48*AB^2-178.04*C^3$

Where, A, B,C and D represent, coconut cake, glucose, $MgSO_4$ and $CuSO_4$, respectively. The significant model terms were chosen for further optimization using response surface methodology. Figure 2 shows contour plots. Interestingly all the four factors were significant and showing good interaction, these interactions are shown in the Figure 2. In the plots there are six figures, which are clubbed and depicting the interactions of the four factors with each other. Interestingly all the four factors were significant and showing good interaction. All the factors are interacting with each other and giving good lipase activity. The optimum values for these parameters from the above model were found to be (g/L): 20, 10, 1 and 1 for coconut meal cake, glucose, $MgSO_4$ and $CuSO_4$ respectively. An experiment was performed in 100 ml flask with filling volume 5 ml with loop full bacteria and kept for 37° C at 200 rpm in an incubator shaker for different intervals 4, 12, 16, 20, 24, 30 and 38 hour. Using the optimum values of factors mentioned above. Figure shows the lipase activity after different intervals. We obtained 1132.6 U/ml lipase activity after 24 hour, which was 90.6% identical to the model predicted value of 1250 U/ml, thus validating the experimental model.

Figure 3 shows the growth and production profile of lipase under the optimized conditions. The specific activity was calculated for the optimized values, and it was found to be 498.5 U/mg. The maximum lipase activity was observed in cultures of *pseudomonas otitidis* (Ramani *et al.*, 2013), *Aspergillus awamori* (Basheer *et al.*, 2011) *and actinomycetes strain* 3 (Selvam, Vishnupriya and Bose, 2011), were 1980 U/ml, 1164.93 U/mg, and 700 U/ml respectively. But all the processes utilized edible oils, which makes the process less sustainable. Whereas, here, the maximum activity of 1132.6 U/ml is obtained using coconut cake and sea water which are cost effective and non-edible which makes the overall process sustainable.

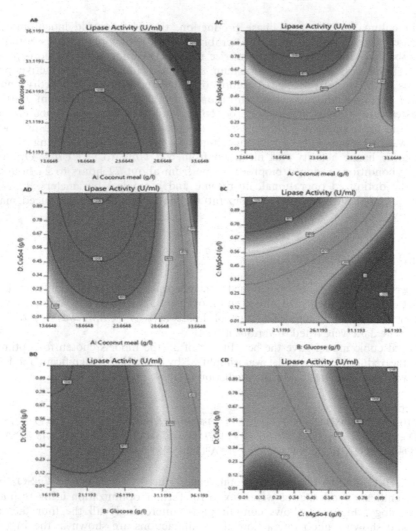

Figure 2.   Contour plot on lipase activity as a function of (AB) coconut meal and glucose (AC) coconut meal and MgSO$_4$ (AD) coconut meal and CuSO$_4$ (BC) glucose and MgSO$_4$ (BD) glucose and CuSO$_4$ (CD) MgSO$_4$ and CuSO$_4$.

5   CONCLUSION

Pseudomonas stutzeri was subjected to the physical and chemical mutagenesis for enhancement of lipase production. These random mutations create diverse clones with different activities, clones were assayed for the lipase activities. The assayed results showed the significant increase in the lipase activity as compared to the parent strain, a mutation was introduced via UV, Sodium azide, and NTG. The activity increased to 13, 56 and 14 fold respectively. After sub culturing NTG mutants obtained stability and were subjected for media optimization. A seawater based sustainable process for the production of lipase is developed and optimized. Moreover, the process used locally available raw (coconut meal cake) materials, which are not competing for the human food chain, which make it an excellent choice for commercialization. After optimization, there are 6.7 folds increased in the lipase activity i.e. 1132.6 U/ml.

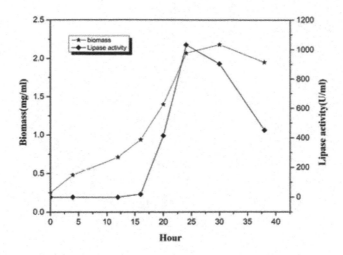

Figure 3. Growth and production profile of lipase in optimized medium at optimized operating conditions.

REFERENCES

Basheer, S. M. *et al.* (2011) 'Lipase from marine Aspergillus awamori BTMFW032: Production, partial purification and application in oil effluent treatment', *New Biotechnology*. Elsevier B.V., 28(6), pp. 627–638. doi: 10.1016/j.nbt.2011.04.007.

Hasan, F., Shah, A. A. and Hameed, A. (2006) 'Industrial applications of microbial lipases', *Enzyme and Microbial Technology*, 39(2), pp. 235–251. doi: 10.1016/j.enzmictec.2005.10.016.

Kaur, J. (2017) 'Studies on Recombinant Lipase Production by E. Coli: Effect of Media And Bacterial Expression System Optimization', *International Journal of Molecular Biology*, 2(1). doi: 10.15406/ijmboa.2017.02.00008.

Kumdam, H., Murthy, S. N. and Gummadi, S. N. (2013) 'Production of ethanol and arabitol by Debaryomyces nepalensis: Influence of process parameters', *AMB Express*. doi: 10.1186/2191-0855-3-23.

Ramani, K. *et al.* (2013) 'Lipase from marine strain using cooked sunflower oil waste: Production optimization and application for hydrolysis and thermodynamic studies', *Bioprocess and Biosystems Engineering*, 36(3), pp. 301–315. doi: 10.1007/s00449-012-0785-2.

Selvam, K., Vishnupriya, B. and Bose, V. S. C. (2011) 'Screening and Quantification of Marine Actinomycetes Producing Industrial Enzymes Amylase, Cellulase and Lipase from South Coast of India', *International Journal of Pharmaceutical & Biological Archives*, 2(5), pp. 1481–1487.

Treichel, H. *et al.* (2010) 'A review on microbial lipases production', *Food and Bioprocess Technology*, 3(2), pp. 182–196. doi: 10.1007/s11947-009-0202-2.

Civil engineering track

Technologies for Sustainable Development – Mahajan, Patel & Sharma (eds)
© 2020 Taylor & Francis Group, London, ISBN 978-0-367-33737-7

Free vibration analysis of laminated composite spherical shells: An analytical solution

Bharti M. Shinde & Atteshamudin S. Sayyad
Department of Civil Engineering, Sanjivani College of Engineering Kopargaon, Savitribai Phule Pune University, Kopargaon, India

ABSTRACT: In the present paper, a new higher order shear deformation theory is developed to study the free vibration response of simply supported laminated composite spherical shells. The present theory includes the effects of transverse normal and shear deformations to predict the natural frequencies. The equations of motion are obtained using the Hamilton's Principle, and solved by using Navier's technique. The numerical results obtained are compared with previously published results to check the accuracy and efficacy of the present theory.

Keywords: spherical, laminated, frequency, shells

1 INTRODUCTION

In the field of aerospace, marine, automobile and civil the structural components are made up of fiber-reinforced composite materials due to their attractive properties like high strength-to-weight ratio, lightweight, high stiffness-to-weight ratio. Also, the parts of the spacecraft, aircraft, and automobiles can be treated as plates and shells. The shell structures are more efficient and gives higher strength than the plates, because of the geometrical coupling of the membrane and flexural forces due to the curvature effects. The failure of these structural components occur mainly due to resonance i.e. when the external applied load frequency is equal to the natural frequency. Therefore, the free vibration study of the shells is made the area of interest of the researchers. Many researchers have carried out the dynamic analysis of the laminated composite spherical shells which can be presented in a review article by Qatu (2002a, 2002b). In past decades, Mindlin (1951) has developed a first order shear deformation theory (FSDT) for the bending and vibration analysis of plates and shells, in which the transverse shear stress is constant across the thickness and it requires the shear correction factor. To overcome limitations of FSDT Reddy (1984b) has developed a higher order shear deformation theory satisfying the zero transverse shear stress condition at top and bottom of the shell. A closed-form formulations to study the free vibration analysis of doubly curved laminated composite and sandwich shells is presented by Garg et al. (2012) using a 2D higher order shear deformation theory. Mantari and Soares (2012) have studied a bending and free vibration analysis of isotropic and multilayered plates and shells by introducing a new 5 degree of freedom higher order shear deformation theory. Sayyad and Ghugal (2019) have recently presented the bending and free vibration analysis of laminated composite and sandwich spherical shells using various refined higher order shear deformation theories.

In this paper, the free vibration analysis of laminated composite spherical shells is presented using a new fifth-order shear deformation theory. The feature of the present theory is that, it considers the effect of shear and normal deformations, to predict the accurate fundamental frequency. The fifth-order polynomial of shape function is considered in the present theory with reference to the Carrera's recommendation. Carrera (2005) recommended that to capture the accurate global response of the beams, plates and shells, minimum fifth-order polynomial of shape function should be considered in the displacement field. The equations of motion are derived using the Hamilton's Principle. The present theory satisfies the zero transverse shear stress conditions at top and bottom surfaces of the shell. Navier's analytical solution is used to solve the equations of motion for a simply supported laminated composite spherical shells. The obtained natural frequency results are compared with the other higher order theories from literature.

2 METHODOLOGY

The simply supported doubly curved spherical shells having a width a in x-direction, breadth b in y-direction and thickness h in z-direction is considered in the present study. The Figure 1 shows the geometry and principle axes in the spherical shell at the mid-plane. The shell is made up of multiple layers of laminas of same thickness.

2.1 Displacement-field

The displacement field of the present theory is as follows,

$$u(x,y,z,t) = \left(1+\frac{z}{R_1}\right)u_0(x,y) - z\frac{\partial w_0}{\partial x} + \varphi_1(z)\phi_x(x,y) + \varphi_2(z)\psi_x(x,y)$$

$$v(x,y,z,t) = \left(1+\frac{z}{R_2}\right)v_0(x,y) - z\frac{\partial w_0}{\partial y} + \varphi_1(z)\phi_y(x,y) + \varphi_2(z)\psi_y(x,y)$$

$$w(x,y,z,t) = w_0(x,y) + \varphi_1{}'(z)\phi_z(x,y) + \varphi_2{}'(z)\psi_z(x,y)$$

$$\text{where,} \quad \varphi_1(z) = \left(z-\frac{4z^3}{3h^2}\right), \quad \varphi_2(z) = \left(z-\frac{16z^5}{5h^4}\right), \quad \varphi_1{}'(z) = \left(1-\frac{4z^2}{h^2}\right), \quad \varphi_2{}'(x) = \left(1-\frac{16z^4}{h^4}\right)$$

$$(1)$$

where, u, v and w are in-plane and transverse displacements and u_0, v_0, w_0 are the displacements in the mid-plane of the shell in x, y and z direction respectively. The parameters $\phi_x, \phi_y, \phi_z, \psi_x, \psi_y, \psi_z$ represents the shear slopes occurred in the mid-plane displacements in x, y, z direction after deformation.

2.2 Strain displacement relationship

Using the general strain relationships the normal and shear strains can be written as,

$$\begin{Bmatrix} \varepsilon_x \\ \varepsilon_y \\ \varepsilon_z \\ \gamma_{xy} \\ \gamma_{xz} \\ \gamma_{yz} \end{Bmatrix} = \left\{ \frac{\partial u}{\partial x}+\frac{w}{R_1}, \quad \frac{\partial v}{\partial y}+\frac{w}{R_2}, \quad \frac{\partial w}{\partial z}, \quad \frac{\partial u}{\partial y}+\frac{\partial v}{\partial x}, \quad \frac{\partial w}{\partial x}+\frac{\partial u}{\partial z}-\frac{u_0}{R_1}, \quad \frac{\partial w}{\partial y}+\frac{\partial v}{\partial z}-\frac{v_0}{R_2} \right\}^T$$

$$(2)$$

The Hooke's law to predict the stress-strain relationship can be written as,

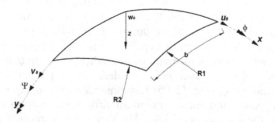

Figure 1.   Geometry and co-ordinate system of spherical shell.

52

$$\bar{\sigma} = \bar{Q}_{ij}\bar{\varepsilon} \tag{3}$$

where, $\bar{\sigma}$ represents a stress vector; $\bar{\varepsilon}$ represents a strain vector; $\bar{Q}_{ij}$ represents a stiffness matrix $(i,j=1,2,3,4,5,6)$.

2.3 Equation of motion

The equation of motion for the free vibration analysis can be obtained using the Hamilton's principle expressed as,

$$\int_{t_1}^{t_2} [\delta U - \delta V + \delta K]\, dt = 0 \tag{4}$$

where, δK is the kinetic energy, δU is the strain energy and δV is the potential energy due to external load applied. δ is variational operator. t_1 and t_2 are the initial and final time respectively. Therefore Eq. (4) leads to,

$$\int_{dv} \left(\sigma_{ij}\delta\varepsilon_{ij}\right)dv - \int_{\Omega} q(x,y)\delta w\, d\Omega + \rho \int_{dv} \left(\ddot{u}_t\delta u + \ddot{v}_t\delta v + \ddot{w}_t\delta w\right)dv = 0 \tag{5}$$

where, ρ is mass density. Therfore, substituting the stress-strain values the equation of motion can be expressed as,

$$\delta u_0 : \dot{N}_{x,x} + \dot{N}_{xy,y} = \left(I_1 + 2I_2/R_1 + I_3/R_1^2\right)\ddot{u}_{tt} - \left(I_2 + I_3/R_1\right)\ddot{w}_{xtt} + \left(I_4 + I_8/R_1\right)\ddot{\phi}_{x,tt}$$
$$+ \left(I_5 + I_9/R_1\right)\ddot{\psi}_{x,tt} = 0$$

$$\delta v_0 : \dot{N}_{y,x} + \dot{N}_{xy,y} = \left(I_1 + 2I_2/R_2 + I_3/R_2^2\right)\ddot{v}_{tt} - \left(I_2 + I_3/R_2\right)\ddot{w}_{ytt} + \left(I_4 + I_8/R_2\right)\ddot{\phi}_{y,tt}$$
$$+ \left(I_5 + I_9/R_2\right)\ddot{\psi}_{y,tt} = 0$$

$$\delta w_0 : \ddot{M}_{x,x}^b + \ddot{M}_{y,y}^b + 2\ddot{M}_{xy,x,y}^b - N_x/R_1 - N_y/R_2 + q = \left(I_1 + I_3/R_1\right)\ddot{u}_{xtt} - I_3\ddot{w}_{xxtt} + I_8\ddot{\phi}_{x,xtt}$$
$$+ I_9\ddot{\psi}_{x,xtt} + \left(I_2 + I_3/R_2\right)\ddot{v}_{ytt} - I_3\ddot{w}_{yytt} + I_8\ddot{\phi}_{y,ytt} + I_9\ddot{\psi}_{y,ytt} + I_1\ddot{w}_{tt} + I_{10}\ddot{\phi}_{z,tt} + I_{11}\ddot{\psi}_{z,tt} = 0$$

$$\delta\phi_x : \dot{M}_{x,x}^{S1} + \dot{M}_{xy,y}^{S1} - Q_x^{S1} = \left(I_4 + I_8/R_1\right)\ddot{u}_{tt} - I_8\ddot{w}_{xtt} + I_6\ddot{\phi}_{x,tt} + I_{15}\ddot{\psi}_{x,tt} = 0$$

$$\delta\psi_x : \dot{M}_{x,x}^{S2} + \dot{M}_{xy,y}^{S2} - Q_x^{S2} = \left(I_5 + I_9/R_1\right)\ddot{u}_{tt} - I_9\ddot{w}_{xtt} + I_{15}\ddot{\phi}_{x,tt} + I_7\ddot{\psi}_{x,tt} = 0$$

$$\delta\phi_y : \dot{M}_{x,y}^{S1} + \dot{M}_{xy,x}^{S1} - Q_y^{S1} = \left(I_4 + I_8/R_2\right)\ddot{v}_{tt} - I_8\ddot{w}_{ytt} + I_6\ddot{\phi}_{y,tt} + I_{15}\ddot{\psi}_{y,tt} = 0$$

$$\delta\psi_y : \dot{M}_{x,y}^{S2} + \dot{M}_{xy,x}^{S2} - Q_y^{S2} = \left(I_5 + I_9/R_2\right)\ddot{v}_{tt} - I_9\ddot{w}_{ytt} + I_{15}\ddot{\phi}_{y,tt} + I_7\ddot{\psi}_{y,tt} = 0$$

$$\delta\phi_z : \dot{Q}_{x,x}^{S1} + \dot{Q}_{y,y}^{S1} - S_{xs1}/R_1 - S_{ys1}/R_2 - S^{S1} = I_{10}\ddot{w}_{tt} + I_{13}\ddot{\phi}_{z,tt} + I_{12}\ddot{\psi}_{z,tt} = 0$$

$$\delta\psi_z : \dot{Q}_{x,x}^{S2} + \dot{Q}_{y,y}^{S2} - S_{xs2}/R_1 - S_{ys2}/R_2 - S^{S2} = I_{11}\ddot{w}_{tt} + I_{12}\ddot{\phi}_{z,tt} + I_{14}\ddot{\psi}_{z,tt} = 0$$

$$\tag{6}$$

where, expressions for stress resultants can be derived from following relations.

$$\left(N_x, N_y, N_{xy}, M_x^b, M_y^b, M_{xy}^b\right) = \int\limits_{-h/2}^{h/2} \left[\sigma_x, \sigma_y, \tau_{xy}, z\sigma_x, z\sigma_y, z\tau_{xy}\right] dz$$

$$\left(M_x^{S_1}, M_y^{S_1}, M_{xy}^{S_1}, M_x^{S_2}, M_y^{S_2} M_{xy}^{S_2}\right) = \int\limits_{-h/2}^{h/2} \left\{ \left[\varphi_1(z)(\sigma_x, \sigma_y, \tau_{xy})\right], \left[\varphi_2(z)(\sigma_x, \sigma_y, \tau_{xy})\right] \right\} dz$$

$$\left(Q_x^{S_1}, Q_y^{S_1}, Q_x^{S_2}, Q_y^{S_2}\right) = \int\limits_{-h/2}^{h/2} \left\{ \left[\varphi_1'(z)(\tau_{xz}, \tau_{yz})\right], \left[\varphi_2'(z)(\tau_{xz}, \tau_{yz})\right] \right\} dz$$

$$\hspace{10cm}(7)$$

$$\left(S^{S_1}, S^{S_2}\right) = \int\limits_{-h/2}^{h/2} \left\{ \sigma_z \left[\varphi_1''(z), \varphi_2''(z)\right] \right\} dz$$

$$\left(S_{xS1}, S_{xS2}\right) = \int\limits_{-h/2}^{h/2} \left\{ \sigma_x \left[\varphi_1'(z), \varphi_2'(z)\right] \right\} dz$$

$$\left(S_{yS1}, S_{yS2}\right) = \int\limits_{-h/2}^{h/2} \left\{ \sigma_y \left[\varphi_1'(z), \varphi_2'(z)\right] \right\} dz$$

2.4 Analytical solutions

The Navier's analytical technique is used to obtain the analytical solutions for the equation of motions stated in Eq. (6). The simply supported boundary conditions are as follows,
along the edges $x = 0$ and $x = a$

$$v_0 = w_0 = \phi_y = \psi_y = \phi_z = \psi_z = M_x^b = M_x^{S1} = M_x^{S2} = N_x = 0 \hspace{1cm}(8)$$

along the edges y = 0 and $y = b$

$$u_0 = w_0 = \phi_x = \psi_x = \phi_z = \psi_z = M_y^b = M_y^{S1} = M_y^{S2} = N_y = 0 \hspace{1cm}(9)$$

The expression for Navier solution can be written as,

$$(u_0, \phi_x, \psi_x) = \sum_{m=1,3,5}^{\infty} \sum_{n=1,3,5}^{\infty} (u_1, \phi_{x1}, \psi_{x1}) \cos\alpha x \, \sin\beta y e^{i\omega t}$$

$$\left(v_0, \phi_y, \psi_y\right) = \sum_{m=1,3,5}^{\infty} \sum_{n=1,3,5}^{\infty} \left(v_1, \phi_{y1}, \psi_{y1}\right) \sin\alpha x \, \cos\beta y e^{i\omega t}$$

$$(w_0, \phi_z, \psi_z) = \sum_{m=1,3,5}^{\infty} \sum_{n=1,3,5}^{\infty} (w_1, \phi_{z1}, \psi_{z1}) \sin\alpha x \, \sin\beta y e^{i\omega t} \hspace{1cm}(10)$$

$$q(x, y) = \sum_{m=1,3,5}^{\infty} \sum_{n=1,3,5}^{\infty} q_0 \sin\alpha x \sin\beta y e^{i\omega t}$$

For free vibration $q = 0.0$

where, $\alpha = m\pi/a$, $\beta = n\pi/b$; $i = \sqrt{-1}$; ω is the fundamental frequency $u_1, \phi_{x1}, \psi_{x1}, v_1, \phi_{y1}, \psi_{y2}, w_1, \phi_{z1}, \psi_{z1}$ are the unknown parameters. By using the foregoing expressions of displacement variables in the equation of motions (6) the resulting equation can be expressed in matrix form as,

$$\{[K] - \omega^2[M]\}\{\Delta\} = \{0\} \tag{11}$$

where $[K]$ is stiffness matrix, $[M]$ is mass matrix and $\{\Delta\}$ is the vector of amplitudes,

3 NUMERICAL RESULTS

In the present study the natural frequency for the three layered laminated composite spherical shells is presented, using the fifth-order shear and normal deformations. The results obtained for natural frequency are compared with Reddy's PSDT (1984b), Sayyad and Ghugal's ESDT (2019) and Mindlin's FSDT (1951). A coding is developed, based on the above mathematical formulation using the software MATLAB 2015. The material properties of the laminated composite spherical shell are given as below (Sayyad and Ghugal [2019]),

$$\frac{E_1}{E_2} = 25, \frac{E_3}{E_2} = 1, \frac{G_{12}}{E_2} = \frac{G_{13}}{E_2} = 0.5, \frac{G_{23}}{E_2} = 0.2, \mu_{12} = \mu_{13} = \mu_{23} = 0.25, \rho = const. \tag{12}$$

The following non-dimensional form to obtain the natural frequency is used,

$$\bar{w} = w\frac{a^2}{h}\sqrt{\frac{\rho}{E_2}} \tag{13}$$

Table 1. Non-dimensional natural frequency for three layered (0°/90°/0°) laminated composite spherical shell at various R/a and a/h ratio. ($m = n = 1$) and ($R_1 = R_2 = R$).

| R/a | Theory | a/h | | | | |
		5	10	20	50	100
5	Present	8.3515	12.0792	15.1567	20.4682	31.4974
	PSDT(1984b)	8.3200	12.0613	15.0499	20.2525	31.2192
	ESDT(2019)	8.3425	12.0412	15.0365	20.2601	31.2189
	FSDT(1951)	9.1605	12.7782	15.3454	20.2981	31.2270
10	Present	8.2908	11.8770	14.4306	16.6907	20.6521
	PSDT(1984b)	8.2593	11.8633	14.3366	16.5276	20.4844
	ESDT(2019)	8.2820	11.8428	14.3225	16.5247	20.4837
	FSDT(1951)	9.1095	12.5953	14.6482	16.5838	20.4962
50	Present	8.2711	11.8111	14.1887	15.2760	15.6296
	PSDT(1984b)	8.2396	11.7988	14.0991	15.1334	15.5166
	ESDT(2019)	8.2625	11.7781	14.0847	15.1302	15.5158
	FSDT(1951)	9.0930	12.5358	14.4164	15.1949	15.5323
100	Present	8.2705	11.8090	14.1811	15.2296	15.4460
	PSDT(1984b)	8.2390	11.7968	14.0916	15.0876	15.3352
	ESDT(2019)	8.2619	11.7760	14.0772	15.0845	15.3343
	FSDT(1951)	9.0924	12.5339	14.4091	15.1493	15.3510
Plate	Present	8.2878	11.8281	14.1997	15.2359	15.4063
	PSDT(1984b)	8.2388	11.7961	14.0891	15.0724	15.2742
	ESDT(2019)	8.2617	11.7754	14.0747	15.0692	15.2734
	FSDT(1951)	9.0923	12.5333	14.4067	15.1341	15.2901

Table 1 shows the non-dimensional natural frequency at various R/a and a/h ratios in laminated composite spherical shells. From Table 1 it is observed that the results obtained using the present theory are in good agreement with those obtained using the other higher order theories published by Reddy (1984b), Sayyad and Ghugal (2019) and Mindlin (1951). As the a/h ratio increases the frequency increases and as the R/a ratio increases frequency decreases.

4 CONCLUSION

In the present study the dynamic analysis of the laminated composite spherical shells is presented. The major contribution of the present study is that it accounts the effect of both normal deformation and shear deformation. It is concluded that, the present theory overestimates the natural frequency due to inclusion of effects of transverse shear and normal deformations. Other theories underestimate the same due to neglect of transverse normal deformation effect.

REFERENCES

Carrera, E. 2005. Transverse normal strain effects on thermal stress analysis of homogeneous and layered plates. AIAA J. 43(10): 2232–2242.

Garg, A.K., Khare, R.K., Kant,T. 2006. Higher-order closed-form solutions for free vibration of laminated composite and sandwich shells. J. Sandw. Struct. Mate. 8: 205–235.

Mantari, J.L. and Soares, C. G. 2012. Analysis of isotropic and multilayered plates and shells by using a generalized higher-order shear deformation theory. Compos. Struct., 94 (8): 2640–2656.

Mindlin, R.D. 1951. Influence of rotatory inertia and shear on flexural motions of isotropic elastic plates. ASME J. Appl. Mech. 18: 31–38.

Qatu, M.S. 2002a. Recent research advances in the dynamic behavior of shells: 1989-2000, Part 1: Laminated composite shells. Appl. Mech. Rev. 55: 325–350.

Qatu, M.S. 2002b. Recent research advances in the dynamic behavior of shells: 1989–2000, Part 2: Homogeneous shells, Appl. Mech. Rev. 55: 415–434.

Reddy, J.N., 1990. Exact solutions of moderately thick laminated shells. J. Eng. Mech., 110 (5): 794–809.

Sayyad, A.S. and Ghugal, Y.M. 2019. Static and free vibration analysis of laminated composite and sandwich spherical shells using a generalized higher-order shell theory. Compos. Struct. 219: 129–146.

Scientometric study and visualization of constructed wetlands (1998-2017)

Nandini Moondra, Robin A Christian & Namrata D Jariwala
Department of Civil Engineering, Sardar Vallabhbhai National Institute of Technology (SVNIT), Surat, Gujarat, India

ABSTRACT: Constructed wetlands is an eco- friendly, green wastewater treatment technology which is an effective substitute to conventional wastewater treatment systems in recent years. The scientometric analysis was conducted in the field of constructed during the years 1998–2017 based on the Science Citation Index Expanded (SCI-Expanded) of Web of Science (WOS)wetlands to investigate its progress and future research trends Co-author analysis, co-citation analysis was conducted. Vymazal J (citation = 992) was the most cited author. The top-ranked journal by citation counts was Ecological Engineering (citation =2521)). Environmental Sciences Ecology (2711 articles) is the top most research area with highest publications. The results of the study helps in developing a comprehensive understanding of constructed wetlands research and further establishes the future research directions in the field of environment to investigate its progress and future research trends.

Keywords: Scientometric, constructed wetlands, treatment, citation, visualization

1   INTRODUCTION

Constructed wetlands are sustainable and green technology to expel pollutants from contaminated water through physical, chemical, and biological mechanisms such as plants, substrates, soils and microorganisms to get rid of various contaminants or boost the water quality(Faulwetter et al., 2009)(Vymazal, 2011) (Wu et al., 2014)(Saeed & Sun, 2012)(Wu et al., 2015). The various processes that help in treating wastewater in wetlands are sedimentation, filtration, oxidation, reduction, adsorption and precipitation (Vymazal, 2013). Defining parameters such as volume of wastewater, HRT, media used, feeding mode and outline of setups has additionally influenced the expulsion effectiveness of contaminants among various investigations(Wu et al., 2014)(Wu et al., 2015). "Scientometrics" is a Russian development which means the study of measuring and analyzing science, technology and innovation. It is an indicator which helps to measure the impact of the research, journals, institutes on the society through mapping, visualizing and understanding of various scientific fields and citations in a particular area (Leydesdorff & Milojević, 2015). It has proven to be an efficient tool in evaluating and assessing the research performance trends in various fields. In recent years it has come to play a major role in the measurement and evaluation of research performance This concept has gained popularity in the year 1990- 2000 leading to increase in the availability and coverage of the citation databases (Mingers & Leydesdorff, 2015).

2   METHODOLOGY

The articles analyzed in this study were taken from the Web of Science center accumulation database, which is the powerhouse of the most vital and influential journals on the planet (Pouris & Pouris, 2011)(Song et al., 2016) and incorporates top productions on constructed wetlands. After pre-investigation and examination, the accompanying recovery code utilized as a part of the WOS center gathering was: TS = (constructed wetlands*). Here, "*" signifies an inquiry and "TS" implies topic search. In this examination only journal articles (review and research) were taken into account as they give maximum complete and higher-quality data than different sorts of distributions. In the study a sum of 3151 bibliographic records were gathered from the Web of Science database for the time traverse of 1998– 2017.

Bibliometric strategies connected in this examination were: (i) co-author analysis which empowers us to distinguish top scientists for constructed wetlands research(ii) co-citation analysis helps in recognizing co-referred to authors, co-referred to articles and co-cited journals through which we can get the idea about the frequency with which various research are cited together by different paper.

3 RESULTS AND DISCUSSION

3.1 Co-author analysis

According to the Web of Science database, the best ten most high -yielding authors were recognized (Table 1).

3.2 Network of co-authorship

The width of the links shows the levels of the valuable links in a given year. The shades of links, e.g., blue, green, yellow, orange and red, relate to various years from 1998 to 2017 respectively (Figure 1). A co-authorship network appears in Figure 2, where authors speak to by nodes and the connections between the writers signify the collaboration built up through the co-creation in the articles. The node size measures the amount of productions. The central authors have strong connections and collaborations with other authors. Many authors have worked with these eminent & most productive authors.

Table 1. Top ten most high -yielding authors.

S. No.	Name of Author	Institute	Country	No. of article Published
1	Garcia J	UniversitatPolitecnica de Catalunya	Spain	59
2	Kuschk P	UFZ Centre for Environmental Research, Leipzig-Halle	Germany	57
3	Vymazal J	Duke University Wetland Center	USA	40
4	Scholz M	The University of Edinburgh	UK	38
5	Zhao YQ	University College Dublin	Ireland	34
6	Brix H	Institute of Biological Sciences	Denmark	33
7	Langergraber G	University of Natural Resources and Applied Life Sciences	Austria	32
8	Molle P	Water quality and Pollution Prevention Research Unit	France	32
9	Wiessner A	UFZ Helmholtz Centre for Environmental Research	Germany	28
10	Tsihrintzis VA	Democritus University of Thrace	Greece	26

1998   2017

Figure 1. Colour coding of network.

58

Figure 2.   Co-authorship network.

3.3   *Co-citation analysis*

Co-citation is being characterized as the recurrence with which more than one articles are referred together by the respective article. In present study author co-citation and journal co-reference have been carried out and presented.

3.4   *Network of author co-citation*

Author co-citation examination was done to distinguish the connections among writers, whose productions were referred to in similar articles and thus helpful in analyzing the evolution of research communities. The node size in Figure 3 demonstrates the number of co-references for a specific author, and the connections between authors to speak about helpful circuitous links set up in view of co-reference checks.

Figure 3.   Author co-citation network.

Thus the top most highly cited authors has been listed as Vymazal J (citation = 992), Kadlec RH (citation = 886), Tanner CC (citation = 210), Langergraber G (citation = 209), Brix H (citation = 205), Garcia J (citation = 199), Mitsch WJ (citation = 186), Faulwetter JL (citation = 176), Zhang DQ (citation = 146) and Calheiros CS (citation = 143). Hence it can be stated that it is not at all necessary that a highly cited author receives high centrality. But it is likely to happen that at the point when the author gets both great references and centrality then that writer would impact the advancement and improvement of constructed wetlands research. Besides, a few authors had reference bursts, with quick increments in reference recurrence over brief periods, including Wu SB (burst = 36.55), Stottmeister U (burst = 33.26), Zhang DQ, (burst=32.24), Saeed T (burst = 31.67),Rousseau DPL (burst = 31.47),Wu HM (burst = 30.30), Reed SC (burst = 27.10), Akratos CS (burst = 22.75), Avila C (burst = 21.56), Knight RL (burst = 21.49).

3.5 *Network of journal co-citation*

The node size signifies the co-citation recurrence of every source journal. The top ranked journal by citation counts and paper publications are Ecological Engineering (citation = 2521, paper published = 723), Water Science Technology (citation = 2224, paper published = 243), Water Research (citation = 2186, paper published = 187), Science and Total Environment (citation = 1518, paper published = 162), Environmental Science and Technology (citation = 1126, paper published = 110), Chemosphere (citation = 1068, paper published = 97), Journal of Environmental Quality (citation = 1056, paper published = 85). and Environmental Pollution (citation = 870). It was worthwhile saying that these journals were one of the better source journals, where articles on constructed wetlands were published. Thus, greater the contribution of the journal to constructed wetland research more is the number of citations. The top-ranked journals by bursts were Environmental Science and Pollution Research (burst = 83.96), Water-SUI (burst = 36.30), Desalination and Water Treatment (burst = 35.63), Chemical Engineering Journal (burst = 34.64), Transactions of the ASAE (burst = 32.22), PLOS ONE (burst = 31.28).These journals are worth following as they represent the significant intellectual turning by receiving strong citations over a short period

Figure 4.   Journal co-citation network.

4 CONCLUSIONS

The expulsion of contaminants in CWs is perplexing and relies upon an assortment of evacuation systems, such as sedimentation, filtration, precipitation, volatilization, adsorption, plant take-up, and different microbial procedures. Constructed wetlands had attracted increasing attention from researchers and practitioners' as these systems are robust and cost-effective with low energy requirement. These attributes of constructed wetlands make it the most suitable engineered green technology which has great acknowledgment among developing nations for decentralized waste treatment. This examination gives a scientometric audit to investigate the status and patterns of developed wetlands inquire about around the world. A sum of 3151 bibliographic records were gathered from the WOS center accumulation database. Co-author analysis, were done so as to visualize and identify the trends and status of constructed wetlands research. The co-authorship and author co-citation analysis were performed to recognize the contributions and influence of the leading researchers. With respect to the distribution of articles on constructed wetlands, most of them originated from the USA China and Spain. This examination gives essential data to the specialists and practitioners in the field of constructed wetlands research. The principal researchers and foundations, the condition of the exploration field, and hotly debated issues on developed wetlands were recognized by analysts. In addition, this investigation will enable experts to get the fundamental discoveries to improve their comprehension of constructed wetlands. The scientometric survey strategy can likewise be utilized to image the exploration incline in different topics.

REFERENCES

Faulwetter, J. L., Gagnon, V., Sundberg, C., Chazarenc, F., Burr, M. D., Brisson, J., Stein, O. R. (2009). Microbial processes influencing performance of treatment wetlands: A review. *Ecological Engineering*, *35*(6), 987–1004. https://doi.org/10.1016/j.ecoleng.2008.12.030

Leydesdorff, L., & Milojević, S. (2015). Scientometrics. *International Encyclopedia of the Social & Behavioral Sciences (Second Edition)*, 322–327. https://doi.org/http://dx.doi.org/10.1016/B978-0-08-097086-8.85030-8

Mingers, J., & Leydesdorff, L. (2015). A review of theory and practice in scientometrics. *European Journal of Operational Research*, *246*(1), 1–19. https://doi.org/10.1016/j.ejor.2015.04.002

Pouris, A., & Pouris, A. (2011). Scientometrics of a pandemic: HIV/AIDS research in South Africa and the World. *Scientometrics*, *86*(2), 541–552. https://doi.org/10.1007/s11192-010-0277-6

Saeed, T., & Sun, G. (2012). A review on nitrogen and organics removal mechanisms in subsurface flow constructed wetlands: Dependency on environmental parameters, operating conditions and supporting media. *Journal of Environmental Management*, *112*, 429–448. https://doi.org/10.1016/j.jenvman.2012.08.011

Song, J., Zhang, H., & Dong, W. (2016). A review of emerging trends in global PPP research: analysis and visualization. *Scientometrics*, *107*(3), 1111–1147. https://doi.org/10.1007/s11192-016-1918-1

Vymazal, J. (2011). Constructed Wetlands for Wastewater Treatment: Five Decades of Experience. *Environmental Science & Technology*, *45*(1), 61–69. https://doi.org/10.1021/es101403q

Vymazal, J. (2013). Emergent plants used in free water surface constructed wetlands: A review. *Ecological Engineering*, *61*, 582–592. https://doi.org/10.1016/j.ecoleng.2013.06.023

Wu, H., Zhang, J., Ngo, H. H., Guo, W., Hu, Z., Liang, S., Liu, H. (2015). A review on the sustainability of constructed wetlands for wastewater treatment: Design and operation. *Bioresource Technology*, *175*, 594–601. https://doi.org/10.1016/j.biortech.2014.10.068

Wu, S., Kuschk, P., Brix, H., Vymazal, J., & Dong, R. (2014). Development of constructed wetlands inperformance intensifications for wastewater treatment: A nitrogen and organic matter targeted review. *Water Research*, *57*, 40–45. https://doi.org/10.1016/j.watres.2014.03.020

Technologies for Sustainable Development – Mahajan, Patel & Sharma (eds)
© 2020 Taylor & Francis Group, London, ISBN 978-0-367-33737-7

Sensitivity of lateral load patterns on the performance assessment of semi-rigid frames

V. Sharma
Ph.D. Student, Malaviya National Institute of Technology, Jaipur, India

M.K. Shrimali, S.D. Bharti & T.K. Datta
Professor, Malaviya National Institute of Technology, Jaipur, India

ABSTRACT: The efficacy of lateral load patterns (LP_L) to estimate the responses of semi-rigid (SR) frames using nonlinear static analysis (NSA) is investigated by comparing their assessment with the standard response, evaluated from nonlinear response history analysis (NRHA). A 5-Story low-rise steel semi-rigid frame is selected as an illustrative example for NRHA, considering the far-field and near-field with forward-directivity earthquakes. The sensitivity of three LP_Ls is compared based on target displacement (T_d) approach by selecting two T_d values, which represent the elastic and elastic-plastic states on the pushover curve of the SR frame. The seismic response parameters included peak-story displacement, maximum inter-story drift ratio, the number of plastic hinges, and their SRSS values. The results show that (i) the estimated error in seismic responses from different LP_Ls relies upon the state (elastic and elastic-plastic); and (ii) the conventional mode based load pattern is found more suitable for low-rise steel frames.

1 INTRODUCTION

The nonlinear response history analysis (NRHA) is preferred to evaluate the seismic responses and design validation of structural members. NRHA anticipates the inelastic seismic demands with excellent precision in the structures subjected to seismic excitations. In spite of its capability, NRHA has not preferred the choice of structural designers due to its intricacies, like proper modeling of structures, more significant computational needs, the need of ample set of response histories, and requirement of hysteresis curve and so on. Thus, the nonlinear static analysis (NSA) or pushover analysis is preferred with sufficient accuracy to estimate the inelastic demand.

Freeman (1975) initially laid the foundation of NSA methods. Further, various researchers modified the NSA procedures, and currently, these methods have become the key methods in performance-based seismic design to achieve the performance objectives (ATC-40, 1996, Fajfar & Gašperšič, 1996, Hasan et al., 2002, FEMA, 2005). The NSA procedure uses the distribution of lateral load pattern (LP_L) along the height to evaluate the inelastic demands. If the structure is vibrating in single mode or its fundamental mode, the load pattern corresponding to fundamental mode or single-mode evaluate accurate results. If the other modes contribute considerably, the single-mode LP_L gives results conservatively. This lacuna inspired the researchers to develop advanced NSA methods. Recently, Bhandari et al. (2018) proposed the two LP_Ls to estimate the inelastic demands in base-isolated structures and compared with standard NRHA procedures. Significant work is carried out to establish the NSA methods for fully rigid steel frames, but fewer studies are available for estimation of seismic demand in semi-rigid (SR) frames. These studies on SR frames did not consider the target displacement (T_d) took as peak story displacement approach for estimating the efficacy of LP_Ls to anticipate the nonlinear behavior.

In this study, the sensitivity of three lateral load patterns, namely, the fundamental mode shape (LP_{L1}) based, the square root of sum of square values of first three modes (LP_{L2}), and the uniform force distribution lateral load pattern (LP_{L3}), are investigated with comparison to seismic response obtained from benchmark NRHA at two target displacements levels (elastic and elastic-plastic). For this, a 5-Story semi-rigid steel frame is designed and analyzed as per Indian standard. An ensemble of three earthquakes, each in far-field and near-field with forward-directivity is employed for NRHA simulations. The response parameters included for comparison are peak-story displacement, maximum inter-story drift ratio, the total number of plastic hinges, and SRSS of maximum plastic hinge rotations.

2 METHODOLOGY AND MODELING OF FRAME

2.1 *Semi-rigid connection modeling*

Two jointed zero-length multi-linear plastic link element is used to model the SR connection in SAP2000. The semi-rigidity in the beam to column connections are dependent on three parameters, namely, stiffness parameter (k); flexural strength parameter (s) and ductility parameter (μ). For the illustrative example, the stiffness and flexural strength parameters are selected as per ANSI/AISC-341 (2016) recommendations and the ductility parameter prescribed by Chan & Chui (2000). The values of parameters k, s, and μ for SR connection are chosen as 15, 1.5, and 0.04, respectively.

2.2 *Analysis*

For this comparative study, two types of analysis, namely, the nonlinear static analysis (NSA) and NRHA, are performed to examine the performance of SR connected frames at two states, defined by the target displacement (T_{d1} = Elastic; T_{d2} = Elastic-Plastic). The target displacement values are based on the pushover curve obtained from the NSA for the three LP_Ls, and the values are shown in Figure 1(a). The first three modes are shown in Figure 1(b). Further, NRHA is executed considering the ensemble of three earthquakes in each FF and NFD types to scale the PGA level corresponding to T_{d1} and T_{d2} (see Table 1). The Hilber-Hughes-Taylor time integration techniques with 5% Rayleigh damping is considered for NRHA simulations.

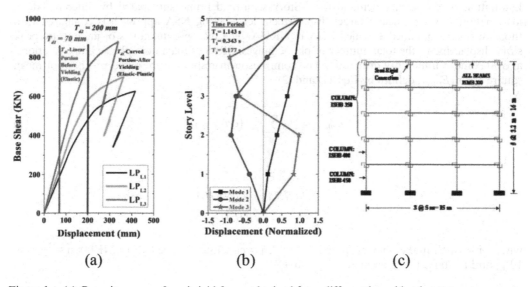

| (a) | (b) | (c) |

Figure 1. (a) Capacity curve of semi-rigid frame obtained from different lateral load patterns representing target displacement levels, (b) Modes, and (b) 5-Story semi-rigid frame with detailing.

Table 1. Scaled PGA (g) values corresponding to target displacements for 5-Story semi-rigid frame.

EQ Type	Earthquake Name	T_{d1}	T_{d2}	EQ Type	Earthquake Name	T_{d1}	T_{d2}
FF	Kobe	0.29	0.851		Erzincan	0.1532	0.4316
	Landers	0.182	0.575	NFD	Northridge	0.1614	0.481
	Superstition	0.1851	0.786		Kocaeily	0.148	0.5144

The geometric nonlinearity (secondary small-displacement P-Delta effects) is also taken into consideration in both types of analysis.

2.3 *Semi-rigid frame modeling*

A 5-story three-bay semi-rigid (SR) frame is designed as per Indian standard provision (IS-800, 2007, IS-1893, 2016) using the SAP2000 software. The frame sections are selected in such a way that it satisfies the strong-column weak beam concept, which means that the plastic moment capacity of columns is 1.2 times more than the plastic moment capacity of the adjoining beam. Figure 1(b-c) shows the modes and detailing of SR frame. The seismic design parameters included the zone factor (Z = 0.36), medium soil condition, importance factor (I = 1), and response reduction factor for special moment frame (R = 5). The gravity load includes the floor dead load of 20KN/m, roof dead load of 15 KN/m and 4KN/m of live load uniformly distributed on each floor. Two different sets of ground motions, namely, far-field and near-field with forward directivity (NFD) are considered for NRHA. The FF consists of (i) 1999 Kobe (FF), 0.51g PGA; (ii) 1992 Landers, 0.42g PGA, and (iii) 1987 Superstition, 0.45g PGA. The other set of NFD includes (i) 1992 Erzincan, 0.5g PGA, (ii) 1994 Northridge, 0.83g PGA and (iii) 1999 Kocaeli, 0.31g PGA. The material nonlinearity in the 2D frame is incorporated in the form of default concentrated plastic hinges as per ASCE-41 (2017) at the end of flexural members.

3 NUMERICAL RESULTS AND DISCUSSIONS

Estimation of the seismic demand for 5-Story semi-rigid frame anticipated by different lateral load patterns at a particular target displacement levels using NSA approach is compared with those responses obtained at scaled PGA in NRHA. The response quantities of interest are peak story displacement, the total number of plastic hinges, SRSS of maximum plastic hinge rotations, and maximum inter-story drift ratio. The comparison in responses are based on the root mean square error (E_{rms}), defined by Eqs. (1) and (2).

$$E_{rms,i} = \sqrt{\frac{1}{3} \sum_{i=1}^{3} \left(error_{ijnt} \right)^2} \tag{1}$$

and

$$error_{ijnt} = \frac{\left((NSA)_{nt} - (NRHA)_{ijnt} \right)}{(NRHA)_{ijnt}} \tag{2}$$

where, i = earthquake number, here i = 1 to 3; j = earthquake type (FF or NFD); n = type to LP_L, and t = target displacement (T_{d1} and T_{d2})

3.1 Estimation of Peak-Story displacement (PS$_d$)

Figure 2 shows the variation of E_{rms} along with height for three LP$_L$s at two target displacement levels in SR frame. The LP$_{L3}$ gives conservative results for both types of earthquakes. It is observed from the figure that the E_{rms} along the height of the building observed for LP$_{L1}$ are very less, the maximum of the order of 12% at T$_{d1}$ and T$_{d2}$. Since the variation in peak story displacement obtained from NRHA and NSA are small for both types of earthquakes, the E_{rms} is also less at T_{d1} and T_{d2}.

As the target values are taken as top peak story displacements, so the difference in E$_{rms}$ values at peak story comes to zero. At the lower story level, the E$_{rms}$ considerably increase, with the maximum of the order of 36% in LP$_{L3}$. This is due to the more inelastic effects exhibited in the NSA with all LP$_L$s as compared to NRHA.

3.2 Estimation of Maximum Inter-story Drift Ratio (MIDR)

Table 2 shows the E_{rms} in MIDR at different story level for SR connected frame at two target displacements for FF and NFD earthquakes. It is observed from Table 2 that the error increases from lower story level to a higher story level. At T_{d1}, the error for FF is negligible in LP$_{L1}$, even at T_{d2} the maximum value reached to near 12%. It is also noticed that the E_{rms} considerable more in NFD type ground motion and reached to near 40%. The NSA produced more inelastic effects at bottom storey level as compared to NRHA. This is the main cause for large error in MIDR values.

3.3 Estimation of inelastic excursion in SR connected frame

The inelastic excursion in SR frames at target displacement level for FF and NFD type earthquakes are captured in the total number of plastic hinges and the SRSS value of maximum plastic hinge rotations. Figure 3 represents the E_{rms} in a number of plastic hinges and SRSS

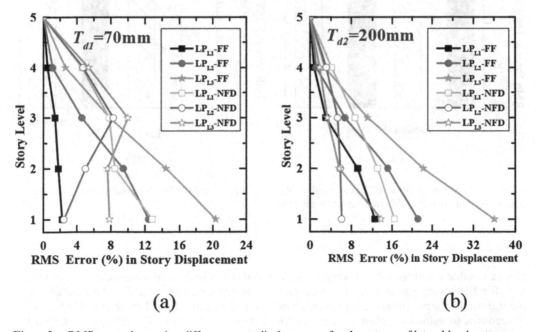

(a)   (b)

Figure 2.   RMS error observed at different target displacements for three types of lateral load patterns.

Table 2. RMS error in Maximum Inter-story Drift ratio (MIDR).

Story Level	LP$_{L1}$		LP$_{L2}$		LP$_{L3}$	
RMS Error at T_d = 70mm						
	FF	NFD	FF	NFD	FF	NFD
0	0.00	0.00	0.00	0.00	0.00	0.00
1	2.26	11.81	12.42	2.43	20.35	7.86
2	2.58	7.37	8.07	6.69	11.62	8.27
3	1.18	15.22	1.93	13.67	0.94	14.32
4	2.44	6.38	8.52	4.54	11.72	6.72
5	2.82	24.36	6.17	22.38	14.87	22.01
RMS Error at T_d = 200mm						
0	0.00	0.00	0.00	0.00	0.00	0.00
1	12.43	15.85	21.16	6.10	36.07	14.30
2	7.92	11.36	12.42	6.27	16.13	4.39
3	8.65	3.42	7.62	5.36	7.56	5.55
4	6.68	12.49	13.31	5.92	18.18	5.65
5	5.08	39.03	6.90	30.29	17.11	17.96

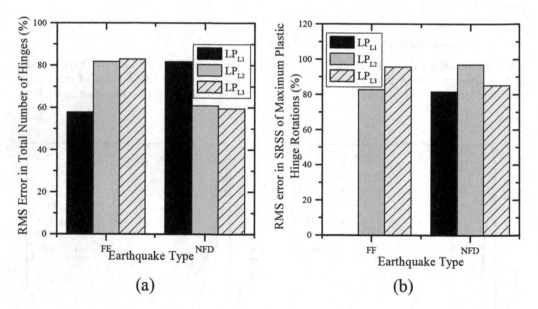

(a)   (b)

Figure 3. RMS error observed in (a) number of plastic hinges, and (b) SRSS of maximum plastic hinge rotations in SR frame at different target displacement.

values obtained at three LP$_L$s as compared with NRHA. The E_{rms} in inelastic excursion measures for FF is very less in LP$_{L1}$ less as compared to LP$_{L2}$ and LP$_{L3}$ at T_{d2}. Thus, the fundamental mode shape lateral load pattern gives better results and found more suitable for NSA procedures to estimate inelastic effects in SR frames. Further, the T_{d1} represents the elastic state, so there are no hinges formed at this level under both types of ground excitations.

4 CONCLUSIONS

The sensitivity of three lateral load patterns (LP_L) used in the nonlinear static analysis (NSA) to estimate the seismic demands is investigated by anticipating the responses obtained from non-linear response history analysis (NRHA) of a 5-Story semi-rigid (SR) frames. The comparative study was carried out considering two target displacements. This study discussed both far-field and near-field with forward directivity earthquakes. The response parameters for comparisons are peak story displacement, maximum inter-story drift ratio, and inelastic excursion in the form of plastic hinges along with their SRSS of hinge rotations. The major outcomes of the numerical study are

The lateral load pattern LP_{L1} produced the best results compared to other two LP_LS, especially predicting the seismic demands viz peak-story displacement and maximum inter-story drift values.

The error in MIDR values is significant in lower stories for T_{d2} and in higher stories at T_{d1} for both suites of earthquakes.

The errors to estimate the inelastic excursion in the form of plastic hinges are more in near-field as compared to far-field earthquakes, and the LP_{L1} turns out to be the best.

REFERENCES

ANSI/AISC-341 2016. Seismic Provision for Structural Steel Buildings. Chicago: American Institute of Steel Construction.

ASCE-41 2017. Seismic Evaluation and Retrofit of Existing Buildings. Reston, Virginia 20191: American Society of Civil Engineers.

ATC-40 1996. Seismic evaluation and retrofit of concrete buildings. Redwood City, California.

Bhandari, M., Bharti, S., Shrimali, M. & Datta, T. 2018. Assessment of proposed lateral load patterns in pushover analysis for base-isolated frames. *Engineering Structures*, 175, 531–548.

Chan, S.-L. & Chui, P.-T. 2000. *Non-linear static and cyclic analysis of steel frames with semi-rigid connections*, Kidlington, Oxford OX5 1GB, UK, Elsevier Science Ltd.

Fajfar, P. & Gašperšič, P. 1996. The N2 method for the seismic damage analysis of RC buildings. *Earthquake Engineering & Structural Dynamics*, 25, 31–46.

FEMA, A. 2005. 440, Improvement of nonlinear static seismic analysis procedures. *FEMA-440, Redwood City*.

Freeman, S. 1975. The Capacity Spectrum Method as a tool for seismic design, Wiss, Janney, Elstner Associates. Inc.

Hasan, R., Xu, L. & Grierson, D. 2002. Push-over analysis for performance-based seismic design. *Computers & structures*, 80, 2483–2493.

IS-800 2007. General Construction in Steel-Code of Practice. Bureau of Indian Standards, New Delhi.

IS-1893 2016. Criteria for earthquake resistant design of structures. *Part 1 General Provisions and Buildings (Sixth Revision)*. New Delhi-110006: Bureau of Indian Standards.

Technologies for Sustainable Development – Mahajan, Patel & Sharma (eds)
© 2020 Taylor & Francis Group, London, ISBN 978-0-367-33737-7

Impact of angle of incidence in rectangular liquid storage tanks

S. Vern
Ph.D. Student, Malaviya National Institute of Technology, Jaipur, India

M.K. Shrimali, S.D. Bharti & T.K. Datta
Professor, Malaviya National Institute of Technology, Jaipur, India

ABSTRACT: The effect of the earthquake motion on the liquid storage tank involves various complex material interactions which makes it a cumbersome problem for the analysis. Generally, all the analysis assumes that the earthquake is always orthogonal to the structural boundary, but in reality, the angle of the incidence can vary. The determination of the critical magnitude of the stresses and moments for a liquid storage tank (LST) is done in the present study by varying the angle of incidence. The nonlinear time history analysis is carried out using ABAQUS which employs Arbitrary Lagrangian-Eulerian (ALE) FEM for solving the Fluid structure Interaction. The response parameters include the top board displacement of the wall, sloshing height, shear stress and overturning moment. The study provides the change in the base shear and overturning moment of order 70-90 %.

1 INTRODUCTION

The study of LSTs requires an updated approach. Due to its ubiquitous presence in the nuclear, petroleum, water supply, chemical industries, etc., this type of structure possesses an integral role in any nations growth. Due to rapid development in liquid storage tanks, various innovations are being employed to tackle the severe difficulty in their overall performance for withstanding any disasters. The common structural failures are elephant foot buckling at the bottom of the tank, uplifting in the base and sliding at the base of the tank.

Housner (1957) represented dynamically the hydrodynamic pressure created by the horizontal component of the earthquake and derived the simple methods in the form of equations. Housner (1963) explained the dynamic behavior of the tank with the help of simplified models for different types of tanks. Liu (1981) presented a nonlinear finite element method which can treat the structural behavior of the tanks in conjunction with fluid, including the dynamics and buckling. Haroun & Tayel (1985) studied the tank responses in both vertical and lateral excitations is to find out the relative importance of the vertical component of ground acceleration. Cheng (2017) examined the massive isolation layer displacement during the earthquake motions for the sliding-based isolators.

The aim of the present study is to understand the effect of the bi-directional behavior of earthquake component on the liquid storage tanks. The FEM approach is used by the ABAQUS software to extract the various parameters of the interests for a three-dimensional liquid storage tank. Along with the bi-directional effect, the study for the changes in the responses of the LST is studied as the angle of the incidence is changed with respect to the principle direction of the tank. To make sure the workability of the tank under sever conditions control responses should be evaluated along with the fragility and reliability analysis of the LSTs under extreme conditions of the earthquakes.

2 METHODOLOGY

The present study takes non-linear behavior of the tank and liquid into the consideration which is developed due to the fluid sloshing. The explicit dynamic analysis option solves the non-linearity in the present study. In an explicit dynamic step, the simultaneous action of the explicit integration rule and the lumped element mass matrices plays an important role. For determining the equation of the motion for the body, the equations of the motions are integrated using the central difference integration rule as given in Equations 1-2.

$$\dot{u}^{\left(i+\frac{1}{2}\right)} = \dot{u}^{\left(i-\frac{1}{2}\right)} + \frac{\Delta t^{(i+1)} + \Delta t^{(i)}}{2} \ddot{u}^{(i)} \tag{1}$$

$$\dot{u}^{(i+1)} = \dot{u}^{(i)} + \Delta t^{(i+1)} \dot{u}^{\left(i+\frac{1}{2}\right)} \tag{2}$$

Here, $\dot{u}$ and $\ddot{u}$ represents the velocity and acceleration. The superscript (i), $(i-1/2)$ and $(i+1/2)$ are the incremental series number the mid incremental numbers of series respectively. The advantage of the mid incremental rule comes handy when the using of the diagonal element mass matrices, as the nature of the acceleration at the beginning of the increment step is triaxial, the mass matrix is calculated easily as shown in Equation 3 (Dassault Systèmes Simulia 2012).

$$\ddot{u}_{(i)}^N = \left(M^{NJ}\right)^{-1} \left(P_{(i)}^J - I_{(i)}^J\right) \tag{3}$$

where M_{NJ} is the diagonal lumped mass matrix, P_J is the applied load vector, and I_J is the internal force vector. The explicit procedure takes lots of the small time increments to integrate through the time. The central-difference operator is conditionally stable. The stable increment is given in the form of the maximum value of eigen value as shown in Equation 4.

$$\Delta t \leq \frac{2}{\omega_{max}} \left(\sqrt{1 + \xi^2} - \xi\right) \tag{4}$$

A small amount of the damping ξ is used for determining the stable time increment of the body. The damping gives two advantages over the analysis stable time increment first by suppressing the high frequency oscillations and secondly by reducing the stable time increment.

2.1 Finite element modelling details

The tank under the present study is taken by considering the flexibility action of the tank wall. To simulate the Fluid Structure Interaction (FSI) a surface to surface interaction is used, which includes default the normal and tangential behavior. The element used in modelling tank is an eight-node linear brick element(C3D8R). this brick element is chosen for its reduced integration technique and hourglass control which helps in convergence of the solution. The element for the fluid is same as that of the tank, but various factors are altered for a smooth transition from the solid to fluid. Those factors are stiffness-viscous weight factor of unity and the use of equation of state. To keep up with the deformation of large magnitude in the fluid domain, an adaptive mesh control is given. Arbitrary Lagrangian and Eulerian (ALE) approach is used which maintains the mesh and results in faster and higher accuracy outputs, as compared to the pure Lagrangian approach.

3 NUMERICAL RESULTS AND DISCUSSIONS

To investigate the behavior of the liquid storage tanks under the action of bidirectional interaction of the earthquake, the tank taken up for the study is flexible and rectangular in cross-section. The base condition is taken as fixed to the ground. The lateral and transverse side being equal to the 60m and 20m respectively resultant geometry comes out to be a simple rectangular tank. The thickness of the tank is 1.2m. The water stored in the tank is up to the height of 12m from the base of the tank. The various material properties taken in the analysis are provided in Table 1. To examine the various response in the LST because of the seismic ground motion, the time history of the 1940s Imperial valley earthquake with El-Centro station is chosen.

The time history analysis of first 23 seconds is taken. The peak ground acceleration of the earthquake is 052g and 0.27g in both horizontal directions respectively. The analysis done for the 10 different orientations of the rectangular tank, with each orientation 10-degree change is added in the last orientation. The sloshing height is extracted at the front right corner of the tank wall. The von-mises stress is calculated at the front corner of the tank base. The top broad displacement of the long and short wall. The values are calculated at the center of the walls.

The validation of the current approach is validated by the Virella, Prato & Godoy, (2008) as shown in Figure 1. In which, a 2D tank model is taken up and a sinusoidal accelerating pulse is given at the base of the tank. The corresponding wave height is extracted of the tank. In the present study the tank is taken up as a flexible tank. The validation of the present study is extended to the 3D tank of rectangular profile. Through the validation of the FSI property the various response quantities of interest are find out by the ABAQUS . The modelling and the computation of the rectangular tank in a FE suite is an expansive and cumbersome process.

The angle of incidence can be a significant factor which can affect the nature of the magnitude of the various critical responses. The sloshing height can be a seen to be affected as the orientation of the tank is changed. The maximum value of the sloshing height is for the 50-degree orientation. The change in the value of the sloshing height can be to have effect on the hydrodynamic pressure. The most critical value of the pressure can be seen at the similar angle of orientation which is 40 degree.

The response quantities such as the base shear and overturning moment are considered by the design engineers for the safety and stability point of view. The change in the base shear at the x-direction is maximum at the starting orientation and as the orientation changes to final position a sequence of the reduction in the base shear is seen. Although the von-mises stress is

Table 1.   Material properties for the tank and fluid.

Concrete	Water
Modulus of Elasticity, E_s = 24.86 GPa	Density, ρ_w = 983.204Kg/m^3
Density, ρ_s = 2450kg/m^3	Equation of state: c_0 = 1450, s = 0, γ_0 = 0
Poisson's ratio, υ = 0.17	Dynamic Viscosity = 0.001 N-sec/m^2

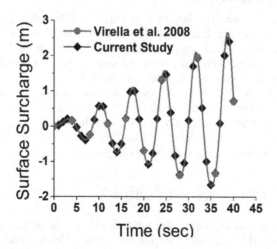

Figure 1.   Validation of current approach.

70

Table 2. Change in the responses with respect to the original orientation.

Angle of Incidence	Wave Ht. (m)	Base Shear (X) MN	Base Shear (Y) MN	Von-Mises (kPa)	Overturning Moment -X (MN-m)	Overturning Moment -Y (MN-m)	Hydrodynamic Pressure (kPa)	TBD (Long Wall) (m)	TBD (Short Wall) (m)
0°	0.13	119.3	38.21	6540	495.2	687.8	217	0.236	0.050
10°	0.13	106.2	43.73	5690	596.9	660.5	222	0.210	0.056
20°	0.14	90.15	52.03	4800	616.4	644.6	255	0.170	0.048
30°	0.16	72.57	48.73	4680	652.4	635.5	263	0.130	0.051
40°	0.18	55.15	46.08	4800	695.14	597.07	319	0.100	0.032
50°	0.19	49.00	40.66	4750	515.83	585.95	236	0.074	0.032
60°	0.18	53.41	47.44	4800	535.71	578.09	268	0.065	0.033
70°	0.17	53.42	65.42	4620	462.87	602.56	259	0.059	0.021
80°	0.14	57.78	82.68	4660	529.62	607.41	222	0.037	0.010
90°	0.13	48.33	97.86	5560	658.92	683.46	229	0.001	0.002

higher at the initial position, the final orientation seems to be most affected in terms of the magnitude. The overturning moment in the y-direction achieves its maximum value in the initial setup. This may be due to the lever arm is highest in the initial shape.

The peak top broad displacements (TBD) in both the short and long wall observed a change due to the angle of incidence. The TBD in both the horizontal directions seems to have decreased as the LST is revolved towards the final position. This may due to the lower value of the stiffness and lesser amount of the flexural rigidity at the center of the walls. This higher amount of the flexibility with hydrodynamic forces with each other resulted in the higher initial values.

The responses quantities in a rectangular tank under the different configurations of angle of incidence have a more in-depth effect. Table 3 gives the representation change with the initial configuration of the rectangular LST. A change of the 42 percentage is present for the 50-degree configuration. This may be due to the increased free surface for the water waves to travel. Base shear in major axis(X) can be changes to an order of the 60 percent at the final orientation. Similarly, the overturning moment in the lateral direction is reduced to a maximum order of the 16 percent. From the analysis it can be concluded that due to the additional unbalanced mass due to the sloshing motion the magnitude of the peak changed overturning moments are at the highest. The sloshing can be major factor in deciding the outcome of the various changed behavior in the LST. The shear forces of the LSTs can be termed as principal factors for the additional induced inertial forces. The aftermath of these unbalanced inertial forces can be seen to result in the almost double magnitude change in the TBD

Table 3. Percentage change in the responses with respect to the original orientation.

Angle of Incidence	Wave Ht. (m)	Base Shear (X) MN	Base Shear (Y) MN	Von-Mises (kPa)	Overturning Moment (MN-m) -X	Overturning Moment (MN-m) -Y	Hydrodynamic Pressure (kPa)	TBD (Long Wall) (m)	TBD (Short Wall) (m)
0°	0.00	0.0	0.00	0.00	0.00	0.00	0.00	0.00	0.00
10°	+2.36	-11.0	14.45	-13.00	+20.54	-03.97	+02.30	-11.02	+12.0
20°	+7.87	-24.4	36.17	-26.61	+24.47	-06.28	+17.51	-27.97	-04.00
30°	+27.2	-39.2	27.53	-28.44	+31.74	-07.60	+21.20	-44.92	+01.0
40°	+40.2	-53.8	20.60	-26.61	+40.38	-13.19	+47.00	-57.63	-36.00
50°	+41.7	-58.9	06.41	-27.37	+4.17	-14.81	+08.76	-68.64	-36.40
60°	+37.9	-49.7	08.48	-15.64	-10.25	-12.48	+20.72	-69.05	-40.89
70°	+20.7	-40.7	25.74	-3.75	-24.91	-06.52	+01.57	-65.29	-56.67
80°	-13.31	-20.4	69.67	-0.43	-18.82	-04.42	-15.59	-71.54	-80.59
90°	-27.08	-12.4	112.4	+15.8	-5.21	+14.47	-28.21	-98.70	-94.38

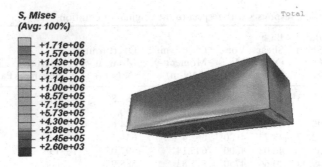

S, Mises
(Avg: 100%)

```
+1.71e+06
+1.57e+06
+1.43e+06
+1.28e+06
+1.14e+06
+1.00e+06
+8.57e+05
+7.15e+05
+5.73e+05
+4.30e+05
+2.88e+05
+1.45e+05
+2.60e+03
```

Total

Figure 2. Von-mises stress in LST for zero-degree orientation.

in both the structural wall. The most interest response quantity behavior is for the TBD in short wall, hydrodynamic pressure and sloshing wave height as in initial orientation up to 40-50 degree the change is either decreasing or increasing but, after the 60-degree orientation this change in percentage changes its approach in opposite direction.

It can be seen from Figure 2 the intensity for the mises stress is maximum at the base of the LST. Due to the larger flexibility in the long wall the critical chances of failure are higher than the shorter wall. This damage at the base can be avoided by placing additional wall support in the form of the braces at the inner walls of the LST.

4 CONCLUSIONS

The effects of the angle incidence in the criticality with respect to the various responses are huge. The response quantity for the comparisons are wave height, base shear and overturning moment in both the horizontal directions, von-mises stress, hydrodynamic pressure in the tank base along with the top board displacement of the long and short wall of the tank. The study yields the following results:

(i) The angle of incidence can be a significant factor when it comes to the most critical arrangement of the orientation of the LST. The sloshing height, overturning moment in the lateral direction and hydrodynamic pressure have a criticality at the 40-50-degree orientation.

(ii) The base shear, von-mises stress and TBD in both walls are most affected when the complete swap of the major and minor axes is present. As the von-mises stress is the damage criteria of a structure, thus it shows that the most critical risk of LST failure is present when the major component of earthquake is normal to the shorter wall.

REFERENCES

Dassault Systèmes Simulia Corp. 2012. Analysis User's Manual Volume 1: Introduction, Spatial modeling, execution and output. Abaqus 6.12, I, 831.
Haroun, M. A., Tayel, M. A. 1985. Response of tanks to vertical seismic excitations. *Journal of Earthquake Engineering and Structure Dynamics*, 13 August 1984, 583–595.
Housner, G. W. 1957. Dynamic pressures on accelerated fluid containers. *Bulletin of the Seismological Society of America*, 47(1), 15–35.
Housner, G. W. 1963. The Dynamic Behavior of Water Tanks. *Bulletin of the Seismological Society of America*, 53(2), 381–387.
Liu, W. K. 1981. Finite element procedures for fluid-structure interactions and application to liquid storage tanks. *Nuclear Engineering and Design*, 65(2), 221–238.
Virella, J. C., Prato, C. A. and Godoy, L. A. 2008. Linear and nonlinear 2D finite element analysis of sloshing modes and pressures in rectangular tanks subject to horizontal harmonic motions. *Journal of Sound and Vibration*, 312(3), pp. 442–460.

Assessing the indoor environmental quality of municipal schools in Ahmedabad

Sneha Asrani* & Dipsha Shah
CEPT University, Ahmedabad, India

ABSTRACT: Humans spend 90 percent of their time indoors, thus, the quality of the built or indoor environment impacts their health as well as productivity. The main aim of an educational building is to satisfy the learning requirements of students and educators by the means of design, architecture and comfort. The Indoor Environment Quality of a space comprises of four comfort parameters which are Thermal Comfort, Acoustic Comfort, Visual Comfort and Indoor Air Quality. In this study, the Indoor Environment Quality of fifty municipal schools was assessed by monitoring Indoor Air Temperature, Relative Humidity, Illuminance, Noise and concentration of pollutants like Formaldehyde, TVOC, CO_2, $PM_{2.5}$ and PM_{10}, twice, once in the month of February (winter – phase 1) and March (spring – phase 2). The monitored results were compared with the relevant standards. It was found that the relative humidity was lower than the standards, noise level higher than the limits and the amount of light within the classrooms was inadequate. The concentration of air pollutants like formaldehyde and TVOC increased during phase 2, while $PM_{2.5}$ and PM_{10} decreased in phase 2. The overall air quality improved in phase 2 of monitoring, which was attributed to better ventilation.

1 INTRODUCTION

1.1 *Overview*

The Indoor Environment Quality is defined by the combined effect of four comfort parameters, namely Thermal, Acoustical, Indoor Air Quality and Visual comfort, which can be quantified by monitoring indoor air temperature, noise level, air velocity, pollutant concentrations and illuminance (ISHRAE standard 10001:2016).

The Indoor Environment Quality impacts the health and overall productivity of occupants. An increase in indoor temperature negatively impacts students' performance. The impact of indoor air quality on students' performance is such that, by increasing the outdoor air supply rate inside the classroom, the speed of work increased by 8% overall, (Wargocki, Wyon (2017)). Infants and young children have a higher resting metabolic rate and higher rate of oxygen consumption per unit body weight than adults, because they are growing rapidly. Thus, indoor air pollution impacts academic performance of children by lowering their concentration levels and increasing absenteeism due to asthma and allergies. The amount of day-light present in the classroom affects the students' performance. Higher amount of day-light has shown to increase test scores. Noise levels can negatively impact the students' learning performance as well as teachers' teaching quality, (Ackley, Donn and Thomas (2017)).

Thus, Indoor Environment Quality needs to be given importance at the time of building design and evaluation. The objective of this study was to assess the Indoor Environment Quality of Ahmedabad Municipal Corporation's Schools by monitoring the comfort parameters.

---

* Corresponding author: sneha.asrani.96@gmail.com

1.2 Literature review

The Indoor Environment Quality is impacted by factors like HVAC use, ambient light, ambient pollution, construction materials and the layout of furniture (M.S.F., de Freitas, Peixoto, Delgado, João M.P.Q (2015)).

Table 1 shows the recommended threshold values for comfort parameters, as taken from ISHRAE standard 10001: 2016, the highlighted cells are the threshold values selected.

1.3 Case study

The study was conducted in the city of Ahmedabad, Gujarat, India. Ahmedabad is divided into 6 zones, which are further divided into 48 wards. There are 374 municipal schools in Ahmedabad, out of which 50 schools were monitored. As far as municipal schools in Ahmedabad are concerned, there were multiple schools that utilize the same building – either during different operational hours or a different floor. Thus, the total no. of schools monitored was 50, but the no. of school buildings monitored were 38, accumulating to 333 class rooms. Figure 1 shows the different zones and wards in Ahmedabad along with the locations of the 38 school buildings monitored.

2 DATA COLLECTION

2.1 Data collected

Table 2 describes the data collected and the devices used. The indoor environment quality of these municipal schools was monitored twice, once between 24th January and 13th

Table 1. Recommended standards as per ISHRAE standard 10001:2016.

Parameter	Unit	Standards		
		Class A	Class B	Class C
Occupant satisfaction	%	90	80	< 80
Formaldehyde	mg/m^3	<0.03	<0.1	NA
TVOC	mg/m^3	<0.2	<0.5	<0.5
PM$_{2.5}$	µg/m^3	<15	<25	<25
PM$_{10}$	µg/m^3	<50	<100	<100
CO$_2$	ppm	Ambient + 350	Ambient + 500	Ambient + 700
Operative Temperature	°C	For summer: 24.5 ± 2.5; for winter: 22.0 ± 3.0		
Relative Humidity (RH)	%		30-70	
Illuminance	lux		300	
Noise	dB(A)		30-45	

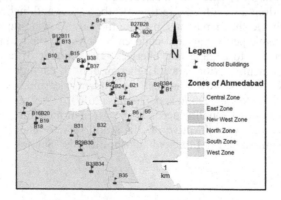

Figure 1. Map of school buildings.

74

Table 2. Data collected, and the devices used.

Comfort parameter	Element	Monitoring device	Unit of measurement
Thermal Comfort (TC)	Indoor air temperature	Airveda Monitor	°C
	Relative Humidity (RH)		%
Indoor Air Quality (IAQ)	Formaldehyde (HCHO)	TVOC and Formaldehyde meter - Dienmern 106b	mg/m^3
	Total Volatile Organic Compounds (TVOC)		mg/m^3
	Carbon Dioxide (CO_2)	Airveda Monitor	ppm
	$PM_{2.5}$		$\mu g/m^3$
	PM_{10}		$\mu g/m^3$
Lighting Comfort (LC)	Illuminance	Kusam-meco KM-LUX-99 Digital Multimeter	lux
Acoustic Comfort (AC)	Noise level	Envirotech SLM 109	dB(A)

February 2019, now onwards referred to as Phase 1, and the second time between 20th February and 13th March 2019, now onwards referred to as Phase 2.

2.2 Data collection methodology

Figure 2 shows the locations at which the data was collected. The monitoring was done at five locations in the classroom, at desk height. The existing conditions of windows, fans and lights were kept the same during monitoring.

3 DATA ANALYSIS

3.1 Relative Humidity (RH)

Figure 3 shows the RH within school buildings during Phase 1 and 2. It was noted that only in 14 buildings during Phase 1, and 18 buildings during Phase 2, the RH was within the standards. The RH increased during Phase 2. The reason for this was the increase in the no. of students in the classroom, since the RH increases with no. of people.

3.2 Visual comfort

Figure 4 shows the illuminance within school buildings during Phase 1 and 2. Only buildings B17, B25, B26, B28, B29 and B34 (i.e. 13% of the buildings monitored) had adequate lighting

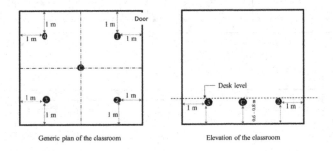

Figure 2. Data collection locations.

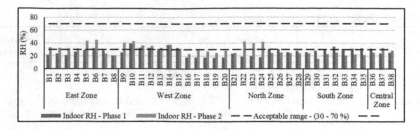

Figure 3. Relative Humidity within buildings during Phase 1 and 2.

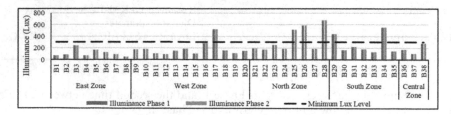

Figure 4. Illuminance within buildings during Phase 1 and 2.

in the classrooms, attributed to more amount of natural light present in the room. It was observed that in majority school buildings the artificial lighting wasn't adequate, and the buildings weren't oriented in order to receive maximum natural lighting.

3.3 *Indoor air quality*

3.3.1 *Formaldehyde and TVOC*

Figure 5 – a and 5 – b show the formaldehyde and TVOC concentration respectively within buildings during Phase 1 and 2. Here, majority of buildings had concentration within the limits, during both phases for both pollutants, formaldehyde and TVOC. The concentration of both pollutants increased during Phase 2 because with the increase in temperature, the

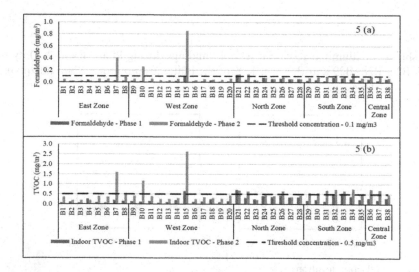

Figure 5. a – b Formaldehyde and TVOC concentration within buildings during Phase 1 and 2.

pollutant emissions from furniture and paints would increase. The buildings B21 and B15 had concentration exceeding the standards during both phases, the reason being the location of the school and higher vehicular emissions around it – B15 was located near a major intersection and a bus stop, while in B21 there was some painting work done a while back, and poor ventilation leading to accumulation of pollutant.

3.3.2 CO_2

Figure 6 shows the CO_2 concentration within buildings during Phase 1 and 2. Here, all buildings had concentration within the limits, except B15. CO_2 concentration would increase with the no. of people in the space. The building B15 had concentration exceeding the standards during Phase 1, the reason being the more no. of students crammed in a room, poor ventilation (windows and fans were kept closed) leading to accumulation of CO_2.

3.3.3 $PM_{2.5}$ and PM_{10}

Figure 7 – a and 7 – b show the PM_{10} and $PM_{2.5}$ concentration, within the buildings during Phase 1 and 2, respectively. The concentration of PM_{10} and $PM_{2.5}$ were higher during Phase 1 than in Phase 2. This can be attributed to the fact that during winter the air is stagnant, which leads to accumulation of the pollutants. Another reason was improved ventilation during Phase 2, winter was receding, fans and windows were kept open.

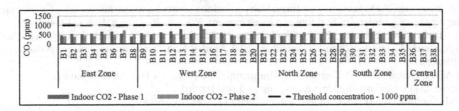

Figure 6. CO_2 concentration within buildings during Phase 1 and 2.

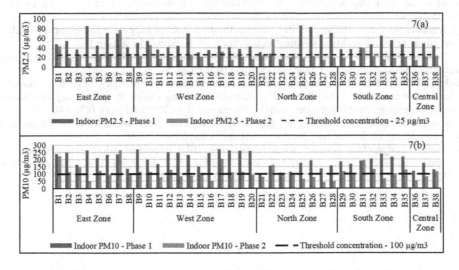

Figure 7.   a – b PM_{10} and $PM_{2.5}$ concentration within buildings during Phase 1 and 2.

4 CONCLUSION

The formaldehyde, TVOC concentration was found to be within the threshold values, for majority of the buildings, during both phase 1 and 2. The concentration of formaldehyde and TVOC increased during phase 2 in majority of the buildings. The probable reason for this can be that with the increase in temperature and RH, the formaldehyde emissions from the furniture increases. CO_2 concentration was within the standards. The pollutants $PM_{2.5}$ and PM_{10} were identified as critical parameters. The concentration of particulate matter exceeded the threshold values in majority of the monitored buildings. The concentration during phase 1 (winter) was higher than that during phase 2 (spring), probable reason being that stagnant air leads to increase in ambient concentration which in turn leads to increase in indoor concentration. The air quality within a school building was influenced by its immediate surroundings and was seen to improve as the season shifted from winter to spring. The relative humidity (RH) for majority of the buildings was found to be less than the prescribed standards during both phases. Only 13% buildings monitored had adequate illuminance. The buildings need to be designed to receive maximum natural light.

REFERENCES

Ricardo M.S.F., de Freitas, Vasco Peixoto, Delgado, João M.P.Q, 2015: Indoor Environmental Quality and Enclosure Optimization.

Blondel, Plaisance,2011: Screening of formaldehyde indoor sources and quantification of their emission using a passive sampler. In: Building and Environment, Volume 46, Issue 6, Pages: 1284-1291.

Ackley, Donn and Thomas, 2017: The Influence of Indoor Environmental Quality in Schools – A Systematic Literature Review. Conference: 51st Annual Conf Architectural Science Association At: Wellington.

ASHRAE Standard 55 2017: Thermal Environmental Conditions for Human Occupancy.

Henderson, 2006: Carbon dioxide measures up as a real hazard.

ISHRAE Standard10001: 2016, Indoor Environment Quality Standard.

Mahmoud M. M. Abdel-Salam, 2015: Investigation of PM2.5 and carbon dioxide levels in urban homes. In: Journal of the Air & Waste Management Association, 65: 8,930–936.

Wargocki, Wyon, 2017: Ten questions concerning thermal and indoor air quality effects on the performance of office work and schoolwork. In: Building and Environment, Volume 112, February 2017, Pages 351–358.

Pontip, Kandar, Sediadi – Multitrait, 2017:Multimethod analysis of subjective and objective methods of indoor environmental quality assessment in buildings. In: Building Simulation 11 (2).

Tess M.Stafford (2015). Indoor air quality and academic performance. In: Journal of Environmental Economics and Management, Volume 70, March 2015, Pages 34–50.

Wargocki, 2009: Ventilation, thermal comfort, health and productivity. In: D. Mumovic & M. Santamouris (Eds.), A handbook of sustainable building design & engineering—an integrated approach on energy, health and operational performance. London: Earthscan (Ch. 14).

Fisk -Estimates of potential nationwide productivity and health benefits from better indoor environments: an update. In: Indoor Air Quality Handbook, Chapter 4.

Al horr, Arif, Katafygiotou, Mazroei, Kaushik, Elsarrag, 2016: Impact of indoor environmental quality on occupant well-being and comfort: A review of the literature. In: International Journal of Sustainable Built Environment, Volume 5, Issue 1, June 2016, Pages 1–11.

Technologies for Sustainable Development – Mahajan, Patel & Sharma (eds)
© 2020 Taylor & Francis Group, London, ISBN 978-0-367-33737-7

Modelling of daytime incoming longwave radiation under cloudy sky over a semi-arid agro-climatic zone of India

Dhwanilnath Gharekhan*
Civil Department, Institute of Technology, Nirma University, Ahmedabad, Gujarat, India

Bimal K Bhattacharya & Rahul Nigam
Space Application Center, ISRO, Ahmedabad, Gujarat, India

Devansh Desai
Department of Physics, Space Science and Electronics, Gujarat University, Ahmedabad, Gujarat, India

P.R. Patel
Civil Department, Institute of Technology, Nirma University, Ahmedabad, Gujarat, India

ABSTRACT: Net surface radiation defines the availability of radiative energy on and near the surface to drives many physical and physiological processes such as latent heat, sensible heat fluxes and evapotranspiration. One of the prime challenges of modelling radiation budget is estimation of net longwave radiation. Incoming longwave radiation (LWin) is one of the key components of net longwave radiation. Estimation of incoming longwave radiation in cloudy conditions has always been a challenge due to the lack of instrumentation and regular measurements at different spatial scale. In this study model for incoming longwave radiation under cloudy sky is developed using half hourly day-time incoming longwave radiation and other meteorological parameters with revised model coefficients. The developed model showed bias of -7.5 W/m^2 over semi-humid climatic zone of India.

Keywords: Incoming longwave radiation, cloud fraction, daytime, modelling

1 INTRODUCTION

The surface radiation (energy) budget of the earth-atmosphere system is currently one of the important research areas pertaining to climate change studies. (Mahalakshmia, et al., 2014). Net surface radiation defines the availability of radiative fluxes for exchange of energy and mass between surface and atmospheric processes such as evapotranspiration and photosynthesis. (Ming, De_Richter, Liu, & Caillol, 2014). Accurate estimation of these components are further helpful for understanding of sustainable development, renewable energy, meteorological systems etc. [(Bo Jianga, 2015), (Male & Granger, 1981), (T M Crawford, 1999)].

Incoming Longwave radiation (LWin) is one of the four key components in the surface radiation budget that is emitted and measured in the range of 4.0 – 100 μm which are affected by trace gases like H$_2$O, CO$_2$ etc. (Idso & Jackson, 1969). While normally longwave radiation in measured using pyro-geometers and all-weather observatories in country generally do not possess them. Simulated LWin can be obtained using Radiative transfer models like **SBDART** or **MODTRAN**. However, information like vertical profiles of temperature and precipitable water is not typically available to simulate LWin

---

* Corresponding author: dhwanilnath@gmail.com

from radiative transfer models. [(Duarte, Dias, & Magiotto, 2006), (Snell, Anderson, J Wang, Chetwynd, & English, 1995)]. Over the years several site-specific models were developed to compensate the limitation and scarcity of instruments. [(Duarte, Dias, & Magiotto, 2006), (Elliott, 2002), (Choi, Jacobs, & Kustas, 2008)]. In the present study an empirical model is developed using high temporal in situ data of incoming long wave radiation and meteorological variables.

2 STUDY AREA AND DATA USED

The ground measurements for air temperature and relative humidity were collected by high response sensor installed with an Eddy Covariance system (22.7969° N, 72.5749° E), installed at Anand Nawagam Station under Indo-UK INCOMPASS (INteraction of Convective Organization and Monsoon Precipitation, Atmospheric Surface and Sea) project. The ground observation site lies under Semi-Arid Climatic zone of India and having a Rice-Wheat cropping system. The mean annual temperature ranges between 9°C to 39°C and the annual precipitation ranges between 700-850mm. In the system, a Net Radiometer (Kipps and Zonen CNR4) is installed along with met assembly. The radiation sensor is a 4-component net radiometer that measures the incoming & outgoing short-wave and incoming & outgoing longwave (Zonen, 2014) .

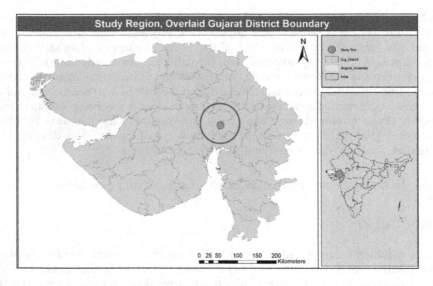

Figure 1.   Ground measurement site over Gujarat, India.

The high temporal data from EC tower and Net Radiometer measurements were averaged at 30 min interval for air temperature (Ta) and relative humidity (RH). The time period of this study was between August -September 2017 and August-October 2018 – 117 days. The study period represents typical monsoon season over Gujarat and having a maximum cloud cover during this period. During computations the days having rain accumulation more than 4 mm were considered as rainy days and excluded from the study. Also, daytime hours have been considered between 8 am to 4 pm (i.e. 16 measurements a day - 1872 total measurement points) coincide with Ta and RH measurements. The dataset was segregated into two section 60% data was used for training the model and rest 40% for validation.

3 CLEAR SKY PARAMETERIZATION

Net surface radiation can be expressed by its components:

$$R_n = SW_{in} - SW_{out} + LW_{in} - LW_{out} \tag{1}$$

$$R_n = SW_{in}(1 - \alpha) + LW_{in} - LW_{out} \tag{2}$$

Where α is the surface albedo, SW_{in} is incoming shortwave radiation, SW_{out} is outgoing shortwave radiation, LW_{in} is incoming longwave radiation and LW_{out} is outgoing longwave radiation. The incoming and outgoing longwave radiation are difficult to obtain due to the lack of instruments. (I. Aladosb, 2003). For years different workers have tried to parametrization in the incoming longwave radiation through its components:

$$LW_{in} = \varepsilon_{ac} * \sigma * T_a^4 \tag{3}$$

Where T_a is air temperature, ε_a is effective clear sky atmospheric emissivity, $\sigma = 5.6705 \times 10^{-8}$ W/m^2K^4 is Stephen Boltzmann's Constant. Overtime there have been several formulas derived which consider ε_a as a function of Air Temperature (T_a) and/or relative humidity (RH) measured near the surface. Six previously published linear models to estimate incoming longwave radiation for clear sky conditions were evaluated. First the models with original coefficients were tested and then with experimental coefficients which were estimated at local conditions.

Table 1.   Incoming longwave radiation estimation empirical formulas for clear sky conditions.

Author	Formula		Constants	
(Brunt, 1932)	$LW_{\downarrow 0} = (a_1 + b_1 e_a^{1/2})\sigma T_a^4$	(8)	$a_1 = 0.55$	$b_1 = 0.0065$ hPa$^{-1/2}$
(Swinbank, 1963)	$LW_{\downarrow 0} = (a_2 T_a^2)\sigma T_a^4$	(9)	$a_2 = 9.36E\text{-}6$ K^{-2}	
(Idso & Jackson, 1969)	$LW_{\downarrow 0} = (1 - a_3 \exp[(b_3(273 - T_a)^2]\sigma T_a^4$	(10)	$a_3 = 0.261$	$b_3 = -7.77E\text{-}4$ K^{-2}
(Brutsaert, 1975)	$LW_{\downarrow 0} = (a_4(e_a/T_a)^{b3}]\sigma T_a^4$	(11)	$a_4 = 1.24$ (K/ hPa)^{b4}	$b_4 = 1/7$
(Idso S., 1981)	$LW_{\downarrow 0} = (a_5 + b_5 e_a \exp[1500/Ta])\sigma T_a^4$	(12)	$a_5 = 0.7$	$b_5 = 5.95E\text{-}5$ hPa^{-1}
(Prata, 1996)	$LW_{\downarrow 0} = (1 - [(1 + w)\exp(-(a_6 + b_6 w)^{1/2})])\sigma T_a^4$	(13)	$a_6 = 1.2$	$b_6 = 3$ cm^2/g

The presence of cloud on the other hand, tend to either increase or decrease the amount of incoming longwave radiation depending on the cloud type and height. Therefore, it is necessary to incorporate their effects into the simplified empirical models. (T M Crawford, 1999) suggested the following:

$$c = 1 - \frac{S\downarrow}{Rext} \tag{4}$$

As a means of calculating cloud correction factor (c), where is insolation (incoming shortwave radiation) and R_{ext} is extra-terrestrial solar radiation for the time period. Furthermore, vapor pressure (e_a) is also required, given as (Allen, Pereira, Raes, & Smith, 1998):

$$e_a = e_s \left[\frac{RH}{100}\right]$$

$$e_a = \left(6.108 exp\left[\frac{17.27T_a}{T_a + 273.3}\right]\right)\left(\frac{RH}{100}\right) \tag{5}$$

Where e_s (in hectopascal) defines saturation vapor pressure, T_a is in degree Celsius while RH is in percentage. Over the years, several cloud correction models were defined to consider the effects of clouds to incoming longwave radiation. Cloudy sky correction formulations generally have one of two basic structures (Duarte, Dias, & Magiotto, 2006)

$$LW_\downarrow = LW_{\downarrow 0}\left(1 + \alpha c^\beta\right) \tag{6}$$

$$LW_\downarrow = LW_{\downarrow 0}(1 - c^\gamma) + \delta c^\zeta T_a^4 \tag{7}$$

Where is incoming longwave under cloudy conditions, α, β, γ, and ζ are the locally calibrated constants determined from cloud types. Seven cloudy sky correction models with original parameters were evaluated in this study, with similar methodology used for clear sky, the initial IDSO1981 results were divided into two groups (3/5 =) 60% for training of local calibration of coefficients and remaining (2/5 =) 40% for validation of the tuned model.

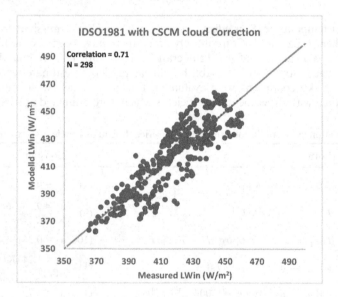

Figure 2.   Regression results for measured against modelled incoming longwave radiation.

In the equations, is incoming longwave radiation (W/m^2), e_a is vapor pressure (hPa), w = perceptible water content – 46.5 (e_a/T_a) g/cm^2. The model based incoming longwave radiation showed best fit with IDSO1981 and in this model cloud correction was integrated from the CSCM2 model (Carmona, Rivas, & Caselles, 2014). The generated results showed instantaneous bias of 7.5 W/m^2. However, being an empirical model, the test parametrizations do tend to show a tendency of overestimating the measurements which leads to a lower correlation value. Table 2 showed the generated calibrated coefficient for the study region using in situ data over monsoon season.

Table 2.   Locally calibrated empirical constants for incoming longwave radiation.

Author	Formula	Constants			
		a_5	b_5	μ	γ
(Idso S., 1981)	$LW_{\downarrow 0} = \left(a_5 + b_5 e_a \exp\left[\frac{1500}{T_a}\right]\right)\sigma T_a^4$	0.7	5.95E-5 hPa^{-1}		
(Carmona, Rivas, & Caselles, 2014)	$LW = LW_{\downarrow 0}(1 - c^\mu) + \gamma c^\mu \sigma T a^4$	0.63 ± 0.002	5.5±0.01 hPa^{-1}	0.85 ± 0.01	0.982 ± 0.02
Our Study	Data -> IDSO1981 ($LW_{\downarrow 0}$) -> $LW_\downarrow$	0.675 ± 0.001	4.5±0.1 hPa^{-1}	0.0585 ± 0.01	0.985 ± 0.01

4 CONCLUSION

In this paper, we customized an empirical model with updated coefficients for semi-arid climatic region to estimate incoming longwave radiation under cloudy sky using in situ data from high response sensor located at Nawagam, Gujarat (India). The updated cloudy-sky model was able to generate incoming long wave radiation with good accuracy however being only limited to one site at the moment. In future the model will be further updated over different agro-climatic regions of India using in situ data and then combined model will be developed for Indian sun-continent with satellite imagery.

ACKNOWLEDGEMENT

The authors would like to thank SAC (Space Application Centre), ISRO, Ahmedabad for providing the datasets and infrastructure to do this study.

REFERENCES

Allen, R., Pereira, L., Raes, D., & Smith, M. 1998. Crop evapotranspiration: guidelines for computing crop water requirements. *Food and Agriculture Organization of the United Nations* (FAO), Rome, 300.

Bo Jianga, Y. Z. 2015. Empirical estimation of daytime net radiation from shortwave radiation and ancillary information. *Agricultural and Forest Meteorology*, 23–36.

Brunt, D. 1932. Notes on radiation in the atmosphere. *Quart J Roy Meteorol Soc.* 58, 389–420.

Brutsaert, W. 1975. On a derivable formula for long wave radiation from clear skies. *Water Resour. Res* 11, 742–744.

Carmona, F., Rivas, R., & Caselles, V. 2014. Estimation of daytime download longwave radiation under clear and cloudy skies conditions over a sub-humid region. *Theor. Appl. Climatol.* 115, 281–295.

Choi, M., Jacobs, J. M., & Kustas, W. P. 2008. Assessment of clear and cloudy parameterization for daily downwelling longwave radiation over different land surfaces in Florida, USA. *Geophysical Research Letters*, 25.

Duarte, H. F., Dias, N. L., & Magiotto, S. R. 2006. Assessing daytime downward longwave radiation estimates for clear and cloudy skies in souther Brazil. *Agric. For. Meteorol.* 139, 171–181.

Elliott, V. S. 2002. On the development of a simple downwelling longwave radiation scheme. *Agric For. Meteorol*, 112, 237–243.

I. Aladosb, I. F.-M.-A. 2003. Relationship between net radiation and solar radiation for semi-arid shrub-land. *Agricultural and Forest Meteorology* 116, 221–227.

Idso, S. 1981. A set of equations for full spectrum and 8 to 10 mm and 10.5 to 12.5 mm thermal radiation from cloudless skies. *Water Resour Res* 17, 295–304.

Idso, S. B., & Jackson, R. D. 1969. Thermal Radiation from the atmosphere. *J. Geophysical Res.* 74 (23), 5397–5403.

Mahalakshmia, D. V., Paula, A., Dutta, D., Ali, M. M., Jha, C. S., & Dadhwal, V. K. 2014. Net Surface Radiation Retrieval using Earth Observation Satellite Data and Machine Learning Algorithm. *ISPRS Annuals of the Photogrammetry, Remote Sensing and Spatial Information Sciences, Volume II-8, 2014 ISPRS Technical Commission VIII Symposium* (pp. 9–12). Hyderabad, India: ISPRS.

Male, D., & Granger, R. 1981. Snow surface energy exchange. *Water Resource Res.* 17 (3), 609–627.

Ming, T., De_Richter, R., Liu, W., & Caillol, S. 2014. Fighting global warming by climate engineering: Is the earth radiation management and the solar radiation management any option for fighting climate change. *Renewabale and Sustainable Energy Reviews*, March, Vol 31, 792–834.

Prata, A. 1996. A new long wave formula for estimating downward clear sky radiaion at the surface. *Quart J Roy Meteorol Soc* 122, 1127–1151.

Snell, H. E., Anderson, G. P., J Wang, J. L. Moncet, Chetwynd, J. H., & English, S. J. 1995. Validation of FASE (FASCODE for the Environment) and MODTRAN3: updates and comparison with clear-sky measurements. *Proc. SPIE Int. Soc. Opt. Eng.*, 2578, 194–204.

Swinbank, W. 1963. Longwave radiation from clear skies. *Quart J Roy Meteorol Soc.* 89, 339–348.

T M Crawford, C. E. 1999. An improved parameterization for estimating effective atmospheric emissivity for use in cacluating daytime longwave radiation. *J. Appl. Meteorol.* 38, 474–480.

Zonen, K. a. 2014. CN4 - Net Radiometer (Instruction Manual). Netherlands: Kipp and Zonen.

Use of BIM and augmented reality in facility management phase

Aathira Satish & Jyoti Trivedi
CEPT University, Ahmedabad, India

ABSTRACT: Augmented reality (AR) is a fast-developing field which provides an interactive experience where-in computer generated perceptual information are projected into real life environment, creating an interactive experience. Application of AR into construction field implies that a 3D BIM model can be visualized at a specific intended location using mobile and other devices specially developed for the same in real time. Application of AR in construction industry can prove to be accurate, saving time and money, enabling better planning of projects and helping in better visualization of the project. This experiment is to understand the potential benefits that can be obtained by implementing Building Information Modelling (BIM) and AR in facility management phase of a building by undertaking a case study approach on the institutional library in ahmedabad city of gujarat. The developed integrated model was applied on a case study and depicted benefits of assets management, better visualization and rectification benefits towards facility management. This project based research work will give evidence to back up the relatively unexplored potential of BIM and AR in facility management.

1 INTRODUCTION

The true potential of Augmented reality (hereon referred to as AR) in construction field, especially for facility management is relatively unexplored. But despite most sectors in the world moving towards digitization, construction industry ranks the lowest as per the industry digitazion index. (Karji, Woldesenbet, &Rokooei, Integration of Augmented Reality, Building Information Modelling, and image processing in Construction Management: A Content Analysis) Focusing on a niche segment integrating three unique techniques of digitization techniques, when applied in construction industry, namely Building Information modelling, augmented reality and Image processing, can bring about potential better project delivery-whereas now all three are being used separately in the sector.

1.1 *Research objective*

o Study the existing facility management processes in the Library building.
o Understand the shortcomings/problems faced by the facility management withthe present manual technique.
o Understand benefits attainable through a BIM-AR sample developed model in terms of possible cost, time and human effort savings.

2 LITERATURE REVIEW

Operation phase of any project constitutes 60% of the total cost on an average. Reactive maintenance and repair have shown to be 3-4 times costlier than proactive maintenance (Sullivian & Mobley, 2010). Asset management market value alone was seen to be near to the value of new constructions in Canada. BIM enabled FM has shown greater ROI in the long run. (Azhar,

Maqsood, &Khalfan, 2012). VR enabled FM would mean a look around the room with the equipment would reveal the facilities hidden behind false ceiling or tiles. The entire facility can be tracked, monitored and upgraded just by sitting at the comfort of the office. (Jurgens, 2016). Interactive VR and AR models can assist the facility managers in case of a confusion.

The applicability of this can be further enhanced if BIM model is converted into an interactive format of visualization aiding faster and informed responses. There are three levels of mixed reality systems according to Milgram's continuum diagram (Milgram, 1994), namely-Augmented reality (AR), Augmented Virtuality (AV) and Virtual Reality (VR). For the purpose of the study, we would be focusing on AR- augmented reality. The task of converting BIM models into AR interface can be daunting work, especially when multiple iterations are to be done to update the model. This manual process can lead to users abandoning the system after the first few cycles. (Jing, Zhengbo, et.al 2017).

3   EXISTING SCENARIO

The BIM model created of the building were Revit architecture and the services considered for the study was of Plumbing and HVAC. The library building is of G+2 with 3 basements below for storing books.

3.1   *Plumbing*

In the institutional building the two main water tanks for the campus are located at the north and south lawns. The library has one main water inlet for the ground floor. All plumbing lines are restricted to the ground level. The drainage is connected to the main campus lines, laid along north gate to SSO (student service office). It drains out to municipal sewer lines. The maintenance for the same is carried out quarterly every year by deputed maintenance team. Present issues faced are compiled in the table below.Wash faucet leakage and clogging of pipe-were the frequent issues observed in plumbing lines.

Following benefits are expected to be obtained by implementing the BIM-AR model.

- Recheck of positions of concealed service lines before drilling or demolition of partsin case of retrofitting.
- Understanding position of concealed service lines in case of leaks.

3.1.1   *Past faults at the library*

Two main plumbing faults were observed in the library since its opening in mid-2017-2018 wash-basin to drain leakage, flooding of basement electrical lines due to improper functioning of rain drain, overflow of urinal flush and clogging of pipes near inspection chamber. Nominal cost range of rework around 1500-2500 INR/.

3.2   *HVAC*

There are both VRV and duct AC systems present in the library, working simultaneously to cater to the demand of the building. AHU is established at first floor level towards FT building east side of the institutional building. There are also 2 split ACs. Problems observed in the building due to HVAC services are AC condenser pipe leak, Trap door clogging and clogging of pipe.Preventive maintenance for the facility is carried out quarterly every year. Agency is also under breakdown compliance by which they need to comply immediately in case of a failure of the systems.Following benefits are expected to be obtained by implementing the model.

- Recheck of positions of concealed service lines before drilling or demolition of partsin case of retrofitting.
- Understanding position of concealed service lines in case of leaks.

4 BUILDING INFORMATION MODELLING

4.1 *Stage 1- Conversion of drawings to BIM model*

The Services drawings were accessed from the Campus office. As received 2D-as building drawings were created as LOD 500 as built drawings for architecture, HVAC and plumbing as given in Figure 1. Based on which paper based AR was created. The data regarding make, model, service dates, guarantee and warranty dates of every HVAC and plumbing component was then added to BIM model as given in Figure 2.

4.2 *Paper augmented reality*

The first set of AR was prepared basing a paper target. The purpose of the same is to create an AR projection of the library HVAC and plumbing systems that will enable an understanding of the ducts and pipes. The model is intended to help visualize the flow of both the services as given in Figures 3-4-5.

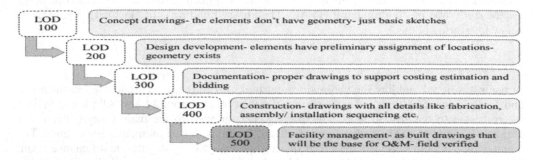

Figure 1. Levels of development.

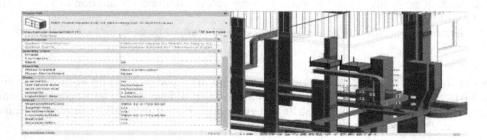

Figure 2. BIM model showing component data.

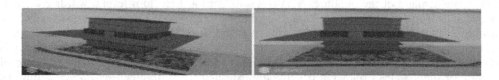

Figure 3. Architecture AR model – paper based.

Figure 4.   HVAC AR model paper based.

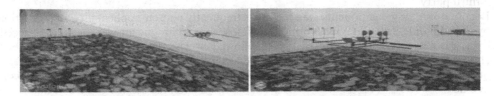

Figure 5.   Plumbing AR model paper based.

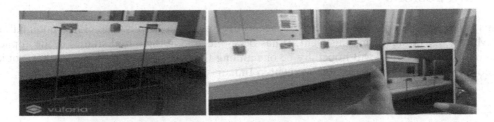

Figure 6.   Plumbing live AR of washroom area.

Figure 7.   Plumbing live AR of washroom area.

4.3   *Live augmented reality*

The next step was to convert select areas of the services into live augmented reality as given in Figure 6, 7. Certain limitations were faced here 1. Lack of skill to do a GPS based AR system. 2. Trackable images could be placed only on vertical faces- thereby making HVAC duct tracking area complex. 3. The viewer should be at 90 degree angle to obtain the correct location.

The final **BIM-AR** model was justified against conventional model in a live-demo mock run to locate a possible pipe line of a selected service of a small scope area. The aim is to understand the benefits obtained by the proposed method in terms of the following aspects:

| Cost benifit | Time benifit | Effort saved | Rework cost avoided |

- 88% of non-users agreed that BIM could be used more heavily in the operations phase.
- Only 55% of users set BIM standards of their own. The rest follow standards put up by a third party.
- Locating building components was the most frequently chosen applicability area criterion by the respondents (91%). Given option to choose more than one answer, facilitating access data was the second most chosen (86%) followed by visualization and marketing. Least chosen option was personnel training and development. The benefits concluded towards quantification of cost (30%), time saving (35%), effort saved (15%), and (20%) rework cost avoided.

5 CONCLUSIONS

This project based thesis work gave evidence to back up the relatively unexplored potential of BIM and AR in facility management. It has been one among the initial studies in the applicability of this field and will hopefully be one among the many foundation papers that would encourage further research on this segment in Indian context.

5.1 *Asset management*

The BIM model has proven to be helpful in maintaining a proper codified list of the plumbing and HVAC assets in the library. The model is updated with the make, model, installation date, warranty, guarantee, last maintenance and next maintenance due date of every element in the building. The list at present is updated with existing plumbing and HVAC data library of the model. This data base of model handed over to the campus archives, the list can be updated and maintained for efficient asset management.Willingness to switch over to the platform was understood from the campus facility managers after a live demo.

5.2 *Benefits of BIM-AR model*

The study has concluded with a live-demo mock drill survey given to a panel of experts including facility managers of the library, the library In charge and faculties of university to quantify the benefits obtained in terms of savings of time, cost, effort taken, rectification cost and time due to erroneous assembly.The responses regarding combined applicability of BIM-AR model showed that 50% of respondents agreed that the platform would be cost beneficial, 62.5% agreed it would be time beneficial, 66.7% agreed it reduced human efforts, and 55.6% strongly agreed and 44,4% agreed it will reduce rework and rework cost.

In total 89% of respondents are willing to implement the proposed system over the existing system where as 11% are confused regarding the applicability of the same as given in Figure 8. (8/9 respondents agreed to shift to the platform, 1/9 respondent was neutral to the question)

The DRP (Directed Research Proposal) is expected to be among the foundation papers for further research in this field. The applicability of BIM-AR in operations and maintenance phase shows wide potential and can be explored to minimize cost, time and human potential wastage.

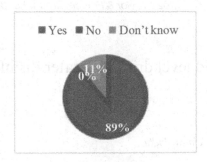

Figure 8.   Willingness to switch over to BIM-AR platform.

REFERENCES

Azhar, Maqsood, and Khalfan. (2012). Building information modeling (BIM): Now and beyond.

Burcin Becerik-Gerber, Jazizadeh, F., Li, N., Calis, Gulben. (2012). Application Areas and Data Requirements for BIM-Enabled Facilities Management. *ASCE.*

Jason Lucas, & Walid Thabet. (2018). Using a Case-Study Approach to Explore Methods for Transferring BIM-Based Asset Data to Facility Management Systems. *Construction Research Congress-ASCE.*

Jing Du, Zhengbo Zou, Yangming Shi, & Dong Zhao. (2017). Simultaneous Data Exchange between BIM and VR for Collaborative Decision Making. *Computing in Civil Engineering ASCE.*

Jurgens, C. (2016, December 20). *Akita Box.* Retrieved from https://home.akitabox.com/blog/integrating-virtual-reality-in-facility-management

Milgram. (1994). Augmented Reality: A class of displays on the reality-virtuality continuum. *Telemanipulator and Telepresence Technologies.*

Sullivian, & Mobley. (2010). Potential utilization of building information models for planning maintenance activities. *International Conference on Computing in Civil and Building Engineering.*

Technologies for Sustainable Development – Mahajan, Patel & Sharma (eds)
© *2020 Taylor & Francis Group, London, ISBN 978-0-367-33737-7*

Developing household level drinking water disinfection unit using copper

Vandit R Shah
Nirma University, Ahmedabad, Gujarat, India

Girivyankatesh Hippargi
NEERI, Nagpur, Gujarat, India

Jaykumar Soni
L.J. Institute of Engineering and Technology, Ahmedabad, Gujarat, India

ABSTRACT: Water is vital part for existence of life and when it comes to drinking water, it should be safe in terms of water borne diseases. In developing countries diarrhoea is major concern for which pathogens in water are responsible. Pathogens enter into water distribution system and groundwater due to various anthropogenic activities even after treatment. In water treatment plant chlorine is commonly used as disinfection agent which is carcinogenic in nature. To prevent pathogens, economic and effective treatment at point of use is required. As per ancient Indian culture and Ayurveda, storing of water in copper vessel prior to consumption provides many health benefits. The brief study was conducted on copper vessels to find out surface area required with respect to optimum time to reach WHO limit of zero E-coli. Physio-chemical parameters such as pH, turbidity, total dissolved solids (TDS), alkalinity, hardness and e-coli concentration were measured at regular time interval before and after storing of water in copper vessel. The bacterial E-coli count was performed using spread plate method and copper concentration was measured using ICP-MS instrument. There was not significant change in water parameters after storage. It was found that E-Coli concentration is mainly responsible for the rate of leaching of copper particles in water. Copper vessel which is having surface area of 640cm^2/litre has achieved 6 log reduction in 7 hours. Further increase in surface area does not cause any reduction in time of disinfection.

1 INTRODUCTION

1.1 *Ground water contamination*

In developing countries like India, people living in rural areas do not have piped water supply. 30.80% of the rural households in India get tap water, and about 70.60% of the Urban Households of the country is covered with tap water supply(Government of India, 2011). In rural areas where piped water supply is not available, people drink water from wells or bore well, which is groundwater.

There is a reason behind using of groundwater as a source of drinking water. Naturally, groundwater ecosystems are well protected by overlying soil and sediment layers. Water from precipitation which recharge groundwater, needs to pass these zone which acts as an effective mechanical and biological filters, therefore providing a natural cleanup of newly generated groundwater. Now a days, groundwater faces increasing threats from anthropogenic impacts of pathogenic micro-organisms(Pedley & Howard, 1997). There is no doubt that pathogenic microorganisms can be found anywhere in the environment. Due to population growth and expanding land use, sources of pathogens contaminated wastes steadily increase hence increasing the potential pollution of ground-water reservoirs with pathogenic agents all around the world. This is mainly true for pathogens originating from human and animal faeces. Improper management of waste-water disposal is the top of the list responsible for contamination of groundwater by pathogens (Willocks et al., 1998). Proper wastewater management, modern sanitation, and groundwater protection strategies prevent contamination of groundwater. Unfortunately, this is not implemented in developing countries. Poverty and over-population are the most noticeable reason of hygiene and sanitation.

1.2 *Necessity of disinfection*

The process of killing pathogens is known as disinfection. The killing of these pathogens is crucial before consumption. Otherwise, it causes diseases such as cholera, typhoid, and diarrhea (Shukla & Mishra, 2015). Among them, diarrhea remains the leading cause of death in children under five years, killing an estimated 321 children every day in 2015 (World Health Organization, 2017). Every year India spends Rs 3.4 billion on treating diarrhea. An estimated 4.1% of the total global burden of diseases is contributed by diarrhea illness (WHO). Despite the active and effective treatment of wastewater, pathogenic microorganisms, and viruses are frequently introduced into the aquatic environment (Feichtmayer, Deng, & Griebler, 2017). Providing drinking water free from microbial contamination is one of the fundamental measures to prevent diarrhea (Walker, Perin, Aryee, Boschi-Pinto, & Black, 2012).

1.3 *Copper as disinfectant*

The Indian Ayurveda suggests storing drinking water in a copper vessel for overnight and consumes it in the morning for health benefits. Generally, people have the practice to store water in a clay pot, but it was not found effective in the destruction of pathogens as copper(Radhar & Susheelap, 2015). Copper has tremendous health and spiritual benefits for humans (Noyce, Michels, & Keevil, 2006). Copper has antimicrobial, antioxidant, anti-inflammatory and anti-carcinogenic properties. Water stored in a copper vessel for overnight does not produce any E-coli colonies. Copper is an ideal material for charging water positively and balancing three doshas of our body, i.e. Vata, pitt, and kapha. Modern studies have demonstrated that copper is a micronutrient that is necessary for the synthesis of Haemoglobin and is an essential component of many enzymes. It is also essential in embryo development, mitochondrial respiration, hepatocyte & neural function and regulation of haemoglobin(Krupanidhi, Sreekumar, & Sanjeevi, 2008). Copperdeficiency is indicated in anaemia, neutropenia (Thorvardur R) and bone abnormalities (Garbreyes). As per WHO guidelines amount of copper in drinking water should not be more than 2000 ppb(Edition, The, & Addendum, n.d.).

1.4 *Destruction of pathogens of copper*

Toxic effect of copper ion on micro-organisms is known as "Oligodynamic Effect." Previous studies on antimicrobial activity of copper have confirmed that ions are responsible for inactivation of bacteria. However, there is no clear explanation has been found for the inactivation mechanism taking place within the cells of microorganisms. Some reports suggest that metal ions bind to DNA, enzymes, and cellular proteins in the bacteria causing cell damage and death (Yamanaka, Hara, & Kudo, 2005). A recent study on the copper surface suggests that hydroxyl radicals in the solution are responsible for their lethal action(Santo, Quaranta, & Grass, 2012). Copper may inactivate bacteria by more than one mechanism(Santo, Taudte, Nies, & Grass, 2008). However, it is not known how they contribute to the actual detoxification mechanism.

2 MATERIALS AND METHODOLOGY

2.1 *Copper fabrication and experimental design*

One of the factors that control the rate of reaction with the pathogen is the surface area of the meta, which is in contact with water. Larger area destroys coliforms at a faster rate(Varkey, 2010). One might think to use the maximum amount of surface area and kill all pathogens within a few seconds. However, at the same time, the volume of water to be treated will be very less and uneconomical. Actually, by experiments, these were essential findings:

(1) Leaching of copper ions requires a minimum time to initiate the reaction
(2) Rate of leaching becomes constant after some surface area.

  Therefore, experiments were conducted with different surface areas by considering the time to reach zero E-coli count with desired Cu concentration. For this experimental study, we have designed and fabricated three different cylindrical vessels which are having a diameter(d) of 6.25cm, 7.5cm, 10cm, 5cm, 3.75c, and 2.25 cm with the same length of 150cm. Surface area is calculated in cm2/litre. Their surface area is 640 cm2/litre, 533 cm2/litre, 400 cm2/litre, 800 cm2/litre, 1066 cm2/litre and 1077 cm2/litre respectively.

2.2 *Framework of experiment*

Experiments were conducted on the water, having a concentration of E-coli of 2.4×10^6 CFU/ml. The reason behind this was to check the amount of copper leached in water. It may be possible that copper leached in water may exceed the WHO limit of 2000 ppb.Bacteria were cultured in tap water, so there was an increment in concentration. Generlly, in the worst case of drinking water, e-Coli is not more than logt 6. In the case of Mohgaon village (Nearby Nagpur), it was of log 5. Physio-Chemical parameters such as pH, turbidity, total dissolved solids (TDS), hardness, and e-coli are measured before andafter storing water. After storing water in a copper vessel, the sample has been collected t the end of 2hs, 4hs, 6hs and 7hs. In those samples E-coli has been checked using spread plate method and copper is checked using ICP-MS (Inductively Coupled Plasma Mass Spectrometry).

3 RESULT AND DISCUSSION

The water sample was taken at 2hs, 4hs, 6hs, and 7hs to check E-coli concentration and amount of copper leached.

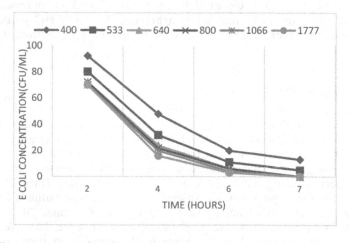

Figure 1. E-coli concentration(CFU/ml) vs time (hours).

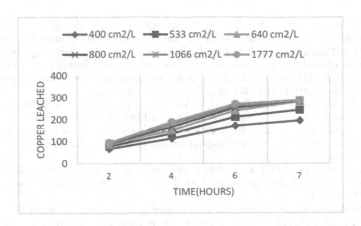

Figure 2.   Copper leached(ppb) vs time (hours).

From the graph it can be seen that the patent of killing bacteria is the same. However, increasing surface area after certain 640 cm^2/litre, there is no change in a number of bacteria killed. It requires 7 hours to kill all bacteria. Moreover, it can be seen that after the surface area of 640 cm^2/litre, there is a negligible increase in leaching of copper and consequently, no change in the rate of killing. It can be said that the surface area about 640 cm^2/litre is optimum.

4 CONCLUSION

Leaching of copper ions and consequently, the rate of killing pathogens depends on surface area and concentration of e-coli. Rate of killing increases up to the surface area of 640 cm2/litre, afterward an increase in the surface area does not increase in the killing of bacteria. Time taken to reach zero e-coli count is 7 hours in a copper vessel of 640 cm2/litre. Rate of copper leaching and killing of bacteria is the same in different concentration of e-coli means a copper vessel of 640 cm2/litre takes 7hours to reach zero e-coli count in different concentration of 2.4×104 and 2.4×106. It is advisable to use copper bottle having a diameter of 6.25cm which is having a surface area of 640 cm2/litre. There is no significant change in other water parameters such as pH, turbidity, total dissolve solids (TDS), alkalinity, and hardness.

REFERENCES

1. Arnone, R. D., & Walling, J. P. (2007). Waterborne pathogens in urban watersheds. *Journal of Water and Health, 5*(1), 149–162. https://doi.org/10.2166/wh.2006.001
2. Barrett, M. H. (2003). A REVIEW OF THE EFFECTS OF SEWER LEAKAGE ON G R 0 U N D WATER QUALITY, (March).
3. Chowdhury, S., & Champagne, P. (2009). The risk from exposure to trihalomethanes during the shower: Probabilistic assessment and control. *Science of the Total Environment, 407*(5), 1570–1578. https://doi.org/10.1016/j.scitotenv.2008.11.025
4. Edition, F., The, I., & Addendum, F. (n.d.). *No Title.*
5. Feichtmayer, J., Deng, L., & Griebler, C. (2017). Antagonistic microbial interactions: Contributions and potential applications for controlling pathogens in the aquatic systems. *Frontiers in Microbiology.* https://doi.org/10.3389/fmicb.2017.02192
6. Gerba, C., & Smith, J. (2005). Sources of pathogenic microorganisms and their fate during land application of wastes. *Journal of Environmental Quality, 34*(1), 42–48. https://doi.org/10.2134/jeq2005.0042
7. Government of India. (2011). *Census of India 2011. State of Literacy.* https://doi.org/10.2105/AJPH.2010.193276

8. Hrudey, S. E., Backer, L. C., Humpage, A. R., Krasner, S. W., Michaud, D. S., Moore, L. E., ... Stanford, B. D. (2015). Evaluating Evidence for Association of Human Bladder Cancer with Drinking-Water Chlorination Disinfection By-Products. *Journal of Toxicology and Environmental Health - Part B: Critical Reviews*, *18*(5). https://doi.org/10.1080/10937404.2015.1067661

9. Kirschner, D. E., & Linderman, J. J. (2009). Multiscale modeling in the immune system. *Cell*, *11*(4), 531–539. https://doi.org/10.1111/j.1462-5822.2008.01281.x.Mathematical

10. Krupanidhi, S., Sreekumar, A., & Sanjeevi, C. B. (2008). Copper & biological health. *Indian Journal of Medical Research*, *128*(4), 448–461. https://doi.org/10.1080/10937400600755911

11. Noyce, J. O., Michels, H., & Keevil, C. W. (2006). The potential use of copper surfaces to reduce survival of epidemic meticillin-resistant Staphylococcus aureus in the healthcare environment. https://doi.org/10.1016/j.jhin.2005.12.008

12. Pedley, S., & Howard, G. (1997). The public health implications of microbiological contamination of groundwater. *Quarterly Journal of Engineering Geology*, *30*, 179–188. https://doi.org/10.1144/GSL.QJEGH.1997.030.P2.10

13. Radhar, & Susheelap. (2015). Comparative Microbiological Analysis of Water Stored in Different Storage Vessels. *Int J Pharm Bio Sci*, *6*(2), 121–128. Retrieved from http://www.ijpbs.net/cms/php/upload/4108_pdf.pdf

14. Santo, C. E., Quaranta, D., & Grass, G. (2012). Antimicrobial metallic copper surfaces kill Staphylococcus haemolyticus via membrane damage. *MicrobiologyOpen*, *1*(1), 46–52. https://doi.org/10.1002/mbo3.2

15. Santo, C. E., Taudte, N., Nies, D. H., & Grass, G. (2008). Contribution of copper ion resistance to the survival of Escherichia coli on metallic copper surfaces. *Applied and Environmental Microbiology*, *74*(4), 977–986. https://doi.org/10.1128/AEM.01938-07

16. Shukla, P. B., & Mishra, M. K. (2015). Antimicrobial Activity of Supported Silver and Copper against E. coli in Water, *1*(1), 11–15.

17. Varkey, A. J. (2010). Antibacterial properties of some metals and alloys in combating coliforms in contaminated water. *Scientific Research and Essays*, *5*(24), 3834–3839.

18. Walker, C., Perin, J., Aryee, M. J., Boschi-Pinto, C., & Black, R. E. (2012). Diarrhea incidence in low- and middle-income countries in 1990 and 2010: a systematic review. *BMC Public Health*, *12*(1), 220. https://doi.org/10.1186/1471-2458-12-220

19. Willocks, L., Crampin, A. Milne, L., Seng, C., Susman, M., Gair, R., Lightfoot, N. (1998). A large outbreak of cryptosporidiosis associated with a public water supply from a deep chalk borehole. Outbreak Investigation Team. *Communicable Disease and Public Health*, *1*(4), 239–243.

20. World Health Organization. (2017). WHO | Diarrhoeal disease. Retrieved from http://www.who.int/mediacentre/factsheets/fs330/en/

Technologies for Sustainable Development – Mahajan, Patel & Sharma (eds)
© 2020 Taylor & Francis Group, London, ISBN 978-0-367-33737-7

Accessibility of BRTS station in Ahmedabad city using GIS: A case study

P.R. Patel
Professor, Civil Engineering Department, Nirma University, India

Fatima Electricwala
P.G. Student, M. Tech by Research (Geomatics), India

ABSTRACT: Accessibility Analyst tool of GIS is employed to analyze the optimum accessibility distance to BRT station around its immediate neighborhood. The accessibility to transit network is a major issue in the planning of efficient bus transit, which affects the ridership significantly. The present study analyses the catchment area, a distance-decay buffer zone with actual street network and interoperates with GIS spatial data. Three corridors with 27 existing stations, field data was collected, which includes the distance between the stations, intersection distance, walking distance of commuter from different modes, age, gender, type of employment, the purpose of the trip, travel time to reach station and income of users. The analysis showed that the station locations is an integral part and resulting in spatial distributions of accessibility over a neighborhood. A GIS analysis gives an opportunity to BRT operators, commuters and authorities to spatial visualization of commuter's accessibility from stations to its neighborhood.

Keywords: BRTS, stations, commuter ridership, GIS, accessibility

1 INTRODUCTION

Commuter's accessibility is an important link between the bus rapid transit (BRT) and land use pattern of the immediate neighborhood (Tiwari et. al. 2016). Accessibility is also an important element in travel demand estimation and planning. The efficient public transit system is one of the significant measures to diminish the expansion in individual motorized dependency. (Gutierrez et.al. 1998). The public transit system is diversified, and every one of its parts must be all around verbalized with a specific end goal to have a lively and efficient public transport. (Alshalalfah et. al. 2007). The accessible public transport services upgrade transit network mobility, spare fuel, reduces congestion, increases commuter ridership. (Gahlot et. al. 2012). Further, the transport demand highly relies on accessibility to the BRT station.

It is necessary to streamline a transit systems interface with other modular alternatives, as system planner amplifies the potential commuters to make the system economically suitable ((NUTP, 2005). The public transport system does not end at the bus station instead of incorporating the entire potential commuter catchment area (Tiwari et. al. 2016). The first contact of the commuter to with the transit service is its bus stop.

Effective transit system incorporates assessment of existing transit station locations. For public bus transport, deciding the maximum accessible station's location is vital. Lack of accessibility to access station location always arises the possibility that they will stop to be a customer for public transit (Gahlot et. al. 2012). The approach accessibility and the state of walking to and from BRT station location decides the ridership. The accessibility analysis should encompass spatial and socioeconomic aspects required to plan public transit system.

The location of stops should be optimally located from its immediate neighborhood (Chien, S. I. et. al., 2004). In measuring the accessibility of an existing system, the tools of GIS such as spatial access to catchment areas presently without adequate access to a public transit stop can be applied. Spatial accessibility through service area analysis feature of Geoinformation System of BRTS stop location is seen as one of the key attributes with coordinate effect on the quality and usage of public transport in urban areas. (Gutierrez et.al. 1998).

Looking into India's urban mode choices, walking mode needs to be enhanced for public transit system, because the vast majority of the urban communities having noteworthy mode share as a walk (Tiwari et. al. 2016). The station location is an essential aspect for effective operational purposes and will guarantee far-reaching accessibility to all classes of the commuters. This research utilizes the apparatuses of Geographic Information System (GIS) in the assurance of the accessibility of the existing BRT stops location. The aim of this study is to find the best location of BRT station by applying Network analyst tools of GIS.

2 A CASE STUDY OF AHMEDABAD BRTS

Ahmedabad city, has expanded area wise by 144% and in accordance with the population it has grown by 58% since 2006. The city grew by about one and a half million people by the last census in 2011. Further, by 2021 AMC has predicted around 9 million people shall be residing in Ahmedabad (AMC, 2018). Ahmedabad experienced a significant increase in the number of motorized vehicles between 2002 and 2009 at a rate of 9.2% per annum (Ministry of Road Transport and Highways 2011: 4)

Public transport in Ahmedabad turns up to only 22% of the total trips of commuters. According to AMC (2018) the count of commuters' trip using the public transportation in Ahmedabad is only 16%. Ahmedabad Janmarg Limited operates the BRTS that covers only 86 km 14 (131 bus stations). The three study corridors were finalized after looking at secondary socio-economic data and conducting an onboard trial survey using BRT commuter as shown in Figure 1.

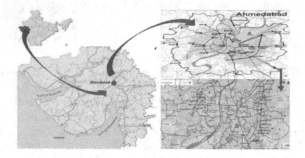

Figure 1.   Study area showing Ahmedabad BRTS route corridor.

3 METHODOLOGY

The three routes comprising 27 stops and all existing geospatial datasets represented by localized data about areas of individual station location, BRT route, the population of occupants of that zone, design of neighborhood and service roads, streets and intersections was collected from the "Janmarg" BRTS service company (AMC, 2018), Ahmedabad, Gujarat. The data related to commuters at the neighborhood, on-board and at the station was acquired by stated preference survey conducted, the respondents were requested to indicate the accessibility distance to the station from origin/to the destination, the purpose of the trip, travel time, gender, occupation, age, travel time and mode of travel. Sample size was selected based on Cochran formulae.

The population density of coverage area represents a spatial coverage buffer. The network area analyst and distance-decay function in ArcGIS 10.1 software were used. The major methodological restriction of the buffer specification tool of GIS is the way that the buffer area doesn't regard any ecological boundaries, (for example, waterworks, street intersections, structures, and so forth.). The significantly summed up and ineffectively corresponds to reality as there were no ecological boundaries found in the study area. The threshold values for station accessibility in the urban area varies from 5 to 7 minutes walking distance, which represents a linear distance within 400 meters (O'Sullivan et. al. 2000). Based on studies, the linear distance of 150, 400, 600 and 800 m maximum was considered for buffering in the service area. In this study, for the sake of simplicity, the population outside the 1200 m range was not considered.

4   RESULTS AND DISCUSSION

The identification of existing station was acquired through primary source data and represented as point features are ArcGIS software. In all, Twenty-Seven (27) stops were identified from selected three different study stretches of Ahmedabad City as shown in Table 1.

All the data layers were converged in ArcGIS 10.1 and were subjected to analysis by Service Area Analyst" function of GIS. All individual BRT station were iterated at the distance of 30 m, at an interval of 150 m. 400 m, 600 m, 800 m, and 1200 m) from its existing location to the proposed location where the maximum percentage of ridership may be observed. Distance decay graphs were plotted by observing the commuters that can easily access the selected transit stops. It was observed from the analysis that commuters increase as the bus stop location is moved towards the catchment area. The descriptive statistics of access distances of corridor-I is shown in Table 2.

Table 1.   Ahmedabad study stretch details.

Details of Route	Name of Urban Route		
	Jasodanagar - Virat Nagar (corridor-I)	Virat Nagar - Naroda Patiya (Corridor-II)	Anjali - Memnagar (Corridor-III)
Corridor lenght (km)	5.4	4.4	6.0
Number station) No.)	10	7	10

Table 2.   Descriptive statistics of access distances.

Variables	Virat Nagar	Lila Nagar	Thakkar Nagar	Hira vadi	Vija park	Krishna Nagar	D.D. Mandir	Naroda Patiya
Mean	1662	1286	1606	1410	1397	1473	1321	1667
Median	1461	1135	1527	1331	1262	1349	1209	1565
Maximum	2337	1823	2184	1998	2033	1986	1867	2377
SD	521	610	489	579	678	558	560	499
85th	2244	1736	2168	1904	1886	1989	1783	2250
R^2	0.796	0.826	0.786	0.720	0.746	0.781	0.778	0.891
P Value	0.00	0.00	0.00	0.00	0.00	0.00	0.00	0.00
Total Sample size = N (1766)	266	186	285	206	167	244	137	275

Independent Variables: Income, Vehicle Ownership, Occupation Pattern, Household size, Age, Gender
Dependent Variables: Access Distance (In metres)
Note: D. D. Mandir = Dhanush Dhari Mandir

Mean values for Thakkarnagar (592m), Vijay Park (588m) and Naroda Patiya (596m) are higher which indicates the people are coming from a longer distance to these stations. The median access distance for five bus stops were near by 450-500m. Amongst them distance of Krishnanagar is 402m. Dhanush Dhari Mandir (415) have lower as compared to Thakkarnagar (513m). It is also observed that Naroda Patiya (539m) have higher median distance value. The 85th percentile (considers as existing service area) for all stations was calculated for the study. Table 2 shows the summary statistics of access distances to transit bus stops as well as mean values or percentage values of independent variables. The 85th percentile of access distance to Thakkarnagar (799m), Vijay Park (794m) and Naroda Patiya (805m), for Viratnagar, Lilanagar, Hiravadi, and Krishnanagar show 700-740m of their present access distance.

Further, to know the relationships between independent variables and dependent variables, the regression results were obtained between predictor variables with the access distance for eight transit bus stops for Corridor-I. The p-values for all eight bus stops is (p = 0.00) were statistically significant results also shown in Figure 2 (a & b). Similarly, another corridor descriptive analysis was performed, and the average distance is mentioned in Table 3. In the analysis, it was observed that six (6) stations out of 27 stations need to be relocated.

The analysis shows that based on Network analysis results for corridor I, only CTM and Expressway BRTS stops needs to be re-located and no other bus stops. The analysis shows the population density increases from 50 to 60%. Except for two stations other stops of stretch 1 are judiciously located, which is near to the intersection and is shown in Table 4. The analysis shows that the relocation of Thakkar Nagar and Krishnanagar stops are necessary. Similarly, Jhansi Ki Rani and University stations need relocation as shown in Figure 3. The distance decay function exponential curves were plotted for cumulative percentages of access distances from origin to the newly relocated transit bus stops. The estimated values for the Six stations show an increase in population after relocation, and the results show R^2 value and equation as mentioned in Table 5.

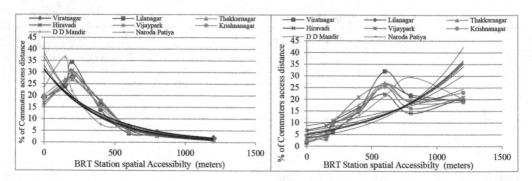

Figure. 2. (a): Accessibility of existing station (Before). (b): Accessibility after the relocated station.

Table 3. Commuter at increment at various distance from bus stop location.

Buffer Radius Distance	0 m	150 m	400 m	600 m	R^2	Exponential Equation
30 m	7	19.59	25.45	32.44	0.7177	y = 54.05e-0.005x
60 m	5	24.45	28.74	37.12	0.8436	y = 67.469e-0.004x
90 m	4	35.45	31.45	28.12	0.8865	y = 76.336e-0.003x
120 m	7	27.45	45.50	27.33	0.9439	y = 88.052e-0.003x
150 m	5	63.45	30.20	28.29	0.9545	y = 97.839e-0.002x

Table 4. Virat Nagar- Naroda Patiya population density in GIS.

Sr No	Stop Location \ Buffer Distance	150 m	400 m	600 m	800 m
1	Lilanagar	-1.1	-11.7	-10.8	-12.8
2	**Thakkarnagar**	**+40.8**	**+30.9**	**+17.5**	**+7.9**
3	Hirawadi	+2.1	-3.5	-0.1	-0.1
4	Vijay Park	-24.1	+12.1	-42.2	-52.2
5	**Krishnanagar**	**+55.2**	**+40.2**	**+23.4**	**+12.4**
6	Dhanushdhari	-0.3	+2.0	-28.2	-30.3
7	Naroda Patiya	+2.1	-7.3	+1.6	+1.8

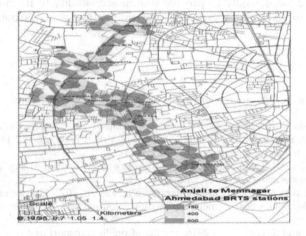

Figure 3. Service area analysis of study Corridor-III.

Table 5. Distance decay function exponential curves & R^2.

S No	Stop Location	R^2	Equation
1	Vadodara Express Way	0.967	$y = 88.57e^{-0.003x}$
2	CTM	0.955	$y = 97.84e^{-0.002x}$
3	Thakkarnagar	0.913	$y = 81.67e^{-0.003x}$
4	Krishnanagar	0.850	$y = 143.43e^{-0.003x}$
5	Jhansi Ki Rani	0.895	$y = 83.21e^{-0.002x}$
6	University	0.962	$y = 90.90e^{-0.002x}$

The results obtained on network analysis shows that the optimizing of bus stop location by relocating it towards commuter catchment area can be very essential and important for increasing ridership. In this analysis increase in commuter ridership in range of 40 to 127 % at network analyst distance of 150m is observed on relocating all the 6 bus stops. Also, increase in ridership was observed at 400m and & 600m in range of 19 to 50 % and 16 to 38 % respectively.

5 CONCLUSIONS

This paper presents a GIS-based tool that has been applied to analyze the accessibility of the stations based on spatial available data. The comparison of accessibility patterns for different locations was made quickly in three corridors. The optimized results have been compared with the existing situation and have shown a good improvement on transit service coverage and increase in population density for six relocated stops. The distance standard (150, 400 m, 600 m) is relevant too. However, as it reflects by due survey, commuters are not willing to walk for a distance more than 200 to 300 m. The distance decay graph results show that the % of the population is reducing with increase in distance from immediate neighborhood to the transit stop. The analysis indicates that the existing average catchment area of all stops have increased up to 600 m. it was observed that when selected stops were relocated at 30m intervals there was increase in commuter ridership by 50 to 60% at buffer distance of 0m to 150m and around 34% at the buffer distance of 150m to 400m and around 21% when stations were relocated to the distance of 400 m to 600m. Results revealed that the optimum location is quite different from the presently located six stations in three stretches. Finally, results suggest that it may be more user-friendly to provide station accessibility to its immediate neighborhood that leads to enlarging station catchment buffer areas around its immediate neighborhood.

REFERENCES

Ahmedabad Municipal Corporation. Janmarg-Ahmedabad Bus Rapid Transit System: Accessible, Ahmedabad 2011.
Alshalalfah, B. & Shalaby, A. 2007. Case Study: Relationship of Walk Access Distance to Transit with Service, Travel, and Personal Characteristics. Journal of Urban Planning and Development, Journal of Urban Planning and Development-ASCE. 133: 2 (114).
Chien, S. l., & Qin, Z. 2004. Optimization of Bus Stop Locations for Improving Transit Accessibility. *Transportation Planning and Technology*, 27 (3): 211–227.
Gahlot, V., & Swami, B.L. & Parida, Manoranjan & Kalla, Pawan. 2012. User oriented planning of bus rapid transit corridor in GIS environment. International Journal of Sustainable Built Environment. 1. 102–109.
Tiwari, G. Deepty, J. and Kalaga, R.R. 2016. Impact of public transport and non-motorized transport infrastructure on travel mode shares, energy, emissions, and safety: Case of Indian cities. *Transportation Research Part* D, 44: 277–291.
Gutiërrez, J., Monzön, A., Pinëro, J. M. 1998. Accessibility, network efficiency, and transport infrastructure planning. *Environment and Planning* A, 30: 1337–1350.
NUTP. 2005. National Urban Transport Policy, Ministry of Urban Development, Government of India.
O'Sullivan, D., Morrison, A. and Shearer, J. 2000. Using desktop GIS for the investigation of accessibility by public transport: an isochrones approach. *International Journal of Geographical Information Systems*, 14 (1): 85–104.

Technologies for Sustainable Development – Mahajan, Patel & Sharma (eds)
© *2020 Taylor & Francis Group, London, ISBN 978-0-367-33737-7*

Shear strength parameters of soil and banana fiber mix

H. Modha*, N. Sharma & H. Chavda
Civil Engineering Department, Institute of Technology, Nirma University, Ahmedabad, India

ABSTRACT: Soil acts as a foundation material for most structure that we see around us. The properties of soil plays an important role in transferring the load from the foundation to subsoil. The soil is a natural and complex material, therefore sometimes the properties of soil required to be modified using various techniques. One of such technique is, the use of natural fibers with soil. The advantages of using natural fibers like coir, jute, sisal, bamboo, hemp, bagasse, straw, banana etc., are freely available in nature, low cost, less in density and relatively high resistance to tension and bending. Due to these advantages, natural fiber material can be used as a reinforcement in soil which enhances the properties of soil. For this study the banana fibers were selected due to its ease availability in large quantity as waste product of banana cultivation in Gujarat. The aim of the present study is to determine the shear strength parameter of soil, with and without randomly mixed banana fiber. The banana fiber is mixed with soil having proportion of 0.5%, 1 % and 1.5% of dry weight of soil. The shear strength parameters, cohesion-c and angle of internal friction -φ values have been increased for soil with 0.5% banana fiber compared to unreinforced soil.

1   INTRODUCTION

Soil is a natural building material, which has number of applications in construction and geo-technical field. In the past soil has been used as a major construction material. Sometimes, the soil properties are not suitable or favorable for the construction of the foundation of structure. Therefore, different soil improvement techniques have been used to stabilize the soil before the construction. The soil can play a major role to fulfil the increasing demand of sustainable material through various application like compressed earth blocks, rammed earth techniques, filling material, mud concrete etc. The properties of soil are modified according to the requirements of the specific application. Sometimes the conventional soil improvement techniques like mechanical or chemical soil stabilization create some geo-environmental issues related to global warming, land, water or soil pollution reported by Sakaray.H et al. (2012), Al-Swaidani A. et al. (2016) & Dilrukshi R.A.N (2016) .Therefore eco-friendly techniques are required to modify the soil properties. Vidal in 1969 invented, the application of fiber in soil, increase the shear resistance of soil which was known as principle of soil reinforcement. The soil reinforcement techniques using different types of natural and synthetic fibres have been already implemented in many countries concluded by Hejazi S.M et al. (2012) .There are variety of fibers available for the reinforcement application like synthetic fibers such as carbon, glass, metal etc., and natural fibers such as coir, sisal, jute, hemp, rice, banana etc. Gowthaman S. et al. (2018) concluded that the fiber reinforment system is divided in two types, such as random orientation of fiber and oriented distribution of fiber sheets. As per the recent requirements of sustainability approach in engineering field, the use of natural fiber in soil has great potential to be used in geotechnical and construction material field. The major disadvantage of natural fiber is, its bio-degradable nature, which can be overcome by treating it with chemicals to increase the resistance to degradation. The natural fibers are treated with various chemical coatings prior to the

---

* Corresponding author: hiral.modha@nirmauni.ac.in

Table 1. Biochemical and physical properties of banana fiber.

Species	Fiber origin	Cellulose (%)	Hemicellulose (%)	Lignin (%)	Reference
Musa indica	leaf, stem	48–59	12-15	14-21	Preneron, A.L et al. and Satyanarayana, K.G et al.
Density (kg/m^3)	Young's Modulus (GPa)	Ultimate Tensile Strength (Mpa)	Elongation at Break (%)		Satyanarayana, K.G et al., Rassiah, K et al., Xu, Y. et al.
1350	27-32	711-799	2.5-3.7		

soil reinforcement applications reported by Gowthaman S. et al. (2018). The coir and jute/bamboo are treated with tetracloromethane, bitumen coatings respectively. Soil reinforcement through fibers can enhance the soil properties like increase in cohesion, decrease in settlement, increase in shear strength of soil etc. The use of natural fibers can prove to be more advantageous as their ease availability, less in cost, good tensile and bending properties. The use of natural fibers for rural development through pavements, embankments, temporary structure, compressed soil fiber blocks for low income housing etc., are the major areas of application.

In the present study banana fiber has been selected as a natural fiber. Banana is a globally important fruit crop with 97.5 million tons of production, In India it supports livelihood of millions of people. With total annual production of 16.91 million tones (vikaspedia.in, 2015). Marwan Mustafa and Nasim Uddin (2016) reported case study of compressed earth block with banana fiber.The use of banana fibers shown better resistance against compression and bending. Moreover due to bio-chemical and mechanical properties of the banana fiber, its potential has been identified for the use of soil reinforcement as suggested by Shivkumar Gowthaman et al. (2018) Banana fibers are a waste product of banana cultivation and available from the fast growing and having high biomass plat. Banana fibers are ecofriendly and have characteristics of low density, low cost, tensile strength, water and fire resistance. Therefore, it has potential to be used as a reinforcement in soil. The bio-chemical and physical properties of banana fibers analysed by researchers are given in Table 1. These properties vary according to types of species and origin of fiber. D.Jagdeesh et al. (2015) have reviewed the banana fiber and their properties, its extraction process and chemical treatment given to banana fibers.

2 MATERIAL AND METHODOLOGY

2.1 Materials

The materials used in the present study are soil and untreated banana fiber. The soil studied in this paper is collected from the Banswara, Rajasthan, India. The soil sample is taken from the 0.5 m depth from the ground. After collecting the soil, it was taken to the laboratory and kept in oven for 24 hours at 105 to 110 ° C to make it dry. The characterization tests were carried out for the soil. Classification tests included visual identification, wet sieve analysis, liquid limit, and plastic limit test. The index properties of soil were found out and given in Table 2. The banana fibers used in the study are shown in Figure1. The fibers are cut approximate in 1 cm length and mixed randomly with soil. The average diameter of banana fiber is 200 μm reported in the literature. The banana fiber is purchased from the commercial vendor from the Maharashtra region.

2.2 Methodology

In the present study series of experiments were done on the soil with and without randomly mixed banana fiber.The direct shear box test was performed to find out the shear strength parameters, c and φ of soil and banana fiber mix. The banana fibers used in the study with proportion of 0.5%, 1% and 1.5% of dry weight of soil. The shear strength was found out with the application of normal stress at 0.5 kg/cm^2, 1 kg/cm^2 and 1.5 kg/cm^2.The rate of displacement was kept at 1.25 mm/minute. The direct shear box test was performed as per IS 2720-13.

Table 2. Index properties of soil.

Sr.No.	Description	Test Results
1.	Color of soil (visual identification)	Blackish
2.	Specific Gravity	2.57
3.	Texture (visual identification)	Fine grain soil
4.	% Sand	30%
5.	Liquid Limit	48%
6.	Plastic Limit	22.2%
7.	Soil classification	CI
8.	Optimum Moisture content	20%
9.	Maximum dry density	1.58 gm/cc

Figure 1. Banana fiber.

Figure 2. Direct shear box test sample failure of banana and soil fiber mix.

3 RESULT AND DISCUSSION

3.1 *Shear strength analysis*

The results of direct shear box test are represented in Table 3. The sample of banana fiber and soil mix after failure is shown in Figure 2. The shear stress vs horizontal displacement of soil with and without banana fiber are presented in the Figure 3.The strength envelope for soil with and without banana fibers are presented in Figure 4. The change in cohesion and angle of internal friction for soil with and without banana fibers of proportion 0.5%, 1% and 1.5% are presented in Figure 5 and Figure 6 respectively. From the results, it can be derived that the inclusion of banana fibers with soil modified the shear strength of soil. Hence the optimum dose of the banana fiber to add in soil, is important to modify the soil properties.

3.2 *Effect on shear stress and horizontal displacement behavior*

Shear stress versus horizontal displacement graph represents a bilinear plot for all the normal stress values 0.5 kg/cm^2, 1 kg/cm^2 and 1.5 kg/cm^2. Plastic behavior starts near the yield shear

Table 3. Results of direct shear box test.

Sr No.	Banana fiber and soil mix	Shear stress (kg/cm^2) with Normal Stress (kg/cm^2) application 0.5, 1 , 1.5			Cohesion-C (kg/cm^2)	Angle of Internal friction φ (degree)
1	BS0 Banana fiber 0% + soil mix	0.69	0.77	0.88	0.59	9.09
2	BS0.5 Banana fiber 0.5% + soil mix	0.74	0.83	0.89	0.70	10.20
3	BS1 Banana fiber 1 % + soil mix	0.62	0.73	0.77	0.53	11.31
4	BS1.5 Banana fiber 1.5% + soil mix	0.43	0.61	0.67	0.38	6.84

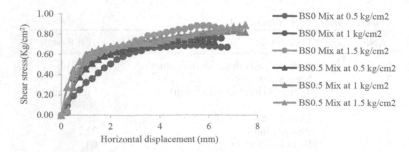

Figure 3.   Shear stress vs horizontal displacement.

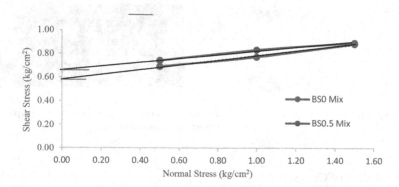

Figure 4.   Shear stress vs normal stress (Strength envelope of banana fiber and soil mix).

value and then after a drop occurs in rate of increase in the shear stress over a wide horizontal displacement compared to the elastic zone as shown in Figure 3.

3.3 *Effect on strength envelope*

The strength envelope of soil and soil with 0.5% banana fibers represents increment in both cohesion-c and angle of internal friction-φ, which indicates that the shear strength increasing with the inclusion of 0.5% of banana fibers with soil. From the results of direct shear box test,

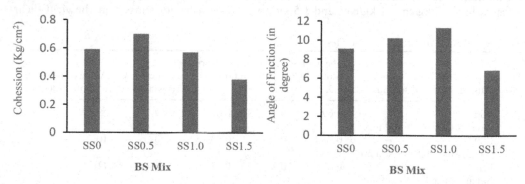

Figure 5.   Cohesion of banana fiber and soil mix.   Figure 6.   Angle of internal friction of banana fiber and soil mix.

as represent in Table 3, shows that the shear strength decreasing at 1% and 1.5% of banana fiber- soil mix. From the behavior of the soil and banana fiber mix, the shear strength can be increased with increasing banana fiber between 0.5% and 1%. From the present study, the optimum dose of the banana fiber is 0.5% of dry weight of soil. However the optimum dose can be identified between 0.5 to 1%. The increment in shear strength at 0.5% banana fiber soil mix is due to the soil reinforcement effect. The randomly placed fibers are subjected to tension, which increase the shear strength of specimen. Random distribution of fiber have adhesive bonding between soil and fibers which create the additional composite strength and helps in increasing the strength of soil structure integrity. The shear strength of soil and banana fiber mix, reduced at 1% and 1.5% proportion because of the friction component is increased between the fibers and soil particles and cohesion is decreased between the soil particles.

4 CONCLUSION

The inclusion of banana fibers in soil increases the shear strength parameters, the cohesion and the angle of friction with 0.5% banana fiber. The cohesion- c and angle of internal friction-φ have increased 18.64% and 12.21% respectively with addition of 0.5%banana fiber compared to unreinforced soil. The shear strength parameters as c and φ values have been increased at 0.5% and deceased at 1% and 1.5 % of banana fiber. Hence the shear strength of banana fiber and soil mix relatively increased compared to the soil without banana fiber. Therefore it can be concluded that optimum dose of banana fiber in soil is between 0.5% to 1%. There is a further scope to determine the optimum proportion of the banana fiber between 0.5% and 1%. However the banana fiber has the potential to be used as a soil reinforcement and the biodegradable nature of banana fiber can be enhanced with some suitable chemical treatment for its long time performance.

REFERENCES

Sakaray H., Togati N.V.V.K., Reddy I.V.R, 2012, Investigation on Properties of Bamboo as Reinforcing Material in Concrete. Int. J. Eng. Res. Appl. 2, 077–083.
Al-Swaidani A., Hammoud I., Mezia A., 2016, Effect of adding natural pozzolana on geotechnical properties of Lime-stabilized clayey soil. Rock Mech. Geotech. Eng. 8, 714–725.
Dilrukshi, R.A.N., Watanabe, J. Kawasaki, S., 2016, Strengthening of Sand Cemented with Calcium Phosphate Compounds Using Plant-Derived Urease. Int. J. GEOMATE, 11, 2461–2467.
Vidal H., 1969, The principle of reinforced earth. Highw. Res. Rec., 282, 1–16.
Hejazi S.M., Sheikhzadeh M., Abtahi S.M., Zadhoush A., 2012, A Simple Review of Soil Reinforcement by Using Natural and Synthetic Fibers. Constr. Build. Mater., 30, 100–116.
Gowthaman Sivakumar, Nakashima Kazunori, Kawasaki Satoru, 2018, A State-of-the-Art Review on Soil Reinforcement Technology Using Natural Plant Fiber Materials: Past Findings, Present Trends and Future Directions, Materials, MDPI, 11,553.
Marwan Mostafa, Nasim Uddin,2016, Experimental analysis of Compressed Earth Block (CEB) with banana fibers resisting flexural and compression forces, Case Studies in Construction Materials 5 (2016) 53–63, http://dxdoi.org/10.1016/j.cscm.2016.07.001.
Preneron, A.L., Aubert, J.E., Magniont C., Tribout, C., Bertron, A., 2016, Plant aggregates and fibers in earth construction materials: A review. Constr. Build. Mater., 111, 719–734.
Satyanarayana, K.G., Arizaga, G.G.C., Wypych, F., 2009, Biodegradable composites based on lignocellulosic fibers-An overview. Int. Rev. J. Prog. Polym. Sci. 34, 982–1021.
Rassiah, K., Ahmad, M.M.H.M., 2013, A Review on Mechanical Properties of Bamboo Fiber Reinforced PolymerComposite. Aust. J. Basic Appl. Sci., 7, 247–253.
Xu, Y., Wu, Q., Lei, Y., Yao, F., 2010, Creep behavior of bagasse fiber reinforced polymer composites. Bioresour. Technol., 101, 3280–3286.
D. Jagadeesh, R.Venkatachalam, G.Nallakumarasamy, 2015, Characterization of banana fiber: A Review, J. Environ. Nanotechnology Volume 4, No.2, pp. 23–26.
IS 2720 (PART 13), 1980, Methods of test for soils: Direct shear test,Bureau of Indian Standards, 1980.
http://vikaspedia.in/agriculture/crop-production/package-of-practices/fruits-1/banana

Technologies for Sustainable Development – Mahajan, Patel & Sharma (eds)
© *2020 Taylor & Francis Group, London, ISBN 978-0-367-33737-7*

Impact of temperature and duration of calcination of metakaolin on the compressive strength of metakaolin incorporated concrete

Sonal Shah* & Satish Desai

Applied Mechanics Department, Sardar Vallabhbhai National Institute of Technology, Surat, India

ABSTRACT: This paper presents the study of impact of temperature and duration of calcination of metakaolin on compressive strength development of metakaolin incorporated concrete. Superplasticized concrete specimens containing metakaolin (20%) in replacement of cement were characterized. Three commercially available metakaolin samples manufactured at different temperature and duration were studied. Two of three metakaolin mixes gave strength higher than normal concrete (NC) mix on all days. The third metakaolin mix showed strength higher than NC in early days, but at 28 days its compressive strength became equal to that of NC mix and did not show any enhancement in compressive strength after 28 days. The SEM images of 28 days hydrated specimens showed that the microstructure of all metakaolin incorporated concrete specimens were dense with a remarkable presence of calcium alumino silicate hydrates. NC matrix showed a clear presence of $Ca(OH)_2$ which was completely absent in the microstructure of metakaolin incorporated concrete.

1 INTRODUCTION

Concrete, due to its mechanical and durability properties along with its versatility, low cost, easy handling and availability is one of the most preferred construction building material (Guneyisi E. et.al, 2008). High performance concrete (HPC) is now a days, extensively used for constructing better performing, durable and sustainable infrastructures. Supplementary Cementitious Materials (SCMs) or Pozzolanas are added to concrete to make HPC (Davidotis et.al 1993, Al Khaja et al 1994). SCMs are those materials which do not possess binding properties in themselves, but when added to cement due to their pozzolanic reactivity, react with the cement hydration products mainly calcium hydroxide (CH) and form strength enhancing compounds such as calcium-silicate-hydrates (C-S-H) gel (Neville AM, 1996). Metakaolin is one such pozzolanic material. Unlike other SCMs such as fly ash, ground granulated blast furnace slag (GGBS), rice husk etc. which are by-products of some other primary products, metakaolin is a manufactured product. Its manufacturing process can be manipulated to produce a highly reactive pozzolanic material. It is well documented in journal papers that the process of manufacturing affects its physical properties such as amorphousness, particle size, loss on ignition (LOI) and surface area and chemical properties which in turn affects the pozzolanicity of a pozzolanic material. Hence an understanding of the performance of metakaolin in terms of these parameters is of paramount importance. In spite of the fact that there are number of research papers available advocating metakaolin as one of the most highly reactive pozzolana with the highest rate of pozzolanic reactivity, it has yet not acquired a place in industry as a preferred SCM. Thus, there is a gap between the research work promoting it to be the most reactive material among all the SCMs and its actual usage as a preferred SCM in concrete. This gap needs to be bridged to make the potential of metakaolin available to the concrete industry.

---

* Corresponding author: sonalshahtripathi@gmail.com

This paper reports the work done to investigate the impact of three different metakaolins varying in their (a) degree of calcination and (b) duration of calcination on the compressive strength development of metakaolin incorporated concrete. The three metakaolins are designated as MK1, MK2 and MK3 here. Compressive strengths were obtained by performing compressive strength tests on concrete specimens prepared from three different concrete mixtures: MCK1, MCK2, and MCK3 having 20% metakaolin in place of cement. The variation in their compressive strength development is discussed in relation to their parameters of process of calcinations and physical characteristics such as amorphousness and particle size.

1.1 *Background*

Two concepts need to be studied to understand the relation between the manufacturing process of metakaolin on its performance in concrete:(a) the working mechanism of microparticles in concrete and (b) the process of calcinations of metakaolin. The following section discusses both the concepts. Reaction Mechanism of microparticles in concrete: Any micro particle when added to concrete as admixture, affects the concrete at (1) physical level as well as (2) chemical level (Philip Lawrence, 2003). At physical level, there are three effects: (a) dilution effect, (b) the filler effect and (c) accelerated cement hydration due to heterogeneous surface nucleation effect. (a)The dilution effect: Microparticles, when added to concrete tend to reduce the strength of concrete in proportion to % of substitution. This reduction in strength that is observed in concrete is called dilution effect. (b) The filler effect: It enhances the strength of the concrete and more seen at the interface zone between the two different ingredients. The micro particles fill the gap between two ingredients in the concrete, improve the packing density and enhance the strength. (c) The heterogeneous nucleation effect: The surface of the added particles acts as sites of nucleation for the cement hydration products leading to the chemical activation of hydration of cement. The impact of this effect is the enhancement in the rate of gain of strength resulting from acceleration of hydration of cement.

The chemical effect is the pozzolanic/chemical reactions that take place between the calcium hydroxide ($Ca(OH)_2$) and the silica and the alumina of the pozzolanic material. They chemically react to form extra C-S-H gel which fill up the pores in concrete, improves the overall health of the microstructure and enhances the strength of concrete. This chemical effect/pozzolanic reaction can occur in the initial days of hydration or at later days of hydration depending upon the reactivity of the material. In highly active pozzolana such as metakaolin this reaction occurs as early as 1 day. In fly ash, ground granulated blast furnace slag (GGBS) which are not highly reactive, this reaction occurs at later days of hydration (after 28 days).

The Process of Calcination: The pozzolanic reactivity of a pozzolanic material depends upon its physical characteristics like its particle size, crystallanicity, LOI and chemical composition. Metakaolin is a manufactured pozzolanic material and its calcination temperature and the duration of calcination affects its physical characteristics. (Kaloumenou M, Badogiannis et al,1995, & M.H. Zang et al. 1995). Thus, for a manufactured product like metakaolin, its manufacturing process is an important aspect which has to be clearly understood to manufacture a highly reactive pozzolanic material. Metakaolin (Al2O3.2SiO2) or AS2, typically consists of alumina and silica. It is obtained by heat treating kaolin clay under controlled temperature conditions (650°C-850°C) for a certain period of time (residence time). During this process of heat treatment, the crystalline structure of kaolinite gets converted into a highly amorphous phase called metakaolin (Kakali G. Perraki T et al. 2001, J.A.K: R.Kostuch, 2000 & B. B. Sabir, 1996). The major reorganization occurs in the Al-O network of the structure, whereas the Si-O network remains intact. The initial days pozzolanic reactivity of metakaolin is due to the highly unstable Al-O atoms of the network. This highly reactive alumina immediately reacts with the cement hydration products mainly CH to form other crystalline products such as calcium aluminate hydrates (C3AH6, C4AH13) and calcium aluminosilicate hydrates (C2ASH8,). The later days pozzolanic reactivity in metakaolin is due to silica network (Si-O) (Bai J. 1999).

2 EXPERIMENTAL DETAILS

2.1 *Materials used for the experiment*

- Cement: Ordinary Portland Cement (OPC) was used for all types of mixes.
- Superplasticizer: A high quality polycarboxylic copolymer superplasticizer perma plast PC-502 was used.
- Aggregates: Good quality aggregates of 20 mm and 10 mm size were being procured for the study. Well graded white river of maximum size 4.75mm was used as fine aggregates.
- Metakaolin: Three types of metakaolin MK1, MK2 and MK3 were used as mineral admixtures. Metakaolin MK1 and MK3 were of the same origin whereas metakaolin MK2 was of different origin. Their calcination temperature and the duration of calcination is shown in Table 2. The XRD profile of the metakaolin samples MK1, MK2 and MK3 are represented in Figure 1. In the figure the metakaolin and quartz peaks are marked letter k and q respectively. Their chemical composition, LOI and particle size were determined using XRF machine, thermogravimetric analysis and dynamic light scattering respectively at SVNIT, Surat and are shown in Table 1.

Table 1. Chemical composition and physical properties of cement, MK1, MK2 and MK3.

Parameters	Cement	MK1	MK2	MK3
SiO_2 (%)	21.42	58.31	50.30	53.48
Al_2O_3 (%)	5.67	37.47	47.3	41.78
Fe_2O_3 (%)	3.16	2.15	1.33	2.3
CaO (%)	63.8	0.17	0.09	0.70
MgO (%)	1.20	-	-	-
Na_2O (%)	0.20	-	-	-
K_2O (%)	0.87	0.23	0.116	0.30
TiO_2 (%)	0.98	1.64	1.11	1.4
SO_3 l(%)	2.4	-	–	–
Loss on ignition (%)	1.18	7.5	0.8	3.5
Particle size (μm)	-	0.65	0.76	0.95

2.2 *Mix details*

Four different concrete mixtures: MKC1, MKC2, MKC3 and NC of M40 grade with water to cementitious material ratio of 0.4 were characterized and cubes of size 150 mm x150 mm x150 mm were prepared from these four mixtures. The mixtures were prepared using IS 10262: 2009 mix design code. The details of the mix proportions are shown in Table 3. MKC2, MKC3 had 20% metakaolin MK1, MK2 and MK3 respectively in them in place of cement whereas NC was prepared without any metakaolin added to it. The % replacement level of metakaolin (20%) was selected based on the results and conclusions of the

Table 2. The parameters of process of Calcination of MK1, MK2 and MK3: Temperature and duration of calcinations.

Type of metakaolin	Calcination temperature	Duration of calcination
	(° C)	(minutes)
MK1	800	45
MK2	1000	60
MK3	800	90

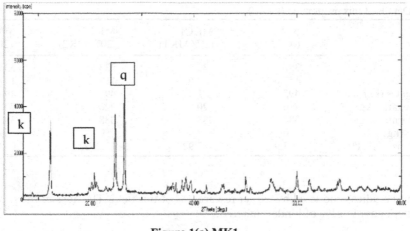

Figure 1(a) MK1

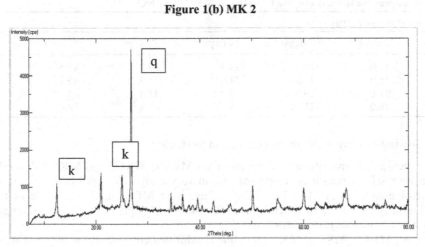

Figure 1(b) MK 2

Figure 1(c) MK 3

Figure 1.   XRD profile of metakaolin samples (a) MK1, (b) MK2 and (c) MK3.

Table 3. Details of mix proportion (kg/m^3).

Ingredients	NC (0%MK)	MKC1 (20%MK1)	MKC2 (20% MK2)	MKC3 (20% MK3)
Cement (kg)	392	333	333	333
Water (kg)	157.6	157.6	157.6	157.6
Fine aggregates (kg)	495	495	495	495
20mm aggregates (kg)	720	720	720	720
10mm aggregates (kg)	388	388	388	388
Metakaolin (kg)	0	59	59	59
Super plasticizer (l) (1%)	3.92	3.92	3.92	3.92

experimental studies conducted by earlier researchers of this field. The dose of the superplasticizer was kept equal to 1% of binder material.

2.3 *Experimental procedure*

The compressive strength test was conducted on Automatic compression testing machine (ACTM). The specimens were removed from water on the day of testing 7-8 hours prior to the test. The rate of loading was kept at 5.2 kN/s. The compressive strength of all the mixes at the age of 3, 7, 14, 28, 56 and 90 days was determined.

3 RESULTS AND DISCUSSION

3.1 *Compressive strength test results*

The compressive strength test results of all mixes: MKC1, MKC2, MKC3 and NC are shown in Table 4. The results shown are the average of results of three specimens taken on the day of testing.

Table 4. Compressive strengths of MKC1, MKC2, MKC3 and NC.

Compressive strength (MPa)						
Mix	3 days	7 days	14 days	28 days	56 days	90 days
NC	12.0	19.0	27.4	39.0	41.5	42.7
MKC1	15.0	26.5	31.6	41.3	41.2	41.9
MKC2	21.0	29.5	36.0	42.3	44.7	53.2
MKC3	18.0	21.0	31.5	41.4	49.0	53.0

The following observations are apparent from the table

– All the metakaolin incorporated concrete mixes MKC1, MKC2 and MKC3 showed a very high compressive strength development rate in very early (3,7,14) days of hydration. The compressive strength of MKC1 at 3 days of age is 15 MPa which is 25% higher than the NC. Similarly, MKC2 and MKC3 show 50% and 75% higher compressive strength than NC at as early as 3 days of hydration.
– At 28 days MKC1, MKC2, MKC3 specimens exhibited compressive strength of 41.4 MPa, 42.3 MPa, 41.4 MPa respectively, which were nearly equal to NC compressive strength value of 39 MPa.
– At 56 and 90 days the MKC2 and MKC3 exhibited higher compressive strength values than NC strength values whereas there is no further enhancement in strength in MKC1 mix.

Thus, it is seen that the compressive strength values of metakaolin added concrete remain higher than the NC for the initial 0-14 days of curing, which then tend to become equal to the compressive strength of NC at 28 days and then again show increasing trend till/at 90 days. Also, it is seen from the Table 4 above that there is variation in the rate of compressive strength development among all the MKC1, MKC2 and MKC3 mixes on all test days (3, 7,14, 56 and 90). To explain this variation in their strength development behavior, the further analysis and interpretation of the test results is done in two parts: (a) Compressive strength development in the initial days (0-28) and (b) Compressive strength development in later days.

3.2 *Analysis and interpretation of the test results*

3.2.1 *Initial (0-28) days compressive strength development in metakaolin incorporated concrete*
Compressive strength development in initial (0-3) days: From the compressive strength Table 4, it is seen that MKC2 specimens represent the highest 3 days compressive strength among all mixtures. It exhibits a compressive strength value of 21 MPa which is 75% higher than the NC (12 MPa), 40% higher than MKC1 (15 MPa) and 16.6% higher than MKC3 (18MPa) specimens. This extremely high rate of gain of strength of MKC2 specimens is attributed to its highly amorphous alumina and smaller particle size. As is seen in their XRD graphs in Figure 1, MK2 exhibits the highest degree of amorphousness among all the metakaolins. The highly unstable alumina ions react mainly with CH to form more stable compounds like calcium aluminate hydrates (C_3AH_6, C_4AH_{13}). Also, the smaller particle size (0.76μm) facilitated the heterogenous nucleation of the hydrated products leading to acceleration of cement hydration process. This in turn produced more strength enhancing C-S-H gel. Thus, the high initial strength of MKC2 is explained by its degree of amorphous and particle size. Comparing the 3 days compressive strength values of MKC1 and MKC3, it is seen that MKC3 has a value of 18 MPa which is higher than 15 MPa of MKC1. This difference is attributed to the higher degree of amorphousness of metakaolin MK3 than MK1. As seen from their XRD graphs in Figure 1 the kaolinite peaks of MK3 exhibit less crystallinity than peaks of MK1. Higher the degree of amorphousness, higher is the pozzolanic reactivity. Thus, the higher strength of MKC3 specimens than MKC1 specimens is attributed to the higher degree of amorphous of MK3 than MK1.

Compressive strength development during the period of 4 to 28 days of hydration: MKC2 continues to exhibit the highest compressive strength among all the mixes on all the 7, 14 and 28 days of testing due to its small particle size and its highest degree of amorphousness among all the metakaolins. But the trend is reversed in MKC3 and MKC2 specimens. At 7 days the MKC1 specimens show compressive strength value of 26 MPa which is higher than the compressive strength value of 21 MPa of MKC3. The reason for this reversal in trend is attributed to the smaller particle size of MK1 than MK3. The particle size of MK3 was much bigger than the particle of MK1. As seen from Table 1, MK1 has a particle size of 0.65 μm which is smaller in size than 0.95 μm size of MK3. Smaller the size of the particle, higher is the surface area available for the cement hydration products to nucleate. Thus, more surface area was available for the hydrates to nucleate in MKC3 specimens than in MKC1 specimens. This in turn accelerated cement hydration process producing more strength enhancing C-S-H gel. Thus, here the 21% higher compressive strength of MKC1 specimens than MKC3 specimens is due to the smaller particle size of MK1 than MK3.

3.2.2 *Compressive strength development in metakaolin incorporated concrete after 28 days (28-90 days)*
Compressive strength development of MKC1 and MKC3: As seen from the compressive strength Table 4, the compressive strength exhibited by MKC1 specimens corresponding to 56 days and 90 days are 41.2 MPa and 41.9 MPa respectively. These values are same as its 28 days strength value of 41.3 MPa, which indicates that there is no enhancement in strength after 28 days. On the other hand, the compressive strength of MKC3 on 56 days and 90 days were 49 MPa and 53 MPa, that are higher than its 28 days value of

41.4 MPa. Thus, although MK1 and MK3 were calcined at the same temperature of 800°C (as seen in Table 2), they exhibited different pozzolanic behavior. This difference in their behavior is attributed to their different duration of calcination. As seen in Table 2, MK1 was calcined for 45 minutes whereas MK3 was heat treated for 90 minutes. Thus, the time period for which MK1was heated was not enough for the crystalline kaolinite to get converted into reactive amorphous metakaolin, while the duration of calcination of MK3 was long enough, that led to partial conversion of crystalline peaks to amorphous form. The same is also confirmed from their XRD profile graphs in Figure 1. The kaolinite crystalline peaks in metakaolin MK1, which showed an intensity of 3031cps and 3251 cps corresponding to 2θ= 12.38 and 24.9 respectively got reduced to 894cps and 974cps in MK3. Thus, the XRD graphs confirm a higher degree of amorphousness of MK3 compared to metakaolin MK1. The later days enhancement in strength seen in MK3 is due to this amorphousness. It is thus concluded that MK3 possessed later days pozzolanic reactivity that led to strength enhancement in MKC3after 28 day whereas MK1 did not have the later days pozzolanic reactivity.

Compressive strength development in MKC2: As seen from the Table 4, the compressive strength value of MKC2 at 28 days (42.3 MPa) is same as its 56 days (44.7 MPa) value, which implies that there is no pozzolanic reaction occurring in concrete during this period. But at 90 days it shows compressive strength value of 53.2 MPa which is 19% higher than its 56 days value. Thus, although there was no pozzolanic reaction in MKC2 from 28 to 56 days, the 90 days higher compressive strength indicates the pozzolanic reactivity during the 56 to 90 days of hydration. The parameter that is responsible for this behavior of MKC2 is its temperature of calcination. MK2 was calcined at 1000°C temperature. At this temperature the amorphous silica again gets converted to highly crystalline cristobalite (Malhotra V.M et al. 1992 & B.B. Sabbir et al. 2001). This highly crystalline silica needs a very long time to diffuse in the solution and react with the hydration products. Thus, MC2 specimens did not show any strength enhancement at 56 days, but later at 90 days the compressive strength increased to 53.2 MPa from 43.2 MPa of 28 days due to the pozzolanic reactivity of the silica network of the pozzolana.

3.3 SEM image analysis

The SEM images are shown in Figure 2. The images were obtained using a XL- 30, Scanning Electron Microscope with EDS, at SICART, Vallabh Vidhyanagar. The SEM images a,c,e and g were taken at 10000x whereas the b,d,f and h SEM images were taken at 35000x. The images revealed that, the matrix of metakaolin incorporated concrete was dense and compact as compared to matrix of normal concrete. The microstructure of MKC1, MKC2 and MKC3 showed a complete absence of calcium hydroxide (CH) crystals in contrast to microstructure of normal concrete that showed a clear presence of CH crystals. In metakaolin incorporated concrete, the highly reactive alumina reacted with the calcium hydroxide liberated during the cement hydration process and produced compounds like calcium alumino hydrates (C_4AH_{13}) and calcium alumino silicate (C_2ASH_8) making the structure dense and compact. Whereas, in NC mix, the calcium hydroxide was left unreacted in NC mix and was clearly seen in their SEM images.

MKC1 SEM images revealed presence of C-S-H gel and C_4AH_{13}. Very little/negligible amount of stratlingite (C_2ASH_8) was found. As it was not properly calcined the later days pozzolanic reaction between silica and the CH did not occur which would have produced stratlingite. The study of MC2 specimen reveals that the microstructure was dense and rich in C-S-H gel and other calcium alumino hydrates (C_4AH_{13}, C_3AH_6). The microstructure of MKC3 had a remarkable presence of crystalline stratlingite. As it was properly calcined for a longer duration of time (1 hr 30 min) its silica (Si-O) network was reactive and produced stratlingite by reacting with the cement hydration products. The SEM images of NC clearly showed the presence of CH crystals and C-S-H gel. Presence of no other compound was traced in the microstructure of the NC specimen.

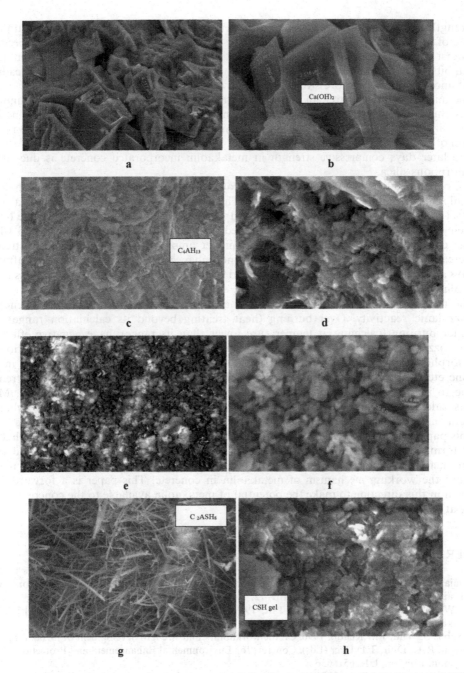

Figure 2.   SEM images of 28 days hydrated samples (a,b) NC. (c,d) MKC1. (e,f) MKC2. (g,h) MKC3.

4   CONCLUSIONS

The conclusions from the study are:

- The study shows that compressive strength development rate in metakaolin added concrete is different in its initial 0-14 days and in its later (28-90) days of hydration. In its initial 0-14 days of hydration, metakaolin incorporated concrete shows a very high rate of compressive strength development as compared to normal concrete. During 14-28 days of hydration the rate of gain of compressive strength decreases substantially and its strength value becomes almost equal to the normal concrete value at 28 days and after 28 days metakaolin again exhibits pozzolanic reactivity due to the amorphous silica content and further leads to enhancement in the compressive strength of concrete.

- The initial (0-14 days) strength imparted by metakaolin is attributed to its highly amorphous alumina which reacts with the CH to form stable C-S-H gel and its extremely fine particle size that accelerates the process of cement hydration and further enhances the rate of gain of strength.

- The later days compressive strength in metakaolin incorporated concrete is due to the amorphous silica.

- It is concluded from the tests results that duration of calcination plays a major role in determining its later days pozzolanic reactivity. This is confirmed from the fact that metakaolin MK1 and MK3 were of the same origin and were calcined at the same temperature but as their durations were different, they exhibited different behavior in concrete. MK3 which was calcined for longer duration showed higher amorphousness and hence its strength increased after 28 days whereas there was no enhancement in strength in MK1 after 28 days as the duration for which it was calcined was not enough to produce amorphous metakaolin with optimum pozzolanic reactivity.

- The temperature of calcination affects the amorphousness of metakaolin and hence its pozzolanic reactivity. Over burning (heat treating beyond its calcination range) and under burning (calcining at temperature lower than its range), have negative effect on its pozzolanic reactivity. MK2 was overburnt, which led to the recrystallization of amorphous silica to crystalline silica that requires a longer duration to react in the concrete. Hence the contribution of the MK2 in enhancing the compressive strength due to its pozzolanic behavior after 28 days was not seen till 56 days. Whereas, MK3 was calcined at appropriate temperature and hence there was a continuous enhancement in strength at all days.

- This paper explains the pozzolanic behavior of metakaolin in concrete and proves that the performance of metakaolin in concrete depends upon its physical properties and their parameters of calcination. More research is required in this direction to explore and understand the working mechanism of metakaolin in concrete. This paper is a forward step, taken in this direction to make the potential of metakaolin available to the concrete industry at large.

REFERENCES

Al-Khaja WA, 1994. Strength and time-dependent deformations of silica fume concrete for use in Bahrain. Construction Building, vol 8: 169–72.

Bai J, Wild S, Sabir B.B and Kinuthia J.M, 1999. Workability of concrete incorporating PFA and metakaolin. Magazine of Concrete Research, vol 12: 207–216.

B. B. Sabir, S. Wild, J.M.Khatib, 1996. On the workability and strength development of metakaolin concrete, in R.K. Dhir, T.D.Dyer (Eds), Concrete for Environmental Enhancement and Protection, E and FN Spon, London, UK: 651–656.

B.B. Sabir, Wild S, and Bai J, 2001. Metakaolin and calcined clays as pozzolans for concrete: A review. Cement and concrete composites, vol 23: 441–454.

Guneyisi E. Gesogl M. Mermerdas, 2008. Improving strength, drying shrinkage, and pore structure of concrete using metakaolin. Material and Structure, vol 41(5): 937–949.

J.A.K: R.Kostuch, V. Walter, T.R.Jones, 1993. High performance concrete incorporating metakaolin: A review, International conference, Infrastructure, research, new applications, vol 2: 1799–1811.

J.davidovits, M. Moukwa, S.L. Sarkar, K. Luke, 1993. Geopolymer cement to minimize carbon-dioxide greenhouse warming, ceramic transaction cement-based material: present, future, environmental Aspect, Westerville: The American Ceramic Society, vol 57: 165–181.

Kakali G. Perraki T, Tsivlis S, Badogiannis E. 2001.Thermal treatment of kaolin: the effect of mineralogy on the pozzolanic activity. Applied Clay Science, vol.20:73–80.

Kaloumenou M, Badogiannis E, Tsivilis S, Kakali G. 1995. Effect of the kaolin particle size on the pozzolanic behavior of the metakaolinite produced. J Therm Analysis Calorimerty, vol 56: 901–907.

Malhotra V.M, 1992. The use of fly ash, slag, silica fume and rice husk ash in concrete: a review, in CBUL CANMET, International Symposium on Use of fly ash, silica fume, slag and of by products in concrete and construction Materials: pp 60–65.

M. H. Zang, V. M. Malhotra, 1995. Characteristics of a thermally activated alumino-silicate pozzolanic material and its use in concrete, Cement and Concrete Research; vol 25(8):1713–1725.

Neville AM. 1996. Properties of Concrete. 4th and final ed. England: Addison Wesley Longman.

Philippe Lawrence, Martin Cyr, Erick Ringot, 2003. Mineral admixtures in mortars Effect of inert materials on short-term hydration. Cement and Concrete Research, vol 33:1939–1947.

Technologies for Sustainable Development – Mahajan, Patel & Sharma (eds)
© *2020 Taylor & Francis Group, London, ISBN 978-0-367-33737-7*

Numerical analysis of shape-memory-alloy-wire based smart natural rubber bearing

Prachi M. Ujalambe
P.G. Student, M.E. Structure, Department Of Civil Engineering, A.I.S.S.M.S College of Engineering, Pune, India

M.V. Waghmare
Assisstant Professor, M.E. Structure, Department Of Civil Engineering, A.I.S.S.M.S College of Engineering, Pune, India

ABSTRACT: Seismic isolation is one of the most effective options for passive protection of structures. In this paper, the effectiveness of Shape Memory Alloy (SMA) based Natural Rubber Bearing (NRB) with Double Cross (DC) SMA wire configuration is presented. Based on the aspect ratio (R) two types of DC-SMA NRBs are considered:(DC-SMA NRB-I has R = 0.22 and DC-SMA NRB-II has R = 0.38). The SMA based bearings show high strength-strain capacity, high resistance to corrosion and fatigue. In this study, FeNiCoAlTaB SMA wire of 2.5mm radius is provided to NRB. The strain in DC-SMA wire based NRB is evaluated through analytical method. The results are compared with available results of straight configuration [Choi et al.2005] and cross configuration [Dezfuli and Alam, 2013]. Results show that smart natural rubber bearing with a cross configuration is more efficient than double cross configuration. ficient. flexibility and wire strain level, the smart rubber bearing with a cross configuration of SMA wires is more efficient.flexibility and wire strain level, the smart rubber bearing with a cross configuration of SMA wires is more efficient.

1 INTRODUCTION

The most uncertain and tough to manage situations are earthquakes, which results in fatal effects to human beings. To get rid of tragic effects of earthquakes, the productive method is use protective systems in structures (e.g. buildings and bridges) as base isolation mechanisms. The Shape Memory Alloys (SMAs) are materials that have super elasticity and shape memory effect that other alloys and metals don't have. The prominent characteristics of SMAs which makes it suitable for seismic protection device are, flag-shape hysteresis curve, large recoverable strain and deformation capability, excellent endurance against fatigue, and corrosion resistance. (Alam and Lagoudas) [1&7]. Wilde et al. [11], suggested that SMA are best material for base isolation devices for earthquake forces with different magnitude and frequencies. Dolce et al. [6] done difficult task of using SMA wires effectively in base isolation. Liu et al. [8] provided the large diameter SMA strand, around rubber bearing, however this system was ineffective in recentering capability as compared to the original rubber bearing. Billah, Alam and Bhuiyan [3] proved that SMA material would regain its original position after the earthquake by restricting the relative motion between the base and the superstructure. Ozbulut & Hurlebaus [9] analyzed that the SMA rubber based (SRB) isolation system for seismic protection of bridge against near field earthquakes can successfully eliminate the seismic response of highway bridges; however, a smart SMA rubber-based isolation system is more useful than the conventional isolation system. Dezfuli & Alam [5] assessed the hysteretic shear response for smart lead rubber bearings (LRBs) equipped with double cross ferrous SMA wires by using finite element method. From analysis it is observed that hybrid SMA-LRB bearing shows a significantly lower shear strain demand and a higher energy dissipation capacity compared to the LRB. Choi et al. [4] noted that, LRB has large horizontal deformation under near field earthquakes which cause harmful problems such as instability of the bearings, pounding and unseating problems of the bridge deck. Hence, in this study, double cross shape memory alloy (DC-SMA) wire configuration is provided to natural rubber bearing (NRB) and its performance is compared to the straight arrangement suggested by Choi et al. [4] and cross arrangement suggested by Dezfuli & Alam [5].

1.1 *Behavior of shape memory alloy*

Shape memory alloys (SMAs) are smart and functional materials. SMAs have mechanical behavior, can recover their pre-determined and original shape after deformation via unloading or by applying a thermal load. Super-elastic and shape memory effects are two exclusive characteristics of SMAs.

When the SMA materials is in austenite phase, it will show super-elastic behavior (Figure 1). In other words, as the temperature of the SMA is above the austenite finish temperature, the strain produced in the SMA will be fully regained. If it is lower than the maximum super-elastic strain, i.e. the generated strain due to the mechanical loading is fully recovered after unloading (Figure 2a). While, in shape memory effect (SME) (Figure 2b), the mechanical deformation should be eliminated by applying a thermal load and increasing the temperature of the alloy. SME have two different shape-memory effects which are one-way and two-way shape memory (Figure 2b).

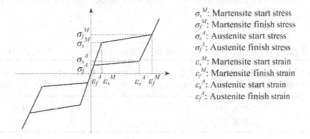

σ_s^M: Martensite start stress
σ_f^M: Martensite finish stress
σ_s^A: Austenite start stress
σ_f^A: Austenite finish stress

ε_s^M: Martensite start strain
ε_f^M: Martensite finish strain
ε_s^A: Austenite start strain
ε_f^A: Austenite finish strain

Figure 1.   Idealized stress–strain diagram of SMA (Auricchio 2001).

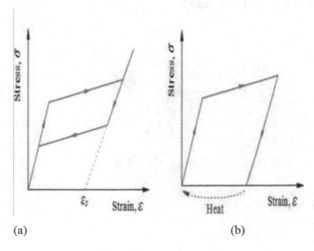

(a)   (b)

Figure 2.   Stress–strain curve for a typical shape memory alloy; (a) super-elastic effect, (b) shape memory effect.

When a shape-memory alloy is in its cold state i.e. below austenite, the metal can be bent or stretched and will hold those shapes until heated above the transition temperature. Upon heating, the shape changes to its original. There is no macroscopic shape change when cooled from high temperature with one-way effect. The two-way shape-memory effect remembers two different shapes: one at low temperatures, and one at the high-temperature. In two-way shape memory effect, a material that shows a shape-memory effect during both heating and cooling.

2 SMART ELASTOMERIC ISOLATORS

Stability, recentering capability, high energy dissipation capacity and long service life are the characteristics of SMA-based smart base isolator. SMAs have excellent performance in terms of fatigue properties and energy dissipation capacity as compared to existing rubber bearings (Suduo and Xiongyan) [10].

Dezfuli and Alam (2013) proposed the model for smart NRBs using SMA wires with the cross configuration. In the procedure of manufacturing SMA-NRB, the laminated pad composed of alternating layers of elastomer and steel plates is fabricated through either hot-vulcanization or cold-bonding process. Before attaching steel end plates to the elastomeric pad, steel hooks should be welded to the plates at locations illustrated in (Figure 3). Based on above procedure, a similar idea can be applied to DC-SMA NRBs.

Table 1.shows geometrical properties of two smart natural rubber bearings and different numbers of elastomeric layers are shown in Figure 4.The schematic plan view of SMA-NRB-1 and SMA-NRB-2 (Figure 4a) consisting of 8 and 14 rubber layers is plotted in Figure 4b and Figure 4c, respectively.

Table 1. Geometrical properties of natural rubber bearings. [Dezfuli and Alam 2013].

Samples	Dimensions of iso-lator (mm x mm)	Dimensions of alternate layers of steel and rubber shims (mm x mm)	t_E (mm)	t_r (mm)	t_s (mm)	n_r	n_s	R
NRB-1	240 X 240	200 X 200	15	4.5	1	8	7	0.2
NRB-2	240 X 240	200 X 200	15	4.5	1	14	13	0.4

Where, t_E: thickness of supporting steel plates; t_r: thickness of rubber layers; t_s: thickness of steel shims; n_r: number of rubber layers; n_s: number of steel shims; R: aspect ratio is the ratio of the height of rubber bearing to the length.

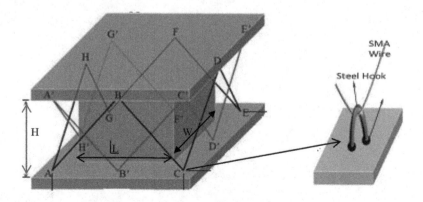

Figure 3. DC- SMA wire-based natural rubber bearing with steel hooks.

118

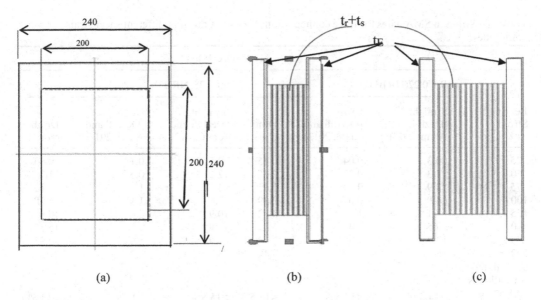

Figure 4. Schematic view of the elastomeric isolator; (a) plan view of NRB-1 and NRB-2, (b) side view of NRB-1, (c) side view of NRB-2.

Table.2. Required initial lengths for three configurations.

| Configuration | Initial Length (L0) mm | |
	R=0.22	R=0.38
Straight	1092	1224
Cross	1871.9	1937.85
Double Cross	982.21	1102.75

$$R = H_{eff}/L \qquad (1)$$

Where, $H_{eff} = t_r * n_r + t_s * n_s$

The Initial length (L_0) required for two aspect ratios (R) with three configurations: straight, cross and double cross (DC) are listed in Table 2.

Accordingly, the induced strain in FeNiCoAlTaB [12], SMA wires (ε_{SMA}) with radius 2.5 mm for double cross (DC) is calculated by using equation.2. The strain in SMA wires (ε_{SMA}) for three configurations: straight, cross and double cross (DC) is listed in Table 3.

$$\varepsilon_{wire} = \frac{LSMA - L0}{L0} \times 100 \qquad (2)$$

Where,

Table 3. Strain in SMA wires for three configurations for aspect ratios at different shear strain amplitudes.

	ε_{SMA} Wire (%)					
	R= 0.22(NRB1)			R=0.38(NRB2)		
Bearing Shear Strain γ (%)	Straight [Dezfuli and Alam, 2013]	Cross [Dezfuli and Alam, 2013]	Double cross	Straight [Dezfuli and Alam, 2013]	Cross [Dezfuli and Alam, 2013]	Double cross
25	0.3	0.04	0.15	0.5	0.1	0.43
50	1.3	0.2	0.6	2.1	0.5	1.7
75	2.9	0.3	1.35	4.4	1	3.82
100	4.8	0.6	2.39	7.4	1.8	6.78
125	7.1	1	3.72	10.9	2.9	10.6
150	9.5	1.4	5.33	14.8	4.1	15.32
175	12.2	1.9	7.23	18.9	5.6	20.95
200	15	2.4	9.42	23.3	7.2	27.5
ε_{SMA} in ferrous alloy is 13.50% (Tanaka et al) [2010]	>13.5%	<13.5%	<13.5%	>13.5%	<13.5%	>13.5%

$$L_0 = \text{Original length of SMA wire}$$

$$L_0 = 4 \times \left[\sqrt{\left(\frac{L}{2} + LE\right)^2 + H^2} + \sqrt{\left(\frac{W}{2} + WE\right)^2 + H^2} \right] \quad (3)$$

$$L_{SMA} = \text{Change in length of SMA wire}$$

$$L_{SMA} = 2 \times \sqrt{\left(\frac{L}{2} + LE + \Delta X\right)^2} + 2 \times \sqrt{\left(\frac{L}{2} + LE - \Delta X\right)^2} + 4\sqrt{\left(\frac{W}{2} + WE\right)^2 + H^2 + \Delta X^2} \quad (4)$$

Where, Δx = Lateral displacement of rubber bearing, H is varying height as aspect ratio increases or decreases, $L_E = W_E = 15$ mm which is a distance between hook position and side face of elastomeric bearing, refer Figure (3).

The strain in SMA wires (ε_{SMA}) is a function of the shear strain amplitude(γ), and the aspect ratio, R. In SMA wires when the shear strain increases, there is increase in strain also.

Where,

$$\gamma\ (\%) = \frac{\Delta X}{(\text{tr} \times \text{nr})} \times 100 \quad (5)$$

3 RESULTS AND DISCUSSION

From the above Table 3 it is observed that, for R = 0.22, the strain in cross SMA and DC SMA wire is 2.4% and 9.42% respectively and for R = 0.38, cross SMA wire has strain 7.2%. Hence it can be concluded that, for higher aspect ratio only cross configuration is suitable.

While, for low aspect ratio (R = 0.22) cross and double cross configurations are suitable. As a result, the effective strain in SMA wires generated due to the shear deformation in the NRB with cross configuration decreases because wires have a longer initial length. The rate of elongation in cross SMA wires is much lower as compared to straight configuration and DC-SMA configuration. Thus, NRB equipped with cross SMA wires reaches its super-elastic strain limit (13.5%) at larger shear strain amplitudes, can carry a higher horizontal deflection. The DC-SMA configuration can be used for low aspect ratio and low shear strain amplitude.

4 CONCLUSION

The induced strain in shape memory alloy wire configuration is function of height, width and length of the rubber bearing. Therefore, following three conclusions are drawn from study:

1) As the height of natural rubber bearing increases, aspect ratio also gets increased. So, the strain in shape memory alloy wire also increases.
2) The strain in shape memory alloy wire increases with increase in shear strain amplitude.
3) To reduce the strain in shape memory alloy wire, increase the initial length of shape memory alloy wire configuration provided to natural rubber bearing.

REFERENCES

1. Alam M S, Youssef M A and Nehdi M 2007.Utilizing shape memory alloys to enhance the performance and safety of civil infrastructure: a review, *Canadian Journal of Civil Engineering.34*, pp 1075–86.
2. Auricchio F 2001. A robust integration-algorithm for a finite-strain shape-memory-alloy, International Journal of Plasticity 17, pp 971–990.
3. Billah A.H.M., Alam M.S., Bhuiyan M.A. 2010. Seismic performance of a multi-span bridge fitted with super-elastic SMA based isolator, *IABSE-JSCE Joint Conference on Advances in Bridge* Engineering-II, *August 8-10, Dhaka, Bangladesh*, pp 303–310.
4. Choi, Nam &Cho. 2005.A new concept of isolation bearings for highway steel bridges using shape memory alloys, *Canadian Journal of Civil Engineering (32)*, pp 957–967.
5. Dezfuli, F.H. and Alam, M.S.,2013. Shape memory alloy wire-based smart natural rubber bearing, *Smart Materials and Structures 22(2013)*, 045013 (17pp).
6. Dolce M, Cardone D and Marnetto R 2000. Implementation and testing of passive control devices based on shape memory alloys, *Earthquake Engineering Structural Dynamic 29*, pp945–968.
7. Lagoudas D C 2008.Shape Memory Alloys, *(New York: Springer Science and Business Media, LLC)*.
8. Liu H, Wang X & Liu J 2008.The shaking table test of an SMA strands-composite bearing, *Earthquake Engineering Vibration. 28*, pp 152–6.
9. Ozbulut, O.E. and Hurlebaus, S. 2011.Seismic assessment of bridge structures isolated by a shape memory alloy/rubber-based isolation system, *Smart Materials and Structures20*,015003(12pp).
10. Suduo X & Xiongyan L 2007.Control devices incorporated with shape memory alloy, *Earthquake Engineering Vibration 6*, pp 159–69.
11. Wilde K, Gardoni P and Fujino Y 2000. Base isolation system with shape memory alloy device for elevated highway bridges, *Engineering Structure 22*, pp 222–9.
12. Y. Tanaka, et al. 2010.Ferrous Polycrystalline Shape-Memory Alloy Showing Huge Super-elasticity, *Science 327,14*, pp 1488–1490.

Technologies for Sustainable Development – Mahajan, Patel & Sharma (eds)
© *2020 Taylor & Francis Group, London, ISBN 978-0-367-33737-7*

Techniques used for structural health monitoring and its application: A review

Gaurav Raj, Nirav Chaudhari, Arihant Jain & Mamta Sharma
IITRAM, Maninagar, Ahmedabad, Gujarat, India

ABSTRACT: In the growing world of infrastructure, Structure Health Monitoring has become an essential subject in Civil Engineering, and is receiving a meticulous review in industry. Structure Health Monitoring is a technique to identify the damage caused to the structure over the due course of time. The damage is referred as the change/modification of material properties, change in boundary conditions or system integrity which predominately defines the system performance. By monitoring and correlating these factors we can predict the expected life and reliability of the structure. Over the past decade significant amount of research has been conducted in this subject due to its associated potential for human life safety and regular monitoring of structures. The chief objective of this paper is to explore and discuss the techniques and measures which are used in order to ascertain and fortify the health of structures and what are their advantages and limitations. A brief overview of wireless technologies being implemented in this field and its advantages over the conventional methods has also been incorporated in the review paper. The paper is thus an amalgamation of different courses of action and exploratory procedures available in the domain of technology on this subject.

1 INTRODUCTION

The fundamental concepts of Structural Health Monitoring (SHM) is to determine and analyze the structural health of infrastructure. The advancement in medical sciences has been aimed at ensuring the health of an individual human being, similarly advancement in this subject is to ensure the safety of structures and human beings. Structural health monitoring concept showcases various techniques to acquire and analyze the complex computational data of a monitored structure using theoretical & traditional technological concepts. The data is utilized to divulge the expected life, dependability and enduringness of an old structure. The health of a structure can be determined predominately in 2 ways. First is using Non-destructive, where a small section of the structure is extracted from multiple locations and laboratory tests are conducted to determine its strength, cracks, internal fractures and other damages. Second is using a sensor based system, where a grid of sensors is installed in structure to capture the response of building and counteract the forces to reduce the damage caused due to such forces and preserves the structural integrity.

2 TECHNIQUES FOR SHM AND ITS APPLICATION

Different techniques have been used in the fields for health monitoring of structures. There applications, advantages and disadvantages have been discussed below.

2.1 *Non- destructive testing*

Non-Destructive Test (NDT) is one of the most widely used concepts to evaluate the strength of a particular structure. In non-destructive testing the specimen is not destroyed or demolished [1]. In this domain of technology extensive techniques are used to examining the properties of a concrete structure without any disfigurement [2]. NDT has various applications and advantages, i.e., seeing through the walls, higher quality of building inspection, safer and faster accumulation of data at site. The main purposes for which NDT is conducted are as listed below:

- Evaluation of uniformity of concrete and its strength at various ages by performing a test.
- To test material components and credibility of insurance without disturbing its state.
- NDT tests provide detailed results of surface hardness and surface adsorption.
- To monitor the history of material properties and long-term variation in structural properties.
- Alkali contents, chloride, sulphate and degree of carbonation can be accessed of structure.
- Elastic Modulus can be measured along with the grouting condition in pre-stressed cables.

Tests performed under NDT: NDT tests consists of various techniques and methods for determining the different properties. Some of the major tests are:

2.1.1 *Rebound hammer test*
The fundamental principle of this method is based on the rebound capacity of an elastic mass which depends on the hardness of the surface. The main objective is determination of the compressive strength of concrete via its correlation with rebound index. It is the most reliable method to examining the grade of concrete [3]. The higher surface hardness of concrete can be interconnected by higher rebound value. This test includes resistivity meters, Windsor probe, and endoscopy with a minimum spacing (20mm increased to 25mm) with acceptance criteria of set's readings. The whole set is discarded if 20% of sets are greater than 6 rebound units from the median. This test is used at five different positions to determine the surface hardness and its strength. Rebound hammer test can be achieved by taking readings vertically upwards, downwards, inclined vertical upwards and downwards.

2.1.2 *Ultrasonic pulse velocity test*
The UPV is carried out to determine the velocity of penetration, which is reliant on the density and elastic modulus of the material. This through-transmission technique primarily comprises of a pair transducers (probes) of discrete frequencies, electric pulse generator, and electrical timings of devices and cables. It helps to determine the homogeneity of concrete and voids/cracks. The velocity pulse in steel is 1.9 times greater than plain concrete. The basic principle of its working is the analogous properties of ultrasonic waves which prevail it to get refracted and focused easily. Having different acoustic properties; reflection and refraction can be achieved via waves of sound interaction with its interfaces. A longitudinal electro-acoustical transducer is utilized for the production of perpendicular vibrations.

2.1.3 *Electromagnetic cover measurement*
The presence of steel is affected by the presence of electromagnet which governs the fundamental principle of this test. Measurement of concrete's cover and the spacing along with the diameter in existing RCC are determined by this method. Allowance for scanning are is 2.0x2.0m/1.0x1.0m/0.5x0.5m. Mostly, it is used to examine the unachievable records of concrete. We can easily detect the allocations of hidden ferromagnetic materials. It is very helpful in the determination of cover of concrete and to ensure the quality control of rebar.

2.1.4 *Half-cell electrical potential method*
In order to detect corrosion in the reinforcement bar of the specimen/structure, we use the fundamentals of electromagnetic concepts through this test. With the help of potential difference and electrical resistance between the surface and reinforcement-corrosion activity can be easily evaluated. The half-cell potential is used in RCC structure for observations of corrosion activities. The apparatus comprises of copper sulphate half-cell connected to a voltmeter of high impedance provided in copper form and sulphate solution within the cell.

2.2 *Sensors, actuators & vibration control device*

Sensors are used in order to determine the external forces being applied to structure or to evaluate the conditions inside a structure. These sensors sense the physical changes including stress (or force), strain (or deformation), crack, damages and change in environmental conditions like temperature and humidity to which a building is exposed over a stretch [5]. Advancements in field of electronics has steered the use of sensors in almost every field from traffic monitoring to defense and military purposes. A wide variety of wired and wireless sensors are readily available in the market for different purposes based on their application. Based on their working principles, Sensors can be broadly categorized in 2 types, i.e. active sensors and passive sensors. A series of instructions/algorithms are fed into these sensors for obtaining the required data and the actuators or control device initiates respond based on the data received by sensors [2]. Two types of methodologies are present for restricting and resisting the external forces, aiming to abate the damage, cracks and collapse of any structure due to external forces.

2.2.1 *Active structural vibrational control*

The main objective of using the active controls is to counter the vibrations caused to the building due to seismic waves and demote the amount of damage caused by these vibrations with the help of engineering techniques. These vibrations can be controlled or countered predominately using 3 techniques:

1. By controlling the source of vibration
2. By improving the structures
3. By changing the path through which seismic waves travel

Active vibration control is a method of applying an equal and opposite force in response to an external force. This helps in canceling the vibrations and providing stability. This technique is very useful as in modern world, the vibrations caused due to seismic waves often lead to enormous loss and have a devastating effect on the humankind, machinery used in industries and Infrastructures [5]. A successful example of AVC can be seen in helicopters. Additionally, this technique also provides more comfort with reduced weight. Components of active vibration control are as following:

- A base with a number of active drivers
- Accelerometer to measure acceleration in all the 3 dimensions
- An electronic amplifier to amplify and convert the signals from accelerometers into electrical signals.
- In the case of large systems, pneumatic or hydraulic components are used as they higher power drive is required.

2.2.1.1 METHODOLOGY OF ACTIVE VIBRATIONAL CONTROL

For the working of active vibration control system, a continuous power source is required by electrohydraulic/electro-mechanical actuators for the analysis [4]. There are two types of active vibration control or cancellation system:

a) Feed forward system: Programmed to counter periodic vibrations.
b) Feedback system: Has a sensing mechanism that reacts every time a vibration is sensed.

2.2.2 *Passive structural vibrational control*

Controlling the structural vibration using passive energy has been an old and economical technique for restraining the damages caused due to the seismic waves. In this type of counter technique the buildings are designed by civil engineering such that the seismic wave gets dissipated and its effect gets reduced over time. The passive system has its own advantages as well as disadvantages that are discussed below.

The catastrophic effects by seismic waves often leaves a deep impact on mankind in almost every aspect. Most of this is caused due to falling of buildings and structures. Therefore, it is indispensable to focus on structural behavior during such event and strengthening the buildings. Passive Energy system is one of the techniques used to achieve this purpose. The passive energy system consists of devices that are already present inside the structure or the building. These devices that are installed inside the structure and helps in dissipating and absorbing the energy from seismic waves.

2.2.2.1 METHODOLOGY

The passive system doesn't require any external source for the dissipation of energy. The most commonly used system in passive energy system are base isolation and energy dissipation. Base Isolation is a type of technique according to which the superstructure is kept completely isolated from the substructure with the help of isolating units and components. Because of this isolation, the ground and the structure are no longer in contact with each other. Therefore, making seismic waves to pass by the building being absorbed by the absorbers. Some examples of base isolation systems are:

Elastomeric System: In this system, rubber bearings and steel plates are used in alternative layers. Rubber to absorb the wave and steel to avoid excessive bending.

Lead-Bearing: It is analogous to the elastomeric system, with an addition lead core to reduce the excessive displacement by inducing damping effect.

Friction system: This type of system uses large, rectangular dampers with high energy dissipation property. It consists of steel plates, specially designed and treated to generate good friction and steel bolts to allow slippage at a fixed load.

Fluid Dampers: Fluid dampers consists of oil filled cylinders attached to the piston type of damper which moves in response to the force exerted due to the seismic waves. Whenever vibration occurs, the liquid moves inside the cylinder leading to dissipation of seismic energy.

Advantages of Passive Structural Vibrational Control:

a) No external power source is needed.
b) It can be installed in a pre-existing structure.
c) Easy maintenance and cost effective.

Limitations of Passive Structural Vibrational Control:

a) Large scale machinery and heavy engineering is required for installation.
b) Highly skilled professionals are required to build up this setup.
c) Design is quite multifaceted and complex.

2.3 *Energy harvesting methods*

Energy harvesting is defined as the process of extraction energy from the surrounding system or environment and its conversion to useable electrical energy. The energy is aimed to be captured from ambient sources as well as other generated energy sources. Rapid advancements in microprocessors and wireless technologies over the last decade has drastic influenced the use of autonomous systems for structural health monitoring [4]. Use of wireless technologies for Structural health monitoring possess multiple advantages over the traditional methods as discussed above. The most substantial advantage is that this system can provide continuous and real time monitoring without enormous investment and complex networks. The Structural Health monitoring system functioning can be broadly classified into 3 (three) categories as explained below:

Operation Evaluation: The objective of operational evaluation is to define basic skeleton of the system. It fundamentally describes the extent upto which damage has to detected, what is

Table 1. Comparison between various techniques.

Methods	Cost	Time	Efficiency	Technical Expertise	Complexity
Non-Destructive Method	Expensive	Quick	High	Intermediate	Simplified
Active Vibration Control system	Expensive	Quick	Low	Advanced	Complex
Passive Vibration Control system	Economic	Prolonged	High	Basic	Complex
Energy Harvesting Method	Expensive	Prolonged	High	Advanced	Simplified

the optimal operational conditions, the methodology to extract the data from sensors, economic feasibility and tailoring monitoring system to obtain the most desired results.

Data Acquisition: It predominately defines the type of sensors to be used, number of sensors and their respective locations, software for acquiring, storing and processing the data. The data acquired is stored and processed continuously for real-time monitoring of the structure. The main purpose lies beyond measuring the damage to structure, rather, it is to sense and measure the response of structure to different conditions and loading. By achieving the latter one we can understand the behavior of system under abnormal conditions and it enables us to predict the health of structure at any particular time.

Feature Extraction and Statistical Modelling: After acquiring and storage of data it has to be analyzed and correlated to obtain the desired outputs. One of the most widely technique is damage-sensitive feature, it involves amalgamation and correlation of sensor's response to structure behaviour and presence of fault/damage in the structure. The response of the sensors changes in consistent pattern with varying damage and identification of feature that can precisely distinguish between damaged and undamaged structure is prime focus of the statistical modelling.

2.3.1 *Solar energy harvesting*

Wireless SHM sensors use solar cells or photovoltaic (PV) cells, it converts the solar energy into the direct usable electricity which harvest solar energy. When the sunlight comes in contact with the PV cell photons, the energy gets absorbed by silicon type semiconducting materials. To harvest electrical energy from light energy, Photovoltaic or solar cell are used which is already an established technology. Thus, for SHM purposes, nodes with solely super-capacitor type of energy storage should be used, as batteries will eventually need replacement, while super capacitors do not.

3 CONCLUSIONS

Structure Health Monitoring is becoming indispensable with every day passing especially for public infrastructure like Metros, Govt. Establishments, and Bridges etc. SHM possess a lot of advantages regarding the safety and reliability of the structure, but it needs extensive research and substantial improvements to establish highly effective and economic techniques. Advancements in the field of wireless sensors and material sciences is essential for its application on a broader scale.

Further, all the methodologies discussed above only identifies the presence of damage/crack in the structure. To resolve this a precise approach has to be developed to record the co-relation between the measured responses and structural parameters of interest. A comparison of various aspects for techniques discussed above can be seen in Table 1.

REFERENCES

1) W. Van Der Lindena, Abbas Emami-Naeini, Yilan Zhang And Jerome P. Lynch, Cyber-Infrastructure Design and Implementation for Structural Health Monitoring, Nondestructive Characterization for Composite Materials, Aerospace Engineering, Civil Infrastructure, And Homeland Security 2013.
2) Sean M. O'connor, Yilan Zhang, et.al., Automated Analysis of Long-Term Bridge Behavior and Health Using A Cyber-Enabled Wireless Monitoring System, Nondestructive Characterization for Composite Materials, Aerospace Engineering, Civil Infrastructure, And Homeland Security 2014.
3) Shinae Jang, Hongki Jo, Et.al., Structural Health Monitoring of A Cable-Stayed Bridge Using Smart Sensor Technology: Deployment, Evaluation & Data Analyses, Smart Structures And Systems, Vol. 6, No. 5-6 (2010) 439–459.
4) Shashank Priya, Daniel J. Inman, Energy Harvesting Technologies, Springer 2009.
5) Rabih Alkhatib and M. F. Golnaraghi, Active Structural Vibration Control: A Review, The Shock and Vibration Digest 2003; 35; 367.

Technologies for Sustainable Development – Mahajan, Patel & Sharma (eds)
© 2020 Taylor & Francis Group, London, ISBN 978-0-367-33737-7

Characterization of air based pneumatic damper for seismic response control

Anukant Jadeja
PG Student, Civil Engineering Department, Institute of Technology, Nirma University, Ahmedabad

Utsav Koshti
Assistant Professor, Civil Engineering Department, Institute of Technology, Nirma University, Ahmedabad

Sharad Purohit
Professor, Civil Engineering Department, Institute of Technology, Nirma University, Ahmedabad

ABSTRACT: In the 20th Century use of lighter material such as steel compared to RCC structure has increased in construction of building. These lighter materials are susceptible to dynamic loading like earthquake, wind etc. Structure made from steel subjected to dynamic loading can result in high response if the natural frequency matches the forcing frequency. To counter this force many strategies are used to control seismic response of the structures. For the present study, development of passive damping device to control seismic response of steel building is targeted. An attempt is made to characterize Pneumatic base passive damper through an experiment. The Pneumatic damper is fabricated and tested for cyclic loading in the developed test-setup. Hysteric behavior of Pneumatic type damper is studied. These dampers are also installed in SDOF building system and analyzed under free and forced vibration. Dynamic properties like damping coefficient and frequencies are ascertained under free and forced vibration. Forced vibration study is carried to evaluate coupling and transmissibility of the SDOF building model. Significant increase in damping is found when Pneumatic dampers are added to structure.

1 INTRODUCTION

Severe damage and poor performance of RCC buildings in India during earthquakes are a matter of serious concern. Earthquakes cause tremendous damage to life and infrastructure, even in countries with modern construction codes and practices. To reduce the economic disruption caused by an earthquake and to create more resilient, sustainable cities, high-performance structural systems that can withstand strong ground shaking with little or no structural damage are required. Seismic response control of buildings can be broadly classified as shown in Passive, Active and Semi-active devices.

2 AIR BASED PNEUMATIC AND METAL-FRICTION BASED DAMPERS

Various passive, semi-active and active damping devices are used to increase damping capacity. Soong and Spencer (2002) Passive damping devices are widely used as it is relatively inexpensive and stable. Metallic dampers were first manufactured in Japan and New Zealand about 50 years ago. Motamedi and Nateghi-A. (2018) have carried out experimental investigations on energy absorbers on accordion metallic damper. Ahmadie Amiri, Najafabadi, and Estekanchi (2018) have conducted experiment on block split type of dampers. Garivani et al. (2016) has conducted experiment on comb teeth shape which transfer the energy through in-plane flexural yielding. Many innovative devices in terms of material and shape such as Piston

Figure 1. Air based pneumatic damper.

Based Self centering (PBSC) Bracing System Haque and Alam (2017), Pre-compressed Rubber Springs and a Flag-shaped Choi et al. (2017) and Self-centering Buckling Restrained Brace Dong et al. (2017) are effectively used to control the energy dissipation in the various structural system. Many experiments and numerical study were conducted on viscous and viscoelastic material to increase passive control of the steel structures Silwal et al. (2015).

In the present study development and fabrication of air based Pneumatic damper is undertaken. Air based Pneumatic damper had two knobs which could be tightened to control the movement of the piston as shown in Figure 1, which increased the damping capacity of the Air based Pneumatic damper.

3 CHARACTERIZATION OF AIR BASED PNEUMATIC DAMPER

Cyclic loading test is conducted on the passive damping devices to understand its behavior. An experimental set-up is developed to conduct cyclic testing. An existing facility of the Civil

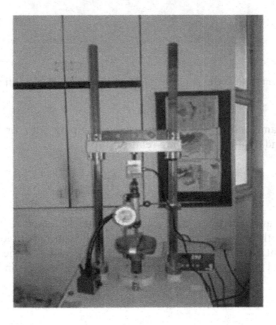

Figure 2. Experimental setup for passive damper for performing cyclic load test.

Engineering Department is used to impart cyclic loading on passive damping devices. The characterization of damper is done by finding damping capacity.

3.1 Test setup and loading protocol

The characterization of air based Pneumatic Damperis carried through cyclic loading of 2.5mm/min. Forces produced in the dampers are measured using S-Type load cell. Dial Gauge is used to measure the displacement at the interval of 1 mm displacement. There are two knobs at each end which are used to control the damping. In the air based Pneumatic damper both the knobs are completely tightened when performing the cyclic loading test. The rate of loading was fixed at 2.5 mm/min for the dampers and cyclic test is done.

3.2 Experimental results and discussion

Load-Displacement curve for 6 cycles is obtained. Displacement at the interval of 1 mm is measured from the dial gauge. One cycle consists of displacement of 12 mm in one direction i.e. the piston moves inwards and then 24 mm cycle in opposite direction i.e. the piston moves outwards. At last piston moves 12 mm inwards which completes the one cycle. Load is measure along the each displacement reading. Load displacement curve is obtained for 6 Cycles. Test was performed on air based pneumatic type damper the type of the dampers.

3.2.1 Equivalent damping and stiffness ratio

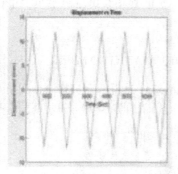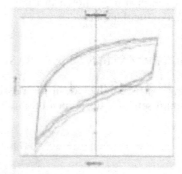

Figure 3.   (a) displacement vs time curve for piston Type damper (b) Force vs time curve for pneumatic type damper.

To evaluate the parameter of the equivalent viscous damping ratio for Air based Pneumatic damper was defined and it has the following relationship

$$\varsigma = \frac{Ed}{2\pi E_s} = \frac{Ed}{2\pi K_{eff} D^2} \tag{1}$$

where Ed is the dissipated energy in each cycle of hysteresis equal to the enclosed area of a complete cycle. Es is the strain energy stored in an elastic spring with an equivalent stiffness K_eff and displacement D, whose equation is $(1/2)K_eff D^2$ Keff is the equivalent stiffness, which can be calculated by the maximum strengths and displacements in both directions as in the formula below;

Table 1. Damping ratio of pneumatic damper.

Sr. No	Ed	Pmax	Pmin	Dmax	Dmin	Keff	z	C_{avg}
1	88.825	3.75	-4.85	12	-12	0.358	0.274	
2	99.4	4.15	-5.8	12	-12	0.414	0.265	
3	93.3	4.25	-5.1	12	-12	0.389	0.265	
4	97.4	4.3	-5.2	12	-12	0.395	0.272	0.267
5	90.85	4.25	-4.7	12	-12	0.372	0.269	
6	86.85	4.25	-4.55	12	-12	0.367	0.257	

$$\varsigma = \frac{P_{\max} - P_{\min}}{D_{\max} - D_{\min}} \tag{2}$$

Here, Ed is the area under the curve of hysteresis curve.

4 STRUCTURAL RESPONSE CONTROL OF BUILDING MODELS

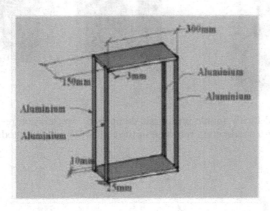

Figure 4. SDOF building model.

The air based Pneumatic damper is attached as single and double bracings in the SDOF building model as shown in Figure 4. These models are subjected to free and forced vibration. Forced vibrations is given using shake table in which the frequency is varying and the response is measured using LabVIEW Software. Response quantities such as displacement, acceleration and the damping are evaluated.

4.1 *Response of SDOF building model*

SDOF building model is prepared and tested under sinusoidal loading with shake table. Response quantities of building models such as acceleration, nature frequency and damping ratio are measured.

Transmissibility Plot is shown in Figure 7. Here,"SDOF 1" is"SDOF model" with Single Pneumatic Damper,"SDOF 2" is SDOF model with Double Pneumatic Damper and "SDOF Forced" is SDOF model without Pneumatic damper subjected to forced vibration.

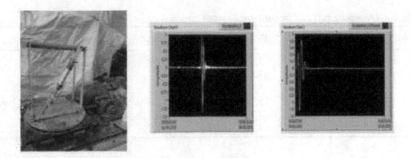

Figure 5. (a) SDOF model with single pneumatic damper (b) Acceleration response of the controlled building model (c) Extracted acceleration response of the controlled building model.

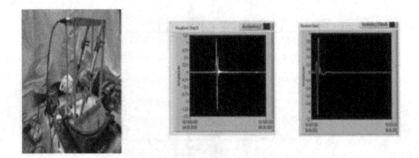

Figure 6. (a) SDOF model with double pneumatic damper (b) Acceleration response of the controlled building model (c) Extracted acceleration response of the controlled building model.

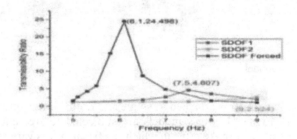

Figure 7. Transmissibility ratio of SDOF building model.

Table 2. SDOF building model.

Cases	Natural Frequency (Hz)	Damping Ratio (Z)
Bare SDOF Model	6.59	.757%
SDOF Model with single Pneumatic Damper	8.49	23.55%
SDOF Model with Double Pneumatic Damper	9.9	59.82%

132

5 SUMMARY AND CONCLUSIONS

The characterization of Air based Pneumatic Damper is performed. Damping Ratio for Pneumatic damper damper is found to be 0.267 from the Hysteric curve obtained by performing cyclic load test on the damper. SDOF system is considered for the present study in which Air based Pneumatic damper is added and the natural frequency of the system and damping ratio are compared. Damping ratio is observed to be very high when SDOF building model attached with single and Double Pneumatic Damper. Natural frequency for SDOF system increases when double Pneumatic damper are attached compared to the single pneumatic damper. The transmissibility plot depicts reduction in response for the SDOF test model when attached with Pneumatic Damper.

REFERENCES

Hossein Ahmadie Amiri, Esmaeil Pournamazian Najafabadi, and Homayoon E. Estekanchi. Experimental and analytical study of Block Slit Damper. *Journal of Constructional Steel Research*, 141:167–178, 2018.

Eunsoo Choi, Heejung Youn, Kyoungsoo Park, and Jong Su Jeon. Vibration tests of precompressed rubber springs and a flag-shaped smart damper. *Engineering Structures*, 132:372–382, 2017.

Huihui Dong, Xiuli Du, Qiang Han, Hong Hao, Kaiming Bi, and Xiaoqiang Wang. Performance of an innovative self-centering buckling restrained brace for mitigating seismic responses of bridge structures with double-column piers. *Engineering Structures*, 148:47–62, 2017.

S. Garivani, A. A. Aghakouchak, and S. Shahbeyk. Numerical and experimental study of comb- teeth metallic yielding dampers. *International Journal of Steel Structures*, 16(1):177–196, 2016.

A. B.M.Rafiqul Haque and M. Shahria Alam. Hysteretic Behaviour of a Piston Based Selfcentering (PBSC) Bracing System Made of Superelastic SMA Bars - A Feasibility Study. Structures, 12:102–114, 2017.

Mehrtash Motamedi and Fariborz Nateghi-A. Study on mechanical characteristics of accordion metallic damper. *Journal of Constructional Steel Research*, 142:68–77, 2018.

Baikuntha Silwal, Robert J. Michael, and Osman E. Ozbulut. A superelastic viscous damper for enhanced seismic performance of steel moment frames. *Engineering Structures*, 105:152–164, 2015.

T T Soong and B F Spencer. Supplemental energy dissipation: state-of-the-art and state-of-the- practice. *Engineering Structures*, 24(3):243–259, 2002.

Generation of wind time history data for nonlinear wind analysis of buildings

Shubhamkumar Patel
Postgraduate Student, Department of Civil Engineering, School of Engineering, Institute of Technology, Nirma University, Ahmedabad, India

Paresh Patel
Professor, Department of Civil Engineering, School of Engineering, Institute of Technology, Nirma University, Ahmedabad, India

ABSTRACT: Cyclone and Hurricanes are one of the natural causes of devastating forces. Many destructive cyclone such as Cyclone Vayu (2019), Cyclone Fani (2019), Cyclone Gaja (2018), Cyclone Nilofar (2014), Cyclone Phailin (2013) and others have affected major cities with heavily occupied residential or business areas. Due to these natural calamities buildings are affected to a great extent causing major loss to individuals and nation. This necessitate development of some reliable method for assessment of the building under expected level of wind storms. Wind Performance Based analysis approach is used to predict the performance of the building under extreme wind events. For performance based analysis primarily wind time history data are required as input parameter. In practice, various procedures like Anemometers, Doppler radar and Laser-based LIDAR or wind tunnel test are used for estimating wind time history data. However, all these procedures are time consuming and expensive. As an alternate to these methods an analytical procedure for generation of wind time history point load data is presented in this paper. Aerodynamic database developed by TPU (Tokyo Polytechnic University) is adopted for generation of wind time history data. TPU aerodynamic database provides wind pressure data of low-rise and high-rise building based upon wind tunnel tests. An example is presented to illustrate the procedure to generate wind time history.

1   INTRODUCTION

Now a days rapid growth of the high-rise buildings in high-density urban areas in or near to coastal areas is observed. Generally, coastal areas are vulnerable to hurricane or wind storms. The increase in population and buildings in coastal areas pose many challenges during disaster management and also causes heavy economic losses. Recently many destructive cyclone such as Cyclone Vayu (2019), Cyclone Fani (2019), Cyclone Gaja (2018), Cyclone Nilofar (2014), Cyclone Phailin(2013) and others have affected major cities of India with heavily occupied residential or commercial areas. Due to these natural calamities buildings are affected to great extent causing major loss globally. Inefficiency of the current wind design approach for computing the performance of the building during extreme wind events. All these facts necessitated use of dependable wind design and building performance assessment approach which describes performance of the building when it is subjected to expected level of wind hazards. Performance based engineering approach results into efficient and economical design for new buildings as well as retrofitting of existing buildings. To carryout wind performance assessment of a building, wind time History data are required as primarily input load. In practice, various procedures like Anemometers, Doppler radar and Laser-based LIDAR or wind tunnel test are used for estimating wind time history data. However, all these procedures are time consuming and expensive. As an alternate to these methods wind time history point load data are generated analytically using Tokyo Polytechnic University (TPU) aerodynamic database. TPU provides online aerodynamic database, based on wind tunnel data for low and high rise buildings. It contains contours of statistical values of local wind pressure coefficients, graphs of statistical values of area averaged wind pressure coefficients on the wall surfaces and time series data of point wind pressure coefficients. These data are used to calculate local wind pressures and area averaged wind pressure coefficient on wall surfaces. Wind pressure coefficients are subsequently used for evaluating dynamic responses of high-rise buildings.

2 LITERATURE SURVEY

Mohammadi et al. (2019) presented wind performance evaluation of an existing 47- storey steel moment frame high-rise building by conducting 3D nonlinear dynamic response history analyses and using an incremental dynamic analysis (IDA) approach. Performance of the building was carried out by evaluating the estimated performance levels as a function of basic wind speed. Authors used both the analytical and wind tunnel procedure for generation of wind time history point load data. Wind tunnel data were obtained by conducting sets of boundary layer wind tunnel tests on 1:400 scale rigid model in the 12-fan WOW facility available at Florida International University. For analytical procedure Pressure coefficient data were obtained from the TPU aerodynamic database and found wind pressure on the prototype building by using Bernoulli's pressure equation. In this paper 3s wind velocity was converted to hourly base mean velocity by using time scale as explained by Simiu and Scanlan (1996). Resampling was considered by four time for increasing the accuracy of analysis. Author converted 64 point load time histories (total 256 time history) at each face of the building by multiplying reference pressure to exposed area around pressure tap locations. Authors found that both the procedures of wind time history generation gives satisfactory results. From the performance assessment approach they concluded that any high-rise building design can be possibly benefited by adopting suitable performance criteria.

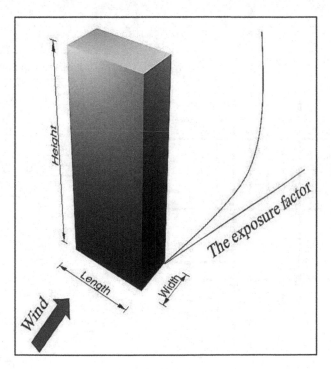

Figure 1. Length, width and height of the building.

3 GENERATION OF WIND TIME HISTORY DATA

TPU aerodynamic database provides wind tunnel data for building corresponding to different aspect ratio of building (B/D) and (B/H) as shown in Figure 1. For the 50 storey building considered in this study has plan dimensions 36 m × 36 m and overall height of the building 180 m. So, aspect ratio B/D is 1:1 and B/H is 1:5. The Alpha parameter to determine exposure factor is selected as 1:4. From these data wind pressure coefficients for different directions from 0 ° to 45 ° are obtained.

3.1 *Tokyo Polytechnic University aerodynamic database.*

TPU Aerodynamic database provides data corresponding to different wind directions and different aspect ratio of the building. Figure 2 shows typical floor plan of the building considered for illustration. Figure 3 shows typical elevation of the building in which pressure tap is located. Figure 4 shows typical data page on TPU website containing various types of raw data required for generation of wind time history. As shown in Figure 3 database provides time series of wind pressure coefficients for different direction of wind. Figure 4 shows a typical positioning of channels around the building surface and the position of pressure taps corresponding to location of channels on different faces of the building provided by the TPU database.

3.2 *Process to generation of wind time history data*

TPU Aerodynamic database gives pressure coefficient time series data. This pressure coefficient data are normalized by multiplying it with square of mean wind speed at reference elevation. Following equation is used to calculate wind pressure corresponding to each pressure tap locations:

$$P_i = 1/2 \times C_{pi} \times \rho \times V^2 \qquad (1)$$

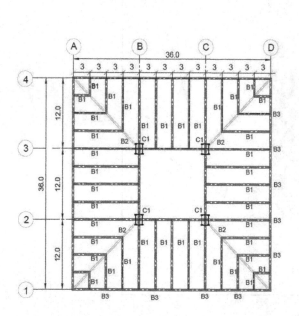

Figure 2. Typical floor plan of the building.

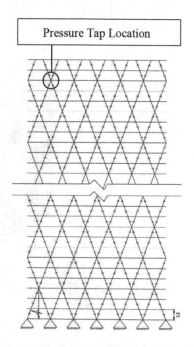

Figure 3. Typical elevation of the building.

136

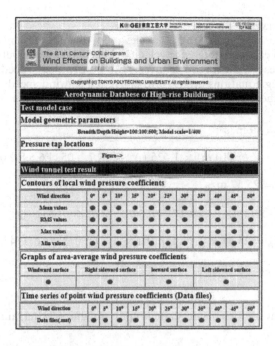

Figure 4.   Typical data page on TPU aerodynamic database.

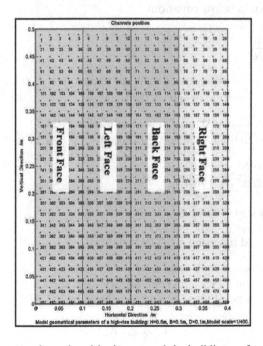

Figure 5.   Typical pressure tap channel positioning around the building surface.

Where,
P_i = Wind Pressure on the surface of the building corresponding to i^{th} pressure tap,
C_{pi} = Wind Pressure Coefficient corresponding to i^{th} pressure tap,
ρ = Density of air (1.2 kg/m^3)

3.2.1 Wind speed conversion

In this paper 3s basic wind speed at 10m elevation used as design wind speed. IS 875 (part 3):2015 is used to calculate reference velocity at each reference level of pressure tap locations. Then 3s basic mean wind speed is converted to hourly mean speed to incorporate wind storms effect. Following equations are used to converted 3s wind speed to hourly mean wind speed. In this paper Exposure condition of building is assumed between B and C.

$$[\bar{V}_{3s}(10 \text{ m})]_c = 39 \text{ m/s} \tag{2}$$

$$[\bar{V}_{3s}(10 \text{ m})]_c = [\bar{V} \, hr \, (10 \text{ m})] \times 1.525 \text{ (A.S.C.E 7 – 05 cl 6.5.8.2 page 27)} \tag{3}$$

$$[\bar{V}_{hr}(10 \text{ m})]_c = 25.57 \text{ m/s}$$

$$[\bar{V}_{hr}(z)]_B = [\bar{V}(10 \text{ m})] \times (Zref/Zc)1/\alpha_c \times (ZB/Zref \, B)1/\alpha_B \tag{4}$$

$$[\bar{V}_{hr}(180 \text{ m})]_B = 25.57 \times (274.32/10)1/6.5 \times (180/365.76)1/4$$

$$[\bar{V}_{hr}(180 \text{ m})]_B = 36.55 \text{ m/s}$$

Where,
α_C and α_B is exponent coefficients dependent upon roughness of terrain

3.2.2 Time scaling

Time scaling is used to determine the equivalent time history duration between the full scale building and model used in wind tunnel test. Time scale λ_T is related to geometric scale λ_L and velocity scale λ_V as shown in below equation:

$$\lambda_t = \frac{\lambda_1}{\lambda_V} \tag{5}$$

$$\lambda_1 = \frac{\text{Scale of Model used in Wind tunned test}}{\text{Actual height of the Building}} = \frac{0.4}{180} = \frac{1}{450} \tag{6}$$

$$\lambda_V = \frac{\text{Velocity in Wind Tunnel test model}}{\text{Actually hourly Wind speed}} = \frac{14.30}{36.55} = \frac{1}{2.55} \tag{7}$$

$$\lambda_t = \frac{1/450}{1/2.55} = \frac{1}{176.47}$$

Database time history step = 0.00195 s
Scaled time step = 0.0056× 176.47 = 0.34 s
Resampling to 4 = Scaled time step = 0.34/4 = 0.085 s

3.2.3 Generation of wind time history point load

After calculation of the Wind Pressure Time history data at each pressure tap locations, the corresponding time history point load data are generated by using following equation:

$$F_i = P_i \times A_i \tag{8}$$

Where,
A_i = Surface area corresponding to i^{th} pressure tap.

Figure 6 shows the typical wind time history point load required for Nonlinear Wind Time History analysis. Ramp-up function is also added to avoid dynamic impact effect, which influencing the dynamic response of the building as shown in Figure 6 (b).

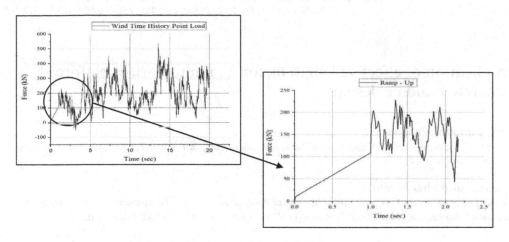

Figure 6.   Wind time-history (a) Typical wind time history point load, (b) Added ramp-up function.

4   CONCLUDING REMARKS

In this paper, step by step analytical procedure for generation of Wind Time History Point load data is presented. Wind time history data is required for carrying out wind performance based assessment of a building. Total 256 time history point load data (64 on each face of the building) are collected for simulation of dynamic wind effect. Main advantage of this method is wind time history data are calculated quickly without any cost of wind tunnel testing. This procedure can be applied for different types of real life buildings for practical acceptance and implementation.

REFERENCES

Mohammadi A., Azizinamini A., Griffis L. and Irwin P. (2019), "Performance Assessment of an existing 47-storey High Rise Building under Extreme Wind Loads", *Journal of Structural Engineering*, Vol-145 (1): 04018232.

Ashcroft J. (1994), "The relationship between the gust ratio, terrain roughness, gust duration and the hourly mean wind speed", *Journal of Wind Engineering and Industrial Aerodynamics*, Vol-53, Issue 3, page no 331–355..

ASCE 7–05 (2005), "Minimum Design Loads for Building and Other structures", *American Society of Civil Engineers*, Reston, Virginia.

IS-875 (part-3): 2015, Design loads (other than Earthquake) for Building and Structures- Code of Practise. Bureau of Indian Standard, New Delhi.

Simiu E, Scanlan R.H. (1996) "Wind Effects on Structures", *3rd Edition, John Wiley and Sons, INC*.

Implication of effective length factor on multi-storied reinforced concrete building design

Sharadkumar Purohit

Professor, Department of Civil Engineering, School of Engineering, Institute of Technology, Nirma University, Ahmedabad, India

Rutvi Ashokbhai Jogani

Student of M. Tech. in Computer Aided Structural Analysis and Design, Department of Civil Engineering, School of Engineering, Institute of Technology, Nirma University, Ahmedabad, Gujarat, India

ABSTRACT: Strength and stability are two key aspects of multi-storied Reinforced Concrete (RC) building design. While strength aspect has been given a due importance and is well understood, stability aspect in design is most underrated. Specially, structural design offices accepts codal based values of Effective Length Factor, k, to ensure stability of designed multi-storied RC buildings. Proper estimation of factor 'k', thus become important and an integral part of structural design of RC buildings. The present study aim towards realistic determination of effective length factor, k, for multi storied RC building. Implication of effective length factor, k, on structural design of a RC building is studied through geometric parameters like Stiffness Ratio, Inter-storey Drift and Maximum displacement and through design parameters like Axial Force and Bending Moment. It has been observed that, most of the commercial soft wares uses the value for effective length, k, as a unity i.e., 1 and warrants P-Δ effect making building frame as a non-sway frame. However, k-value varies largely and depends on joint stiffness ratio between column-beam.

1 INTRODUCTION

Stability of any multi storied RC building is represented by buckling of columns. Euler's buckling theory is widely accepted to estimate buckling load on the columns, however, theory has its own limitations. Major limitation being, Euler's theory treats an individual columns with different end conditions and other allied assumptions which cannot be comparable to practical point of view. Determination of effective length factor become critical. Most practically sought method for determining effective length factor k, for multi storied RC building is to use Alignment chart and Wood's chart. Wood's chart primarily depends on joint stiffness i.e., column-beam junction stiffness at the top and bottom ends of the column to estimate effective length factor. Current professional structural design practises heavily involves use of professional commercial software or customised spreadsheets for the design of compression member, column in the multi-storied building. Common tendency on use of effective length factor as 0.85 for non-sway frame and as 1.2 for sway frame for designer is not well thought off, and may lead to either conservation or unsafe design of the building. The black box type use of commercial software creates larger challenges as effective length factor value used is unknown to the user. Apart, effective length factor, k, is by enlarge treated as a geometric parameter and considers no impact from internal forces, percentage of longitudinal steel etc.

The present study is an attempt to study implications of effective length factor, k, on the structural design of a compression member, column, in the multi-storied building. The work is divided into three parts; (i) development of Wood's chart from the first principle to understand the impact of parameters on effective length factor and to validate Wood's chart provided in IS:456-2000; (ii) to consider effective length factor, k, as a geometric parameter and study it's impact through real-life multi-storied building on response parameters like maximum displacement and maximum drift with respect to stiffness ratio; and (iii) to conduct a parametric study in terms of internal forces (Axial Force, Shear Force and Bending Moment) and percentage of longitudinal steel for compression member of the real-life building.

2 DEVELOPMENT OF WOOD'S CHART

A simplified method for frame stability, using stiffness distribution imitated from moment distribution method, was developed by Wood, in the year 1974. Subsequently, Wood has developed a chart to estimate effective length of compression member, column, for non-sway and sway building frames. These charts finds place in most country codes, including Indian Design Code of practice IS: 456-2000. Using-slope deflection relationship of column-beam frame elements, equation in terms of effective length factor k, for non-sway and sway frame are obtained as mention in Equation (1) and Equation (2), respectively. Using Equation (1) and Equation (2), customised spreadsheet has been developed and Wood's charts are prepared for, both, non-sway and sway frames in the present study. When using commercial software like ETABS, compression member design invokes effective length factor, k, based on Wood's chart through equation (3) as follows. Figure (1) shows Wood's chart developed using customised spreadsheet using Equation (2) and Equation (3) and charts developed using ETABS software for comparison and validation is given in Figure (1) for sway frame.

$$\frac{s}{4}k\left[1 - c^2\left(\frac{\frac{ks}{4}}{\frac{ks}{4} + \sum^k}\right)\right] + k\left(\frac{1}{\eta_i} - 1\right) = 0 \tag{1}$$

Where,
η_i = Relative Rotational Stiffness Distribution Factor at Top of Column
η_i = Relative Rotational Stiffness Distribution Factor at Bottom of Column
k = Effective Length Factor
s = Stability Function for Non-sway Frame
c = Moment Carry Over Factor for Non-sway Frame

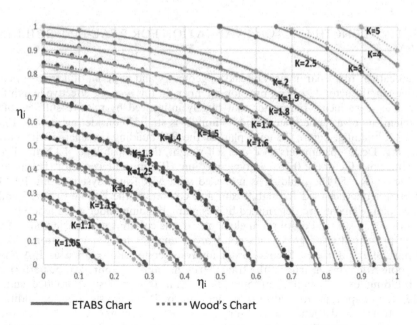

Figure 1.   Wood's chart for sway frame obtained through equations and ETABS software.

141

$$4\left[\eta_i(2\eta_j - 1) - \eta_j\right]\left(\frac{\pi}{k}\right) + \left[\eta_i\eta_j\left(\frac{\pi}{K}\right)^2 - 16(\eta_i - 1)(\eta_j - 1)\right]\sin\left(\frac{\pi}{k}\right) = 0 \qquad (2)$$

Where,
η_i = Relative Rotational Stiffness Distribution Factor at Top of Column
η_i = Relative Rotational Stiffness Distribution Factor at Bottom of Column
k = Effective Length Factor

$$k = \left[\frac{1 - 0.2\left(\eta_i + \eta_j\right) - 0.12\eta_i\,\eta_j}{1 - 0.8\left(\eta_i + \eta_j\right) + 0.6\eta_i\,\eta_j}\right]^{0.5} \qquad (3)$$

Where,
η_i = Relative Rotational Stiffness Distribution Factor at Top of Column
η_i = Relative Rotational Stiffness Distribution Factor at Bottom of Column
k = Effective Length Factor

It is evident from Figure (1) that Wood's charts plotted shows good agreement with each other and establishes that commercial software ETABS uses Wood's chart to determine effective length factor for a compression member. Wood's chart have been widely adopted world-wide for the design of buildings. However, Corrections to stiffness distribution factors of frame used while developing charts for accurate estimation of effective length factor value (Duan and Chen, 1988). Wood's charts based value of effective length factor may be inaccurate as value of k is estimated considering column as an individual element which may not be true as influence of other framing members to the column is dominating to buckling behaviour in most of practical cases (Gante and Mageirou, 2005). It has been suggested that value of 'k' used in code may be conservative due to following assumptions; (i) all column of the storey buckles simultaneously and (ii) negative stiffness induced in the column due to adjoining columns have not been considered (Webber et al, 2015). Further non-linearity, both, geometrical and material, for reinforced concrete column for developing modified charts (Tikka and Mirza, 2014). It is clear that determination of effective length factor for frame system is an active area of research.

3 EFFECTIVE LENGTH FACTOR EVALUATION FOR REAL-LIFE BUILDING SYSTEM

As discussed earlier, many parameters related to geometry and material of a frame system influence effective length factor, k, of compression member. In this section, effective length factor 'k' is considered as a geometric parameter that hugely influenced by relative stiffness of column and beam elements of a frame system and its impact is studied considering real-life Reinforced Concrete (RC) building. Figure 2 shows plan of the RC building considered with gravity and seismic loading. Design parameters used are: self-weight, floor finish load: 1 kN/m^2, live load: 2 kN/m^2, seismic zone factor (z): 0.16, soil type: medium (II), importance factor (I):1 and response reduction factor (R): 5. The building is modelled with MR frame system using ETABS software. Material properties used for slab, beam and column of the building is M25 and size of beam and column elements are determined by designing building under given loading.

Slab is modelled as a rigid diaphragm element. All the columns are considered as fixed support for the present study. Invoke of second order analysis for determination of k value is a critical part of the analysis. Second order analysis, commonly known as P-Δ analysis, is related to geometric non-linearity of the frame structure and is required to be performed when any RC building exceeds height of 50m as per an Indian design practice standard IS 16700:2017. It is important to realize that inclusion of P-Δ analysis produce additional end moment due to lateral deflection of column. Thus, total end moments for which column has to be designed increases leading to either higher section size or higher percentage of

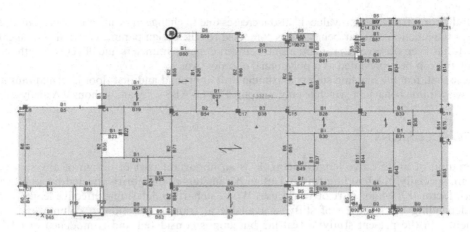

Figure 2.   Plan of real life building modelled in ETABS.

longitudinal steel. However, keeping practical aspects in mind, cracks develops at end of column which is proved through experimental investigation on the column. Development of crack results in to reduction in stiffness of the column and thus reducing bending moment. It has been found that, on invoking P-Δ analysis in ETABS software, all the columns of the RC building becomes non-sway irrespective whether building frame is sway or non-sway as defined through stability index (Q) as per IS 456:2000. Thus, value of 'k' considered for the analysis becomes one and thus reducing end moments on column that further reduces percentage of longitudinal reinforcement required. This might be a common design practice followed by professionals to keep size and percentage of longitudinal steel in practical range, but ETABS software enforces frame to behave as non-sway.

In the present study, effective length for columns of real life building as shown in Figure (2) is evaluated without invoking P-Δ analysis in the ETABS software. Apart, two important design parameters - axial force and bending moment in the columns are considered to study impact of an effective length factor on the same. A ground floor column, highlighted by circle, is identified for the study which is framed by two orthogonal beams. Design parameters at each floor level of real life building are determined and its relation with effective length of the column is represented in Figure (3). It has been observed that, with increase in axial load while moving from roof floor to

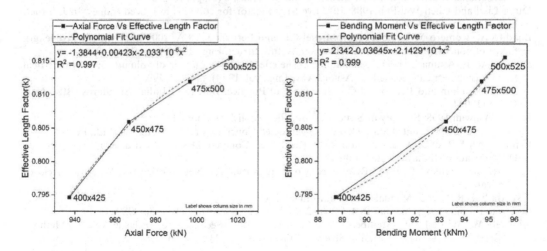

Figure 3.   Effect of change in column size on k value with respect to various design parameters.

ground floor effective length value 'k' also increases due to change in beam size so as relative stiffness at the junction. Similar behaviour is computed for the design parameter bending moment as well. It has been observed that relationship between design parameters and effective length factor of column can be represented as a polynomial of order two.

Note that, real life building suffers maximum drift at ground and first floor level and thus indicate sway frame behaviour which is only computed by ETABS software without P-Δ analysis.

4 CONCLUSIONS

Estimation of an effective length factor 'k' is an important part of design of any structural system, specially, the building frame. Design parameters are majority affected by an effective length factor 'k'. Major part of world uses Wood's chart to determine effective length factor 'k', depending upon degree of stiffness available at framing junction of column and beam elements. In the present study, a real life building is considered and is modelled in ETABS software. Effective length factor 'k' is calculate for a selected column, frame by two orthogonal beams, at each floor using ETABS software. A P-Δ analysis is not considered in the present study, as it enforce frame system to be non-sway and value of effective length factor as '1'. This is mostly followed in practice as it reduces size and percentage longitudinal steel in the column member. It has been found that effective length factor 'k' increases with design parameter - axial force and bending moment and has polynomial relationship with them. Real life building shows that frame structure sways at ground floor and first floor level and hence results in to higher percentage of steel. Use of P-Δ analysis warrants careful attention for determining effective length of column, while using ETABS.

REFERENCES

Wood R.H. 1974. Effective Lengths of columns in multi-storey buildings. In The Structural Engineering, 52(8), pp. 235–244.
Webber A., Orr J.J., Shepherd P., Crothers K. 2015. The effective length of columns in multi-storey frames. In Engineering Structures 102, pp.132–143.
Charis J. Gante and Georgios E. Mageirou 2005. Improved stiffness distribution factors for evaluation of effective buckling lengths in multi-storey sway frames. In Engineering Structures, 27, pp. 1113–1124.
Tikka T. K. and Mirza S. A. 2014. Effective length of reinforced concrete columns in braced frames. In International Journal of Concrete Structure and Materials, 8 (2), pp.99–116.
Duan Lian and Chen Wai-Fah 1988. Effective length factor for columns in braced frames. In J. Struct. Eng., 114, pp. 2357–2370.
Bendito A., Romero M.L., Bonet J.L., Miguel P.F. and Fernandez M.A. 2009. Inelastic effective length factor of non sway reinforced concrete columns. In J. Struct. Eng., 135 (9), pp. 1034–1039.
Slimani A., F. Ammari, and R. Adman 2018. The effective length factor of columns in unsymmetrical frames asymmetrically loaded. In Asian J. Civ. Eng., vol. 19 (4), pp. 487–499.
IS 456:2000, Plain and Reinforced Concrete Code of Practice, Bureau of Indian Standards (BIS), New Delhi 110002.
Kumar Ashwini 1998. Stability of Structures. In Allied Publishers Limited.
Varghese P.C. 2008. Limited State Design of Reinforced Concrete, PHI Learning Pvt. Ltd., Ed.2.
Unnikrishna S.P. and Devdas Menon 2009. Reinforced Concrete Design, 3rd Ed., Chennai: McGraw Hill Education (India) Private Limited.
Park R. and Paulay T. 1974. Reinforced Concrete Structures New York: The Wiley Interscience Publications.
CSI Analysis Reference Manual 2016. California: USA.
Concrete Frame Design Manual IS 456:2000 2015. For ETABS 2016, California, USA.
King B.S.W., White D. W., Member A, and Chen W. F. 1992. Second-order Inelastic Analysis Methods for Steel-Frame Design. Journal of Structural Engineers, vol. 118 (2), pp. 408–428.

Computer science and engineering track

Technologies for Sustainable Development – Mahajan, Patel & Sharma (eds)
© 2020 Taylor & Francis Group, London, ISBN 978-0-367-33737-7

Enhancement of policer algorithm on fast-path using VPP

Dhrumil Thakkar

PG Student, Institute of Technology, Nirma University, Ahmedabad, Gujarat, India

ABSTRACT: The Vector Packet Processing (VPP) is a framework which provides high-speed packet processing in user space. In VPP, policer algorithms use classical token bucket algorithm which checks the availability of tokens. If any time huge numbers of packets are received for processing and enough tokens are not available, many packets may be dropped. This paper proposes enhanced algorithm called Hierarchical Token Bucket (HTB). In the proposed methodology, VPP is used for fast packet processing and DPDK (Data Plane Development Kit) as a driver. The HTB is basically a tree structure in which various bandwidths are assigned to tree nodes and ranked according to levels of importance. While processing the packets, these nodes are traversed in order to check for the availability of tokens. It helps in efficient and dynamic shaping of traffic among the hierarchy of assigned bandwidths.

Keywords: Traffic shaping, hierarchical token bucket algorithm, VPP, policer, fastpath

1 INTRODUCTION

Traffic shaping and traffic policing are two components of Quality of Service (QoS) which are used to control traffic flow in different services [1]. By observing performance elements of the network to have optimum bandwidth and network performance, QoS is preferred [2]. Traffic flow and queuing plays a vital role in understanding QoS. Major factors that affect the quality are bandwidth, jitter, delay and packet loss [2]. Received traffic is verified using the traffic policy to confirm whether the received traffic is according a predefined traffic profile or not. It helps to prevent unpredicted or undesired behaviour in the network by marking or dropping non-predefined traffic.

In congestion control and quality of service, the main focus is on data traffic [3]. In congestion, we try to ignore traffic congestion and in quality of service, we try to create proper environment for traffic [3]. To avoid congestion, we use congestion control techniques such as Open Loop and Closed Loop. Open Loop Congestion Control technique is used to prevent or avoid congestion and ensur that the system never enters in a congested state. Open Loop Congestion Control techniques such as Leaky Bucket Algorithm and Token Bucket Algorithm are used. Closed Loop Congestion Control technique is used to treat congestion after it happens. One technique use the Backpressure system [2].

Considering the in Linux, Hierarchical Token Bucket algorithm helps to implement traffic control mechanism for classful queuing. HTB works on network layer (L3). HTB is basically a Tree structure where admin can configure hierarchy of assigned bandwidths. It provides more efficient and dynamic shaping of traffic and works with better efficiency. This algorithm also provides better traffic sharing bandwidth management [4].

Vector Packet Processing (VPP) framework is used for building high-speed data plane functionalities in software. VPP is a network stack which utilizes different application domains, ranging from Virtual Switch in data-centre to Support Virtual Machines. Also it is used in 4G/5G technologies [14]. The VPP is fast, scalable and deterministic software that runs on commodity CPUs [7]. It process graphically which provides high-speed processing [3]. VPP has a rich feature set [7] and it is the most efficient software packet processing engine [8]. It is a framework which allows anyone to "plug in" new graph nodes without the need to change core/kernel code [5]. It can be used as a basis for the Load Balancer, Firewall, Virtual switches, etc [6]. In this paper, use of Data Plane Development Kit (DPDK) for network Input and output is discussed [6]. DPDK is a set of data plane libraries & network interface controller driver for fast packet processing [10].

The paper is organized as follows: Section 2 holds VPP architecture and packet handling. Section 3 holds existing policer algorithms in VPP like 1r2c, 1r3c, 2r3c-2698 and 2r3c-4115. Section 4 details the proposed methodology (HTB). Section 5 explains the experimental setup and section 6 discusses the result analysis and conclusions.

2 VPP ARCHITECTURE

VPP is an open source software which was released under Fast Data IO (FD.io) project by Linux Foundation [8]. It is a networking stack [14]. VPP framework is written in C language. It runs in user space on a different architecture (x86, ARM) and embedded device with high-speed packet processing. Due to rich features VPP technology is used in Cisco router products since long. Here, the layer-3 functionalities are focused. Currently Layer-4 functionalities are in the development phase at the product developer site [8]. VPP runs with different software like DPDK, Netmap and so on. Notifications or polling plays an important role in increasing the processing throughput in high traffic conditions. VPP supports the Poll-mode [14].

The VPP follows a "run-to-completion" model [16]. Run-to-completion scheduling model in each task runs until it either finishes or explicitly turns out to get control back to the scheduler. VPP works through batch processing to increase throughput which is its unique feature. VPP consists a set of low-level libraries as well as high-level library for custom packet processing applications. In VPP framework, user can also define extensions called plugins which may define some additional functionality or replace existing ones. The main core and plugins together form a forwarding graph, which shows the possible paths for packets which can be followed during its processing.

2.1 Packet handling

As shown in Figure 1, firstly data arrives at NIC (Network interface card) then after DPDK takes data from NIC and provides to the userspace application. VPP takes data from DPDK and vice versa. Here, DPDK means Data Plane Development Kit which was created in 2010 by Intel [9]. It is a collection of set of data plane libraries and network interface controller drivers for fast packet processing [10]. It is an open source program [9]. It is used as drivers for bypassing the kernel and data transfer between user space applications (i.e. VPP) and NIC. Also it is used for fast packet processing [9].

The important thing about VPP is that it processes the packets systematically, efficiently and in "vectored" form. Each node processes all packets in a batch instead of each packet traversing the entire forwarding graph. It enhances performance.

In VPP, there are set of nodes available like input nodes, internal nodes and process nodes. Input nodes are responsible for handling data of NIC and pushing them into remaining or needed portions of the graph for further process. Input nodes and other internal nodes explicit call the internal nodes for traverse. Internal nodes get their packets from another source nodes and it has responsibility for processing incoming data in the graph. Process nodes behave like thread. These nodes are related with the time and schedule. These nodes are very useful for sending periodic notifications or polling some data that is managed by another node.

In VPP graph, total 253 nodes and 1479 edges are available. Different nodes have different functionalities and different processes. VPP's process flow starts with a node devoted to packet reception like dpdk-input and policer-by-sw-if-index. Then after, full vector is passed to next node which deals with parsing like l2 input node. After parsing, vector can be divided into multiple protocols for process which depends on type of vector. After this step, vector enters into the routing procedure and refer node like IPv4, IPv6. The process is finally done

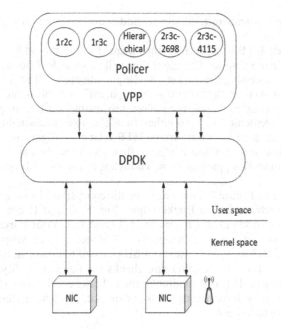

Figure 1. System architecture.

with forwarding decision. In the whole process, if any condition stays false, it decides to drop the packet at that time and refer error-drop node.

3 EXISTING ALGORITHM

In VPP, there are 4 different policer algorithms available such as 1r2c-Single Rate two Color Marker algorithm [15], 1r3c-2697 - Single Rate Three Color Marker algorithm [12], 2r3c-2698-Two Rate Three Color Marker Algorithm [11], 2r3c-4115-Differentiated Service Two-Rate, Three-Color Marker with Efficient Handling of in-Profile Traffic Policer algorithm [13].

4 PROPOSED METHODOLOGY (HTB)

The Hierarchical Token Bucket algorithm (HTB) is basically a tree structure in which band-widths are assigned to tree nodes and ranked according to levels of importance. It works on L3 layer [4]. It provides rate and ceil rate so that user can control the absolute bandwidth to particular classes of traffic as well as it indicates the ratio of distribution of bandwidth upon availability of extra bandwidth (up to ceil) [4]. HTB is proposed as improvement over the existing classical token bucket algorithm in which different policers are independent of each other. While in HTB, the same policers are maintained in hierarchy of tree nodes. The leaf nodes represent actual policers which are configured. The parent nodes provide a way to relate few of the leaf nodes belonging to same class of bandwidth allocation. The tree structure as a whole provides a view of the bandwidths distributed in hierarchy of different classes.

When packet arrives, it checks for availability of required tokens (equivalent to packet length) for a particular matching policer node. If tokens are available the packets will be transmitted. In case tokens are not available, it will traverse the tree nodes to check if it is related class of nodes have enough extra tokens. Firstly it will traverse the siblings of the node. If needed further, it will check next level of related nodes having common parent node. Finally, if extra tokens are found, the packet will be using tokens from those related nodes' tokens and packets will be allowed to transmit. If related nodes do not have enough tokens, the packet will be dropped.

The above mentioned HTB algorithm is implemented inside the VPP framework. New node structure is created as parent node. The algorithm will traverse leaf nodes and parent nodes to check for tokens while processing the packets. To help configure HTB inside VPP, few commands are newly implemented such as "configure policer_parent" is implemented to configure parent node and its related child nodes. "configure policer_htb_enable" and "configure policer_htb_disable" commands are implemented to help either enable/disable functionality. "show htb_status" and other commands are used to print current HTB related configuration. Implementation will work with existing policer creation and configure functionalities. Similar to existing policer HTB also can be used in both modes pps and kbps. Though it can work with both modes, our focus is on pps for this paper.

Example: According to Figure 2 There are five different IP addresses which are configured at L3-layer with Hierarchical Token Bucket algorithm. F, G and H are virtual nodes. Nodes A, B and C's parent is F, node D and E's parent is G and F, G node parent is H. It configured predetermined bandwidth like A = 10 Mbps, B = 3 Mbps, C = 12 Mbps, D = 10Mbps, E = 15Mbps and a total of 50 Mbps. If A wants 14 Mbps extra bandwidth, it checks sibling nodes (B, C) if available, then gives to A otherwise checks for the availability of parent node (F). Let's checks their sibling node (G) if available, then checks child nodes (D, E) for availability and to check whether they have enough tokens or not. So here E has enough tokens, so deduct from E and give to node A.

5 EXPERIMENTAL SETUP

This section shows the experimental setup used for experimenting proposed approaches. Our hardware setup includes Intel i5 core processor with 2.50GHz. We have used three different VMs and all VMs run Ubuntu 16.04.1 LTS version with two network interfaces NAT and Bridge. Each VM consists of 2GB RAM. In these VMs, VPP (Version VPP-17.10) and DPDK (Version 18.05) are installed. Other two VMs consist of 1GB RAM. The CPU 2.49 GHz with scale equal 1 is fixed for experiment in VPP. The packets are sent using two different VMs to other VMs which have installed VPP. The VPP code was altered for implementation of the HTB as explained in previous sections. The resultant compiled version of VPP was deployed on VM. Different configurations were applied on VPP to achieve different tree patterns representing various bandwidths. Packet transmission and drops were captured for these configurations. More over the achieved results are explained in next section.

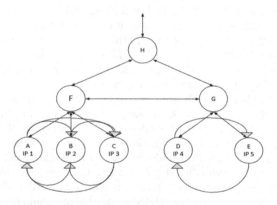

Figure 2. Proposed model.

6 RESULT ANALYSIS

It is a challenging task to reduce the number of packet drops using an efficient methodology. Here, in the implementation shown above we take different scenarios like CIR 200, 100 and CB 50, 10 with different combination and generate the result. If we pass 500 packets at a time with no siblings then according to different analysis figures, we must say that the numbers of packets dropped are 417. When the same scenario is executed with one sibling node and two sibling nodes the packet dropped are 345.5 and 278.7 respectively on an average. So it can be said that when multiple sibling nodes are used, the number of packets dropped will be less.

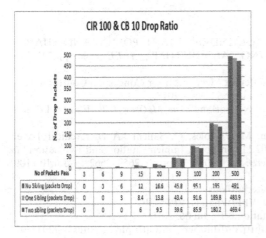

CIR 100 & CB 10 Drop Ratio

No of Packets Pass	3	6	9	15	20	50	100	200	500
No Sibling (packets Drop)	0	3	6	12	16.6	45.8	95.1	195	491
One Sibling (packets Drop)	0	0	3	8.4	13.8	43.4	91.6	189.8	483.9
Two sibling (packets Drop)	0	0	0	6	9.5	39.6	85.9	180.2	469.4

Figure 3. Scenario one.

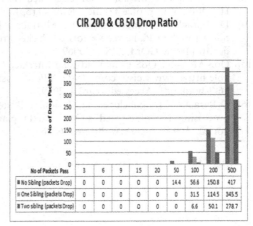

CIR 200 & CB 10 Drop Ratio

No of Packets Pass	3	6	9	15	20	50	100	200	500
No Sibling (packets Drop)	0	0	3	8	13.1	42.4	90.6	188.5	483.2
One Sibling (packets Drop)	0	0	0	2.7	6.9	35.8	80.9	180.8	466.8
Two sibling (packets Drop)	0	0	0	0	2	27.2	77.4	166.4	451.7

Figure 4. Scenario two.

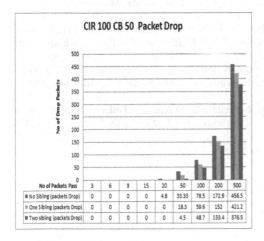

CIR 100 CB 50 Packet Drop

No of Packets Pass	3	6	9	15	20	50	100	200	500
No Sibling (packets Drop)	0	0	0	0	4.8	33.33	78.5	172.9	456.5
One Sibling (packets Drop)	0	0	0	0	0	18.3	59.6	152	421.2
Two sibling (packets Drop)	0	0	0	0	0	4.5	48.7	133.4	376.5

Figure 5. Scenario three.

CIR 200 & CB 50 Drop Ratio

No of Packets Pass	3	6	9	15	20	50	100	200	500
No Sibling (packets Drop)	0	0	0	0	0	14.4	56.6	150.8	417
One Sibling (packets Drop)	0	0	0	0	0	0	31.5	114.5	345.5
Two sibling (packets Drop)	0	0	0	0	0	0	6.6	50.1	278.7

Figure 6. Scenario four.

7 CONCLUSIONS

To cater to low number of packet drops and dynamic configuration of the network software, this paper introduces an enhanced algorithm called Hierarchical Token Bucket (HTB) in VPP. The hierarchical token bucket algorithm provides an organizational structure in which the bandwidths are ranked according to levels of importance. The implementation of HTB in VPP provides fast packet processing with better handling of the packets. Proposed methodology will help to get more efficient and dynamic shaping of traffic among predetermined bandwidth and also it provides help to get lowest packets drop adhering to configured hierarchy. Implementation of this algorithm provides better user-friendliness in terms of QoS provisioniry and data experience.

REFERENCES

[1] C.Hu, H.Saidi, P.Yan, P.S.Min, "A PROTOCOL INDEPENDENT POLICE AND SHAPER DESIGN USING VIRTUAL SCHEDULING ALGORITHM", *IEEE*, 2002,pp.791–795, doi:10.1109/ICCCAS.2002.1180731

[2] P.P.Kotian, V.K.Shetty, S.Begum, "Study on Different Mechanism for Congestion Control in Real Time Traffic for MANETS", *IRJET*, Volume: 04 Issue: 11 Nov- 2017.

[3] B.A.FOROUZAN "Data communication and Networking", *in McGraw-Hill*, Fourth Edition, 2007.

[4] J.L. Valenzuela, A. Monleon, I. San Esteban, M. Portoles, O. Sallent "A Hierarchical Token Bucket Algorithm to Enhance QoS in IEEE 802.11: Proposal, Implementation and Evaluation" *in IEEE 60th vehicular technology conference*, 2004, pp. 2659–2662, doi:10.1109/ VETECF.2004.1400539

[5] Fast data project 2018. *Available in* https://fd.io/technology

[6] Vector Packet Processing. https://wiki.fd.io/view/VPP/What_is_VPP%3F.

[7] J.DeNisco,"Vector Packet Processor Documentation 0.1", Jul 06, 2018.

[8] FD.io-vector packet processing available in https://fd.io/wpbontent/uploads/sites/34/2017/07/ FDioVPPwhitepaperJuly2017.pdf

[9] Intel. 2010. Data Plane Development Kit. Available in http://dpdk.org/.

[10] Data Plane Development Kit. Available in https://en.wikipedia.org/wiki/Data_Plane_Development_Kit.

[11] J.Heinanen, T.finland, R. Guerin, "A Two Rate Three Color Marker", RFC 2698. Sept-1999.

[12] J.Heinanen, T.finland, R. Guerin, "A Single Rate Three Color Marker ", RFC 2697, Sept-1999.

[13] O.Abou-Magd, S.Rabia." Differentiated Service Two-Rate, Three-Color Marker with Efficient Handling of in-profile Traffic", RFC 4115, July-2005.

[14] D. Barach, L. Linguaglossa, D. Marion, P.Pfister, S. Pontarelli, and D. Rossi "High-Speed-Software Data Plane via Vectorized Packet Processing" *IEEE, 2018*, pp:97–103, doi: object identifier 10.1109/MCOM.2018.1800069

[15] QoS Modular QoS Command-Line Interface Configuration Guide, Cisco IOS XE Fuji 16.7 Available:https://www.cisco.com/c/en/us/td/docs/iosxml/ios/qos_mqc/configuration/xe-16-7/qos-mqc-xe-16-7-book.pdf. Accessed: April 4, 2019.

[16] D.Barach, L.Linguaglossa, D.Marion, P.Pfister, S.Pontarelli, D.Rossi, J.Tollet. "Batched packet processing for high-speed software data plane functions" *IEEE INFOCOM 2018*, April2018, doi:10.1109/INFCOMW.2018.8406826

Technologies for Sustainable Development – Mahajan, Patel & Sharma (eds)
© 2020 Taylor & Francis Group, London, ISBN 978-0-367-33737-7

Efficient occluded face detection and classification using artificial intelligence

Ketan Bhavsar & Foram Makwana

Electronics and Communication Department, L.J.Institute of Engineering & Technology, Ahmdabad, Gujarat, India

ABSTRACT: In Current Scenario the increased rates of crime associated in house, road, parking place, ATM Security built up by surveillance techniques have been main agent for industrial and academia. Camera are already install in ATM, road, parking place, public place for capturing real time movement with facial image of users but the function in imitated for criminal investigation which could be useless when a criminal's face is to be occluded .Occlusion can be glass,mask,scarf, beard etc. also to detect face in unconstrained environment challenging due to various poses, illuminations. For solving the current issue the feature extraction method named as local patterns of gradients (LPOG) is introduced for robust face recognition. Based on LPOG descriptor, we have proposed a novel face recognition system which exploits Whitened Principal Component Analysis (WPCA) for dimension reduction and weighted angle based distance for classification. Experimental results on user define database prove that LPOG+WPCA with SSR Enhancement robust against have a wide range of challenges such as: illumination, poses, occlusion, pose, time-lapse variations and low resolution and classify face with SVM classifier.

Keywords: WPCA, LBP, LPOG, occlusions, enhancement, svm etc.

1 INTRODUCTION

Automated teller machines (ATMs) are now goals for criminals of all types. For example, fraudsters can reap card details and PINs the usage of a wide range of tactics. Physical attacks have also elevated. Without powerful protect measures, ATM crime has probably to come to be even extra attractive as the extra offerings are furnished from new ATMs. Among the feasible strategies to be define in opposition to ATM crime, real time automated alarm structures appears to be the maximum straightforward technical solutions to maximize protection. This is due to the fact the surveillance cameras are installed in almost all ATMs. However, present day video surveillance for ATMs requires consistent body of workers monitoring, which has the obvious risks of human error as a result of fatigue or distraction. Face occlusion detection has been studied for several years with some of techniques posted. The algorithms can be more or less divided into the subsequent classes. (i) Classifier based, (ii) component presence-based totally, and (iii) skin color area ratio (SCAR)-primarily based [1]. Component presence-based totally technique detects facial elements and determines face occlusion based at the presence of facial components at suitable locations. SCAR-based totally technique extracts skin color pixels in face vicinity and determines face occlusion via the percentage of area and skin coloration location. However, these techniques are particularly problem based. Most of them adopt a two-step framework: (i) a handmade function is extracted from uncooked pix (ii) a classifier is skilled based on the calculated capabilities. Such a succession of separated modules can't reach to most useful solution for the general device. Recently, the development of deep mastering has attracted plenty of attention in laptop vision research as increasingly more promising consequences are posted on a range of different vision responsibilities. Deep gaining knowledge of commonly refers to representational studying hierarchical functions from records, that is in marked contrast with hand crafting features as SIFT or HOG. The most vital attributing factor for the fulfillment is the end-to-end framework which integrates function extraction and category or different computer imaginative and prescient obligations [1].

2 EASE OF USE

In today's world of digitization face detection, classification and recognition have a huge application in mall, roads, society etc. but still we face problems like partial occlusion, low-intensity images, side view images, etc. Hence, thieves, local criminals, and terrorists take advantage of it. Government is doing a lot of work to identify thieves and criminals, which needs manual interference. Our objective to implement an automated system which can detect and recognize human face with occlusion. Different types of face occlusion are mentioned in the Figure 1 below [4].

3 RELATED METHODS

Applications of Face Recognition Person Identification: Face recognition system is used to identify people to provide authorization similar to secret identification key or password. This is mainly used to remove duplications in voter registration system, adhaar etc. thus providing security to the nation. Access Control: Face recognition system is mainly used in organizations to control the computer login and to protect the office access, thus providing security to organizations information. Security: Face recognition system is mainly used in public areas such as bus station, railway station, airport etc. to

Figure 1.   Different types of occlusion [4].

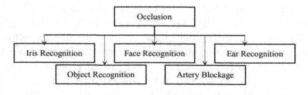

Figure 2.   Types of occlusion [4].

identify terrorist, criminals etc. Database investigation: Face recognition system is mainly used for searching licensed drivers, missing child and police booking. Video Surveillance: Face recognition system is mainly used in large organizations and malls to provide protection against terrorist attack.

3.1 *Wighted Principal Component Analysis (WPCA)*

WPCA is more efficient than Principal issue evaluation (PCA) for face recognition beneath the condition of the schooling set which has one sample according to person. Compared to PCA, the extra benefit of WPCA is that it normalizes the contribution of every important element by using the usage of whitening transformation which divides the essential additives by way of trendy deviations. The columns of U are composed of the eigenvectors of the covariance matrix. A high-dimensional vector can be compressed into a low-dimensional vector y by using projecting it on U, this is: The 2nd step of WPCA is to convert the U into W by the usage of whitening transformation [5]:

$$y_1 = U^T X \tag{1}$$

$$W = U \, (D)^{-1/2} \tag{2}$$

Where $D = \text{diag}\{\lambda_1, \lambda_2, \cdots \lambda_n\}$. Then the projected WPCA Features y_1 are

$$y_1 = W^T X = \left(U(D)^{-1/2}\right)^T X \tag{3}$$

However, WPCA may suffer performance degradation problem when the eigenvalues of the covariance matrices are very small or close to zero. If the eigenvalues of covariance matrix are too small and we use WPCA to whitening the features, it will over-amplify the influence of the small eigenvalues in feature matching. To address the issue, we standardize the WPCA features with z-score standardization (ZSCORE), which is denoted as:

$$Z = \frac{y_1 - \text{MEAN}\,(y_1)}{\text{STD}\,(y_1)} \tag{4}$$

Where MEAN (y_1) and STD (y_1) denote the mean and standard deviation of y, respectively. In face recognition,
standard Euclidean distance is used as the similarity between two face images [5].

3.2 *Local Patterns of Gradients (LPOG)*

Feature extraction method named as Local Styles of Gradients (LPOG) for strong face recognition. LPOG uses block-wised elliptical neighborhood binary styles (BELBP),a polished variation of ELBP, and local section quantization (LPQ) operators directly on gradient pix for taking pictures nearby texture patterns to build up a function vector of a face picture. From one input images, two directional gradient images are computed. A symmetric pair of BELBP and a LPQ operator is then separately implemented upon every gradient picture to generate local patterns images. Histogram sequences of local pattern images' non-overlapped sub-areas are eventually concatenated to form the LPOG vector for the given images Based on LPOG descriptor, LPOG method is faster than many advanced feature extraction algorithms and can be applied in real-world applications [2].

3.3 *Local Binary Pattern*

This technique is extraordinarily effective to give an explanation for the image texture options. LBP has blessings comparable to high-velocity computation and rotation changelessness that facilitates the huge utilization within the fields of image retrieval, texture exam, face reputation, photo segmentation, etc. Disadvantages: It is a planned technique which isn't touchy to little modifications inside the Face Localization, and victimization large native regions will increase the errors. It is inadequate for nonmonotonic illumination adjustments and is purely used for binary and grey pictures. The LBP operator is used to explicit the feel of photograph Evaluation of Histograms Local Features and Dimensionality Reduction for 3-D Face Verification patches. LBP has been broadly implemented with numerous algorithms of face reputation structures as a neighborhood feature extraction method. The LBP description of a pixel photograph is produced through thresholding the 3×3 community with the important pixel and devolving the end result as a binary code [3].

Table 1. Comparision between classifier.

Classifier	Advantage	Disadvantage
Artificial Neural network	• It is a non- parametric Classifier. • It is an universal Functional approx. with arbitrary accuracy. • capable to present funs such as OR, AND, NOT • It is a data driven self adaptive technique • Computation rate is high	• It is Semantically poor. • The training of ANN is time taking. • Problem of over Fitting. • Difficult in choosing the type network architecture.
Decision tree	Can handle nonparametric training data • Does not required an extensive design and training.	The usage of hyper plane decision boundaries parallel to the feature axes may restrict their use in which classes are clearly Distinguishable.
Support Vector Machine	It gains flexibility in the choice of the form of thethreshold. • Contains a nonlinear transformation. • It provides a good generalization capability. • The problem of over fitting is eliminated. • Reduction in computational complexity. • Simple to manage decision rule complexity and Error frequency.	Result transparency is low. • Training is time consuming. • Structure of algorithm is difficult to understand • Determination of optimal parameters is not easy when there is nonlinearly separable training data.

4 PROPOSED MODEL

4.1 *Implementation steps*

1) Apply query image from the system.
2) After that apply Pre-processing using Median filte for remove noise and other redundent data.
3) After Preprocess image apply feature Extraction methodology WPCA normalizes the contribution of every important element by using the usage of whitening transformation which divides the essential additives by way of trendy deviations.

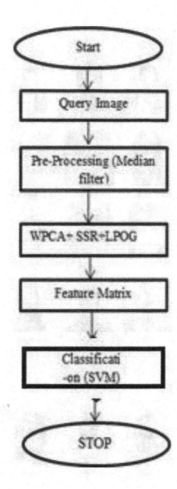

Figure 3.   Proposed model.

4) Using LPOG method for characteristic extraction, we advocate a unique FR frame paintings called LPOG whitened PCA (LPOG WPCA), whose steps are mentioned in Figure 7. From input pics and their eyes' coordinates, a face cropping system is carried out to have aligned face images. Next, an illumination normalization method is employed for face photo preprocessing. We use a retinal filter out or the histogram equalization set of rules for this step. After that, function vectors are generated from normalized images by way of using the LPOG approach. The resulting LPOG feature vectors are then projected into WPCA subspace for size reduction.

5) After that,use SSR Enhancement technique which represent Illumination containg geometric properties of the scene and Reflectance contains information about the object. Based on the assumption that the illumination varies slowlyacross different locations of the image and the local reflectance may change rapidly across different location, the processed illumination should be drastically reduced due to the high-pass filtering, while the reflectance after this filtering should still be very close to the original reflectance. The reflectance can be also finding by dividing the image by the low pass version if the original image, which is representing illumination components.

6) After that apply svm classifer on WPCA+SSR+LPOG Feature matrix and classify it.

7) Result analysis using Acuuracy,SSIM and FSIM parameters.

5 EXPERIMENTAL RESULT

Figure 4.   Occlusion image database.

Figure 5.   Testing data.

Figure 6.   Training data.

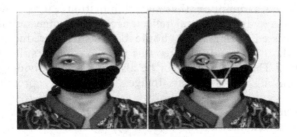

Figure 7.   Occluded face detection.

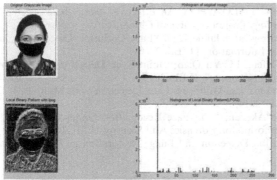

Figure 9.   Final equivalent image.

Figure 8.   LPOG histogram.

Table 2.   Result analysis.

Image	FSIM	SSIM	ACCURACY
1.JPG	0.6179	0.7366	95%
2.JPG	0.6323	0.5278	94.95%
3.JPG	0.9022	0.8254	98%
4.JPG	0.9350	0.8953	99%
5.JPG	0.9001	0.9311	96%

6   CONCLUSION

After successfully implementig proposed methodology it can be concluded that Face recognition system is very important in our daily life for security number of challenges face when detecting real time face like occlusion, intensity, resolution, skin color, shape. Occlusion, disguise, with makeup face is big challenge in face recognition. The face recognition is having very low accuracy and difficulties when performed on occluded images like occlusion detection and recovery So, our proposed approach solve existing issue like occlusion, low intensity with high accuracy as seen in result analysis table with machine learning approach.

7   FUTURE WORK

In Future work will be based on other types of database and will also try to solve other types of issue, like shape, monkey cap occlusion, and also try to improve accuracy using other classification technique.

REFERENCES

[1] Yizhang Xia, Bailing Zhang, Frans Coenen, "Face Occlusion Detection Based on Multi-task Convolution Neural Network",12th International Conference on Fuzzy Systems and Knowledge Discovery (FSKD)-IEEE, 2015.
[2] Huu-Tuan Nguyen and Alice Caplier, "Local Patterns of Gradients (LPOG) for Face Recognition", Transactions on Information Forensics and Security- IEEE, 2015.
[3] Riddhi H. Pancholi, Saket Swarndeep, "Efficient Face Recognition Method For Occluded Images", IJARIIE-ISSN(O)-2395-4396, Vol-3 2 2017.
[4] Krupa A. Patel, Prof. BarkhaBhavsar, "A Survey on Face Detection and Classification for Partially Occluded images", IJARIIE-ISSN(O)-2395-4396, Vol-5 1 2019.

[5] Chaorong Li, Wei Huang Huafu Chen, "LGLG-WPCA: An Effective Texture-based Method for Face Recognition", Science and Technology Project of Yibin of China, 2016.

[6] Gahyun Kim, Jae Kyu Suhr, Ho Gi Jung, and Jaihie Kim, "Face Occlusion Detection by using B-spline Active Contour and Skin Color Information", IEEE, 2010.

[7] Kaipeng Zhang, Zhanpeng Zhang, Zhifeng Li, Yu Qiao, "Joint Face Detection and Alignment using Multi-task Cascaded Convolutional Networks", IEEE, 2015.

[8] Cha Zhang and Zhengyou Zhang, "Improving Multiview Face Detection with Multi-Task Deep Convolutional Neural Networks", IEEE, 2014.

[9] Nese Alyuz, Berk Gokberk, and Lale Akarun, "3-D Face Recognition Under Occlusion Using Masked Projection", Transactions on Information Forensics And Security-IEEE, 2013.

[10] A.Srinivasan, Balamurugan V, "Occlusion Detection and Image Restoration in 3D Face Image", IEEE, 2014.

Technologies for Sustainable Development – Mahajan, Patel & Sharma (eds)
© 2020 Taylor & Francis Group, London, ISBN 978-0-367-33737-7

Image Super-Resolution through quick learning from self-examples

Rutul Patel

Assistant Professor, Institute of Technology, Nirma University, Ahmedabad, India
Research Scholar, Gujarat Technological University, Ahmedabad, India

ABSTRACT: Image Super-Resolution (SR) is an ill-posed problem with an objective of constructing High Resolution (HR) image from given Low Resolution (LR) image. The existing popular SR approaches include deep neural networks which require large dataset for training and also they are computationally expensive. In this paper, a novel algorithm to achieve SR which utilizes examples taken from LR itself for training have been proposed. The proposed algorithm fastens the training process via selectively fetched examples (patches) from given LR image. For the training purpose, Simultaneous Codeword Optimization (SIMCO) based dictionary learning algorithm is used along-with Least Absolute Shrinkage and Selection Operator (LASSO) for sparse representation. The trained dictionary is further utilized for constructing an HR image. Experimental results show that the proposed algorithm outperforms in terms of computational complexity and quantitative measures also with respect to deep learning algorithms.

Keywords: Image Super-Resolution, self-example, Sample Mean Square Error (SMSE), dictionary learning, sparse coding

1 INTRODUCTION

Image Super-Resolution (SR) tries to estimate High Resolution (HR) image from given one or a set of Low Resolution (LR) images. Considering this, SR algorithms are extensively used in Remote sensing, Medical Imaging, and Security Surveillance applications. Image SR is an extremely ill-posed problem since many HR images can be mapped to a single LR image. Therefore, finding the optimum HR image corresponding to a given LR image with predefined reconstruction constraint is a challenge.

Various algorithms for image SR exist in the literature, which is primarily classified into Single Image-based SR (SISR) and Multiple Image-based SR (MISR) (Capel and Zisserman 2005; Yuan, Zhang, and Shen 2012). In the case of MISR, multiple acquired LR images are merged through image registration in order to obtain the HR image. In many applications, multiple LR images are not available and therefore, SISR is much of interest nowadays. The simplest SISR algorithms are interpolation-based methods (Gao et al. 2012; S. Dai et al. 2009) which are extremely fast to obtain HR image for given LR image. However, these methods are unable to estimate high-frequency textures which yield smooth HR image. Therefore, Example-based learning methods (Baker and Kanade 2000; Freeman, Jones, and Pasztor 2002; Zhang et al. 2012) are much of interest among researchers for SISR algorithms. These learning-based methods are mainly dictionary learning-based or deep neural network-based which require large dataset for training. These optimum trained models or dictionaries through extensive dataset are further used to construct the HR image. Since each image exhibits its own texture information for which utilizing the model or dictionary which is optimum for that large dataset is not a feasible idea. Therefore, training from given LR image itself to predict HR image is more reasonable reconstruction constraint. In addition, deep neural networks are computationally expensive with respect to the dictionary learning method. Therefore, in the proposed framework, the HR and LR dictionaries are trained from the patches sampled from given LR image and its down-sampled Bicubic LR

(BLR) image respectively. Any dictionary learning methods are two-step iterative process of sparse coding and dictionary update until specified convergence criteria achieved. For sparse coding, many popular algorithms like Basis Pursuit (BP) (Chen, Donoho, and Saunders 1998), Matching Pursuit (MP) (Mallat and Zhang 1993), Orthogonal Matching Pursuit (OMP) (Tropp and Gilbert 2007) and Least Absolute Shrinkage and Selection Operator (LASSO) (Tibshirani 1991) are used. In the proposed SR algorithm, LASSO is used in the sparse coding. Following this, dictionary update can be done with widely used approximations like Maximum Likelihood (ML) (Olshausen and Field 1996), Maximum a Posteriori (MAP) (Kreutz-Delgado et al. 2003), Method of Optimized Directions (MOD) (Engan, Aase, and Hakon Husoy 1999), K-means Singular Value Decomposition (K-SVD) (Aharon, Elad, and Bruckstein 2006). However, the majority of these algorithms suffers from singularity problem which is a major bottleneck in dictionary update. In order to overcome this issue, Simultaneous Codeword Optimization (SIMCO) based dictionary learning method is proposed by Dai et al. (W. Dai, Xu, and Wang 2012), which is used in the proposed dictionary update stage. To summarize, the proposed algorithm is developed to satisfy the following objectives:

- To reduce the computational complexity of the training phase by choosing selective patches which exhibits texture information.
- To implement the training phase without the usage of the exhaustive training dataset to acquire textures from given **LR** image only.

Furthermore, to achieve these objectives, the major contribution of this paper is mentioned below:

- In most SR algorithms, patches are sampled randomly for training. However, the proposed algorithm, patch selection strategy for coupled-dictionary learning is based on Sample Mean Square Error (SMSE) (Wang and Bovik 2009) for effective (to extract meaningful patches) and efficient (computationally) training.
- For quick convergence of dictionary learning algorithm, the dictionary is initialized as a subset of selective training patches. Due to this, with respect to the randomly initialized dictionary, the previously initialized dictionary converges with five to ten folds lesser iterations.

2   PROPOSED ALGORITHM

This section describes how SMSE based chosen examples extracted from the LR image itself can be used for training the dictionaries. Furthermore, these trained dictionaries are used to obtain the HR image using sparse representation.

2.1   *Training phase*

First, for the given input LR image, obtain its down-sampled (factor s) version using bicubic interpolation refereed as BLR. From both these LR and BLR images, extract L patch-pairs each of size $sN{\times}sN$ and $N{\times}N$ respectively. Next, convert each patch into column vector $s^2N^2{\times}1$ and $N^2{\times}1$ to create training dataset as $\boldsymbol{I_{LP}} \in R^{s^2N^2 \times L}$ and $\boldsymbol{I_{BLP}} \in R^{N^2 \times L}$ respectively. Considering $\boldsymbol{I_{LP}} = \left\{ I_{LP}{}^{(1)}, I_{LP}{}^{(2)}, \ldots, \cdot\cdot I_{LP}{}^{(L)} \right\}$, SMSE for each patch of $\boldsymbol{I_{LP}}$ is computed using Equation 1 as,

$$ SMSE\left(I_{LP}^{(i)} \right) = \frac{\Sigma_{j=1}^{s^2N^2} \left(I_{LP}^{(ij)} - \Sigma_{j=1}^{s^2N^2} I_{LP}^{(ij)} / s^2N^2 \right)^2}{s^2 N^2 - 1} \tag{1} $$

Out of these L patches, patches having SMSE greater than threshold δ, are only fetched. Assuming L_{eff} patches satisfies SMSE threshold criteria, they are used to prepare effective training dataset as $\boldsymbol{I_{LPeff}} \in R^{s^2N^2 \times L_{eff}}$ and $\boldsymbol{I_{BLPeff}} \in R^{N^2 \times L_{eff}}$. Next, the objective is to train corresponding HR and LR dictionaries $\boldsymbol{D_h} \in R^{s^2N^2 \times m}$ and $\boldsymbol{D_l} \in R^{N^2 \times m}$ respectively which gives a common sparse representation of training data. Here m refers the size of a dictionary which needs to be at least larger than four times of the size of patch vector. To maintain correlation among HR-LR patch pairs, corresponding dictionaries are normalized and concatenated. To learn these dictionaries, first sparse representation is obtained using widely used LASSO as formulated in Equation 2 below,

$$\underset{\{Z\}}{argmin} \; \frac{1}{2}\|I_c - D_c Z\|_2^2 + \lambda\left(\frac{1}{M} + \frac{1}{P}\right)\|Z\|_1 \tag{2}$$

where, $I_c = \begin{bmatrix} \frac{1}{\sqrt{M}} I_{LPeff} \\ \frac{1}{\sqrt{P}} I_{BLPeff} \end{bmatrix}$, $D_c = \begin{bmatrix} \frac{1}{\sqrt{M}} D_h \\ \frac{1}{\sqrt{P}} D_l \end{bmatrix}$, M and P are dimensions of HR and LR patch respectively.

Following this, dictionaries are updated based on the SIMCO algorithm by solving the following optimization problem as per Equation 3 below:

$$\underset{\{D_h, D_l\}}{argmin} \; \frac{1}{2}\|I_c - D_c Z\|_2^2 + \mu\|D_c\|_2^2 \tag{3}$$

These updated dictionaries are further refined with a number of iterations between sparse coding and dictionary update. However, for quicker convergence, dictionaries are initialized with randomly selecting patches from training sets which satisfied SMSE criteria. After convergence, these trained dictionaries are used for reconstruction of HR patch.

2.2 Reconstruction phase

Previously learned dictionaries are now used to obtain HR image for given LR image using patch-based processing. Patch based processing refers to obtain HR patch for each LR image using the sparse coding algorithm. In this reconstruction phase, widely used ScSR algorithm (Yang et al. 2010b) is used which is summarized in Algorithm 1 as below:

Algorithm 1 Single Image Super-Resolution via sparse representation

---

Input: Learned dictionaries D_h, D_l and LR image I_{LR}.
For each extracted 5×5 patch i_{LR} of I_{LR} starting from the upper left corner with stride 1 scanning as raster-scan order,

- Convert the extracted patch i_{LR} to be zero mean by subtracting mean $\overline{i_{LR}}$ from each pixel of the patch i_{LR}
- Compute sparse vector which shares same sparse representation for HR and LR patch through,

$$\underset{Z}{argmin} \; \|D_c Z - i_{LR}\|_2^2 + \lambda\|Z\|_1$$

- Obtain HR patch $i_{HR} = D_h z$
- Add mean $\overline{i_{LR}}$ into HR patch i_{HR} and put in I_{HR0}

end
Through global reconstruction constraint, obtain the closest image to I_{HR0} which satisfies,

$$I_{SR} = \underset{I_{HR}}{argmin} \; \|SHI_{HR} - I_{LR}\|_2^2 + c\|I_{HR} - I_{HR0}\|_2^2$$

Output: Super-Resolved image I_{SR}

The results are obtained to validate the effectiveness of the proposed algorithm, the proposed algorithm is compared with popular SISR algorithms using quality metrics PSNR and SSIM as discussed in the next section.

3 EXPERIMENTAL RESULTS

All experiments are performed on a windows machine with Intel® core™ i3-5005U CPU having a 2GHz clock and 4GB of RAM. In order to demonstrate effectiveness of the proposed algorithm, popular SISR algorithms Glasner (Glasner, Bagon, and Irani 2009), SRCNN (Dong et al. 2016), ScSR (Yang et al. 2010a) along with interpolation based method is compared (for upscale factor 2) based on quality metrics Peak Signal to Noise Ratio (PSNR) and Structured Similarity (SSIM) index for widely used dataset *Set14* and shown in Table 1. To obtain these results, in the sparse coding stage, regularization parameter λ is chosen to be 0.20, and for dictionary update stage, regularization parameter μ is chosen to be 0.05 through cross-validation. Results show that average PSNR and SSIM for the proposed algorithm is comparable with SRCNN. The SRCNN is utilizing 3,95,909 images to train CNN based deep learning model where the proposed algorithm utilizes average 15k patches of size 5 × 5 extracted from LR image itself. Furthermore, the proposed algorithm outperforms with respect to ScSR which uses 100k patches for training. Hence, the proposed algorithm is computationally efficient without sacrificing quality metrics. In addition, a visual comparison is made for existing SISR algorithms and shown in Figure 1 and 2.

Table 1. PSNR and SSIM comparison of *Set14* dataset for SISR algorithms (x2).

Image	Bicubic	Glasner	SRCNN (~395k images)	ScSR (100k patches)	Proposed (~15k patches)
baboon	24.6606/0.6368	25.1119/0.6688	**25.3626/0.6932**	25.239/0.6774	25.3481/0.6908
barbara	27.9346/0.8221	28.5427/0.8415	28.5021/**0.8553**	28.527/0.8467	**28.5448**/0.8524
bridge	26.4965/0.7923	27.1901/0.8245	25.8107/0.8458	25.529/0.8337	**27.4993/0.8459**
coastguard	29.1379/0.7758	29.8068/0.8087	**30.457**/0.8357	30.2921/0.8228	30.4292/**0.8372**
comic	26.0551/0.8437	26.658/0.8638	**28.3004/0.8988**	27.6679/0.888	27.6068/0.8859
face	34.8348/0.8012	35.2177/0.8105	35.5806/0.8214	35.5411/0.8183	**35.5977/0.8229**
flowers	30.4185/0.883	31.4789/0.8893	**33.0583**/0.8987	32.3753/**0.9004**	32.3755/0.8971
foreman	32.6673/0.9428	34.1581/0.9559	33.7996/0.9581	34.4633/0.9589	**34.6546/0.9595**
lenna	34.7126/0.8521	35.7744/0.8576	**36.4613/0.8646**	36.2026/0.8623	36.3429/0.8642
man	29.26/0.8322	30.3145/0.8573	**30.808/0.8721**	30.4663/0.8642	30.6522/0.8705
monarch	32.9571/0.9509	36.2158/0.9606	**37.1023/0.9629**	35.9167/0.9612	36.3445/0.9611
pepper	33.0587/0.8361	35.0775/0.8398	33.9433/0.8403	34.1208/**0.8417**	**34.2974**/0.8416
ppt3	26.8521/0.9379	29.6587/**0.964**	**30.2398**/0.9605	28.9818/0.9612	29.9382/0.9622
Zebra	30.6785/0.9032	31.1288/0.9115	**33.2304/0.934**	32.9928/0.9296	32.94/0.9327
Avg. PSNR/SSIM	29.9803/0.8436	31.1667/0.861	**31.6183/0.8744**	31.3082/0.869	**31.6122/0.8732**

Figure 1. SISR for upscale (x2) and quantitative measures PSNR and SSIM. Left to Right: Original, Bicubic (26.0551, 0.8437), Glasner (26.658, 0.8638), SRCNN (28.3004, 0.8988), SCSR (27.6679, 0.8880), Proposed (27.6068, 0.8859).

Figure 2. SISR for upscale (x2) and quantitative measures PSNR and SSIM. Left to Right: Original, Bicubic (34.7126, 0.8521), Glasner (35.7744, 0.8576), SRCNN (36.4613, 0.8646), SCSR (36.2026, 0.8623), Proposed (36.3429, 0.8642).

4 CONCLUSION AND FUTURE WORK

Experimental results show that the proposed algorithm outperforms in terms of quantitative parameters (PSNR and SSIM) and visual quality as well over other existing algorithms. The training stage has made computationally efficient using SMSE strategy along with dictionary initialization from training patches. The threshold for SMSE is chosen as 400 empirically which selects an average 30% of maximum possible patches. This in turn, significantly fastens the training process since sparse coding iterations are saved by 70%. As shown in results, the proposed algorithm able to retrieve high frequency textures at significantly reduced training patches (~15k) compared to existing SISR algorithms which is proven quantitatively and qualitatively as well.

Results so far have been encouraging and despite this, the proposed algorithm could be computationally efficient in the reconstruction phase. Furthermore, the performance of the algorithm primarily depends upon SMSE threshold, a stochastic model needs to be developed for effective choice of threshold.

REFERENCES

Aharon, Michal, Michael Elad, and Alfred Bruckstein. 2006. "K-SVD: An Algorithm for Designing Overcomplete Dictionaries for Sparse Representation." *IEEE Transactions on Signal Processing* 54 (11): 4311–22. https://doi.org/10.1109/TSP.2006.881199.

Baker, Simon, and Takeo Kanade. 2000. "Hallucinating Faces." In *Proceedings - 4th IEEE International Conference on Automatic Face and Gesture Recognition, FG 2000*, 83–88. https://doi.org/10.1109/AFGR.2000.840616.

Capel, D., and A. Zisserman. 2005. "Super-Resolution from Multiple Views Using Learnt Image Models." In, 627–34. https://doi.org/10.1109/cvpr.2001.991022.

Chen, Scott Shaobing, David L. Donoho, and Michael A. Saunders. 1998. "Atomic Decomposition by Basis Pursuit." *SIAM Journal on Scientific Computing*. https://doi.org/10.1137/S1064827596304010.

Dai, Shengyang, Mei Han, Wei Xu, Ying Wu, Yihong Gong, and Aggelos K. Katsaggelos. 2009. "Soft-Cuts: A Soft Edge Smoothness Prior for Color Image Super-Resolution." *IEEE Transactions on Image Processing*. https://doi.org/10.1109/TIP.2009.2012908.

Dai, Wei, Tao Xu, and Wenwu Wang. 2012. "Simultaneous Codeword Optimization (SimCO) for Dictionary Update and Learning." *IEEE Transactions on Signal Processing* 60 (12): 6340–53. https://doi.org/10.1109/TSP.2012.2215026.

Dong, Chao, Chen Change Loy, Kaiming He, and Xiaoou Tang. 2016. "Image Super-Resolution Using Deep Convolutional Networks." *IEEE Transactions on Pattern Analysis and Machine Intelligence* 38 (2): 295–307. https://doi.org/10.1109/TPAMI.2015.2439281.

Engan, K., S.O. Aase, and J. Hakon Husoy. 1999. "Method of Optimal Directions for Frame Design." In *1999 IEEE International Conference on Acoustics, Speech, and Signal Processing. Proceedings. ICASSP99 (Cat. No.99CH36258)*, 2443–46 vo l.5. https://doi.org/10.1109/ICASSP.1999.760624.

Freeman, William T., Thouis R. Jones, and Egon C. Pasztor. 2002. "Example-Based Super-Resolution." *IEEE Computer Graphics and Applications*. https://doi.org/10.1109/38.988747.

Gao, Xinbo, Kaibing Zhang, Dacheng Tao, and Xuelong Li. 2012. "Image Super-Resolution with Sparse Neighbor Embedding." *IEEE Transactions on Image Processing* 21 (7): 3194–3205. https://doi.org/10.1109/TIP.2012.2190080.

Glasner, Daniel, Shai Bagon, and Michal Irani. 2009. "Super-Resolution from a Single Image." In *Proceedings of the IEEE International Conference on Computer Vision*, 349–56. https://doi.org/10.1109/ICCV.2009.5459271.

Kreutz-Delgado, Kenneth, Joseph F. Murray, Bhaskar D. Rao, Kjersti Engan, Te Won Lee, and Terrence J. Sejnowski. 2003. "Dictionary Learning Algorithms for Sparse Representation." *Neural Computation*. https://doi.org/10.1162/089976603762552951.

Mallat, Stephane G., and Zhifeng Zhang. 1993. "Matching Pursuits With Time-Frequency Dictionaries." *IEEE Transactions on Signal Processing*. https://doi.org/10.1109/78.258082.

Olshausen, Bruno A., and David J. Field. 1996. "Emergence of Simple-Cell Receptive Field Properties by Learning a Sparse Code for Natural Images." *Nature*. https://doi.org/10.1038/381607a0.

Skretting, K, and K Engan. 2010. "Recursive Least Squares Dictionary Learning Algorithm." *IEEE Transactions on Signal Processing* 58 (4): 2121–30. https://doi.org/10.1109/TSP.2010.2040671.

Tibshirani, Robert. 1991. "Regression Shrinkage and Selection via the Lasso." *Journal of the Royal Statistical Society. Series B (Methodological)*. https://doi.org/10.2307/2346101.

Tropp, Joel A., and Anna C. Gilbert. 2007. "Signal Recovery from Random Measurements via Orthogonal Matching Pursuit." *IEEE Transactions on Information Theory* 53 (12): 4655–66. https://doi.org/10.1109/TIT.2007.909108.

Wang, Zhou, and Alan C. Bovik. 2009. "Mean Squared Error: Lot It or Leave It? A New Look at Signal Fidelity Measures." *IEEE Signal Processing Magazine* 26 (1): 98–117. https://doi.org/10.1109/MSP.2008.930649.

Yang, Jianchao, J Wright, T S Huang, and Yi Ma. 2010a. "Image Super-Resolution Via Sparse Representation." *IEEE Transactions on Image Processing* 19 (11): 2861–73.

Yang, Jianchao, John Wright, Thomas S. Huang, and Yi Ma. 2010b. "Image Super-Resolution via Sparse Representation." *IEEE Transactions on Image Processing* 19 (11): 2861–73.

Yuan, Qiangqiang, Liangpei Zhang, and Huanfeng Shen. 2012. "Multiframe Super-Resolution Employing a Spatially Weighted Total Variation Model." *IEEE Transactions on Circuits and Systems for Video Technology* 22 (3): 379–92. https://doi.org/10.1109/TCSVT.2011.2163447.

Zhang, Kaibing, Xinbo Gao, Dacheng Tao, and Xuelong Li. 2012. "Multi-Scale Dictionary for Single Image Super-Resolution." In *Proceedings of the IEEE Computer Society Conference on Computer Vision and Pattern Recognition*, 1114–21. https://doi.org/10.1109/CVPR.2012.6247791.

Technologies for Sustainable Development – Mahajan, Patel & Sharma (eds)
© 2020 Taylor & Francis Group, London, ISBN 978-0-367-33737-7

Multimodal mini U-net for brain tumor segmentation

Rupal Kapdi
Computer Science and Engineering Department, Institute of Technology, Nirma University, Ahmedabad, India

ABSTRACT: The paper explores the ensemble approach for the segmentation of brain tumors from Magnetic Resonance Images. For an automated brain tumor segmentation, well-known U-net architecture has been modified for the number of layers and input image type. In this modified architecture, the total number of layers is reduced to three, which have a reduced number of parameters required for tumor segmentation. Moreover, an ensemble of three architectures is used to segment the tumor. From three different views i.e., axial, coronal and sagittal, segmentation result is combined considering the mean of the output of these networks. The highest dice similarity coefficient achieved by this approach is 0.78.

Keywords: Brain tumor segmentation, Fully Convolutional Neural Networks (FCNN), Magnetic Resonance Imaging (MRI), ensemble approach

1 INTRODUCTION

Glioma is one of the most life-threatening brain tumors. It occurs in the glial cells of the brain. Depending on the severity, it is divided into four grades ranging from grade I to grade IV. A brain tumor can further be divided into constituent parts like -Necrosis, Enhancing tumor, Non-enhancing tumor, and Edema. As brain tumor affects the soft tissues of the brain, MRI is the most advisable imaging technique, it is non-invasive imaging technique, which uses radio signals and magnetic field to capture the soft tissue functioning and it provides 3D brain images. It is time-consuming for human experts to analyze an entire image which leads to the development of semi-automated/automated tumor analysis. Since the 1990s, the researchers are working on the devising methods for brain tumor segmentation. Methods for automated brain tumor segmentation are divided into three categories: basic, generative, and discriminative methods [1]. With the growth of deep learning, most state-of-the-art methods use Convolutional Neural Network (CNN) for pixel-wise classification [1]. Similarly, a Fully Convolutional Neural Network (FCNN) focuses on semantic segmentation and segments the entire object from an image [2].

In [3], the authors are using five-layer U-net architecture. 2D axial images are augmented with flip, rotation, and transformations. Such augmented images are then supplied to U-net for segmenting the entire tumor from the given image. Authors in [4] are using four 3D U-nets to segment the tumor and fine-tune the result of the segmentation. Authors in [5] are applying to pre-process to the images then they are divided into 16x16x16 patches considering the foreground (tumor) and background (everything else). Once the segmentation is completed, erosion and dilation are applied to remove the false positives of the tumor. In [6] two 3D U-nets are used: one is for coarse segmentation of tumor which gives the entire tumor from the image and second to extract, tumor core and enhancing tumor. In [7] three different U-net architectures of depth three are proposed with pre-processing on the 2D axial images. Authors in [8] consider multi-modal images taken from the x-axis, y-axis and z-axis to train 3D U-nets. Different U-nets are given these images and majority voting is done to label the voxel.

This paper focuses on modified design of the U-net architecture to improve the results of segmentation. The report is organized as follows. Section II provides the proposed method, section III gives results followed by a conclusion and future work.

2   PROPOSED METHOD

3D MR images can be captured from three different views: axial, coronal and sagittal. 3D MR image is a 2D image stack created by capturing the image in any of the three views. Figure 1 and 2 show the images in three different views. The actual capture view of MR image gives proper information terms of voxel relationship. To take advantage of the exact view of the MR image for segmentation, three networks are derived from U-net with three-layer architecture [3]. Each network takes as an input 2D images of axial, sagittal and coronal views respectively. Figure 3 shows the proposed network architecture. The dimension of the input image for the network is specified in Table I. After training these networks, segmentation results are combined using the majority vote to get improved segmentation results.

3   EXPERIMENTAL SETUP

The multimodal brain tumor segmentation (BraTS) challenge invites researchers to develop robust brain tumor segmentation techniques from magnetic resonance imaging scans [9]. In order to pinpoint the clinical relevance of the segmentation task, the BraTS 2017 challenge also focused on the prediction of patient overall survival. The BraTS 2017 challenge comprised of two tasks: segmentation of the Gliomas, and prediction of patient overall survival. Annotations comprise the GD-enhancing tumor (ET label 4), the peritumoral edema (ED

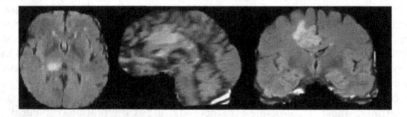

Figure 1.   (a) Axial (b) Sagittal (c) Coronal.

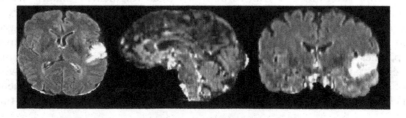

Figure 2.   (a) Axial (b) Sagittal (c) Coronal.

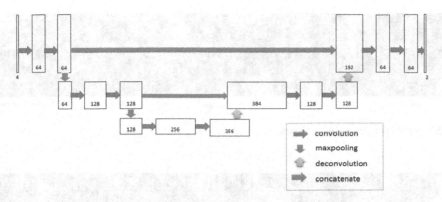

Figure 3.   Proposed network architecture.

Table 1.   Network layer dimensions.

Network Dimension	Layer 1	Layer 2	Layer 3	Layer 4	Layer 5
Axial	240x240	120x120	60x60	120x120	240x240
Sagittal	240x156	120x078	60x39	120x078	240x156
Coronal	156x240	078x120	39x60	078x120	156x240

label 2), and the necrotic and non-enhancing tumor (NCR/NET label 1), as described in the [9]. The dataset was distributed after pre-processing, i.e., co-registered to the same anatomical template, interpolated to the same resolution (1mm3) and skull-stripping [9]. The dataset has 210 HGG samples and 75 LGG samples, with each sample having four 3T MRI modalities (T1, T2, T1C, and FLAIR) along with the ground truth. Each sample has 155 slices with 240x240 pixels per slice. Also, the overall survival, defined in days, is included in a comma-separated value (.csv) file with 'Patient ID', 'Age' for 163 samples. The suggested classes for classification based on the prediction of Overall Survival were long-survivors (e.g., >15 months), short-survivors (e.g., <10 months), and mid-survivors (e.g. between 10 to 15 months).

The dataset is divided into 210 training images and 75 test images. The test set has 40 HGG images and 25 LGG images. Images are augmented using flip horizontal, flip vertical, rotation, shift, shear, zoom, brightness, and elastic distortion. All the networks are trained for a minimum of 75 epochs. Once the weights are achieved, test images are segmented on each network and output is the mean of all the outputs. These mean segmentation results are post-processed using the 3D K-nearest approach (KNN) approach with the value of K is 15 in the 26-neighborhood approach. The result of segmentation is shown in Figure 4, 5, 6 and 7.

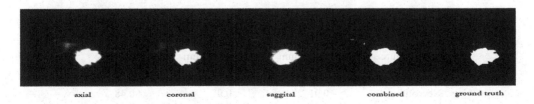

Figure 4. Whole tumor segmentation.

Figure 5. Necrotic substructure segmentation.

Figure 6. Enhancing substructure segmentation.

Figure 7. Edema substructure segmentation.

3.1 *Dice Similarity Coefficient (DSC)*

Achieved DSC is as shown in Table II. As shown in the box plot of Figure 8, when the network is trained for the entire tumor(as indicated with 1), it gives better results compared to when it is trained on individual subcomponents due to complicated voxel relationship.

Results are zero in most LGG cases as there is not much difference between brain tissues and tumorous tissues. But if only HGG cases are considered then DSC improves as shown in Table III. The box plot of the HGG cases is shown in Figure 9, it is shown that the network better understands the relationship of voxels in substructure when the network is trained on the only HGG compared to when the network is trained on HGG and LGG cases.

Table 2. Dice similarity coefficient for all cases.

Substructure	Maximum	Minimum	Average
Whole tumor	0.94	0.08	0.76
Necrotic	0.91	0.0	0.48
Enhancing	0.92	0.0	0.55
Edema	0.87	0.0	0.67

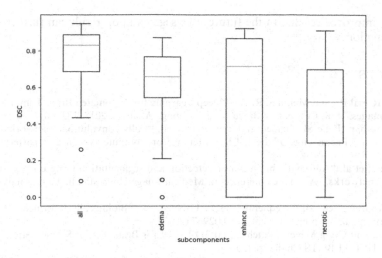

Figure 8. Variation of results for all test cases.

Table 3. Dice similarity coefficient for HGG cases.

Substructure	Maximum	Minimum	Average
Whole tumor	0.94	0.26	0.78
Necrotic	0.88	0.0	0.46
Enhancing	0.92	0.0	0.66
Edema	0.86	0.24	0.66

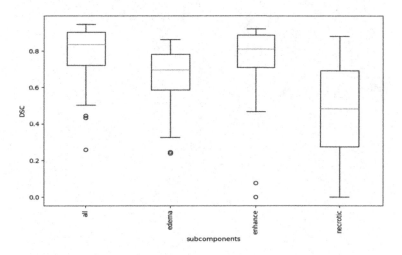

Figure 9. Variation of results for HGG test cases.

4 CONCLUSION

In this paper, U-net like architecture is used to train the fully convolutional network in three different views: axial, sagittal and coronal. As the architecture is shallow, the number of parameters is very less compared to the original U-net architecture. But by considering three view it gives improved results. In the future, this segmentation result can further be used for survival prediction.

REFERENCES

[1] Agravat, Rupal R., and Mehul S. Raval. "Deep Learning for Automated Brain Tumor Segmentation in MRI Images." Soft Computing Based Medical Image Analysis. 2018. 183–201.

[2] Long, Jonathan, Evan Shelhamer, and Trevor Darrell, "Fully convolutional networks for semantic segmentation." Proceedings of the IEEE conference on computer vision and pattern recognition. 2015.

[3] Dong, Hao, et al. "Automatic brain tumor detection and segmentation using u-net based fully convolutional networks." Annual Conference on Medical Image Understanding and Analysis. Springer, Cham, 2017.

[4] Beers, Andrew, et al. "Sequential 3D U-Nets for Biologically-Informed Brain Tumor Segmentation." arXiv preprint arXiv:1709.02967 (2017).

[5] Feng, Xue, and Craig Meyer. "Patch-Based 3D U-Net for Brain Tumor Segmentation." 2017 International MICCAI BraTS Challenge (2017).

[6] Colmeiro, RG Rodrguez, C. A. Verrastro, and T. Grosges. "Multimodal Brain Tumor Segmentation Using 3D Convolutional Networks." International MICCAI Brain lesion Workshop. Springer, Cham, 2017.

[7] Kim, Geena. "Brain Tumor Segmentation Using Deep Fully Convolutional Neural Networks." International MICCAI Brain lesion Workshop. Springer, Cham, 2017.

[8] Li, Yuexiang, and Linlin Shen. "Deep Learning Based Multimodal Brain Tumor Diagnosis." International MICCAI Brain lesion Workshop. Springer, Cham, 2017.

[9] Menze BH., et. al. "The Multimodal Brain Tumor Image Segmentation Benchmark (BRATS)", IEEE Transactions on Medical Imaging 34(10), 1993–2024 (2015) DOI: 10.1109/TMI.2014.2377694

Electrical engineering track

Technologies for Sustainable Development – Mahajan, Patel & Sharma (eds)
© 2020 Taylor & Francis Group, London, ISBN 978-0-367-33737-7

Testing of modern distance protection scheme on 12-pulse FCTCR compensated transmission line

Abhishri Jani
GTU PhD Student, Electrical Engineering, Ahmedabad, India

Vijay Makwana
Electrical Department, G H Patel College of Engineering & Technology, GTU, Vallabh Vidyanagar, India

ABSTRACT: The conventional distance protection scheme, used for the protection of transmission line, is affected by various factors, such as fault location, value of fault resistance, power angle etc. Recent research proves that the presence of compensating devices affects the performance of conventional distance relays as well. All these factors lead to change of impedance seen by distance relay, thereby affecting the overall performance of the protection scheme. This paper presents the issues faced by conventional protection scheme considering the above specified problems been encountered in the line. It also presents modern scheme to protect a doubly-fed transmission line compensated by shunt connected 12-pulse FCTCR provided at the mid-point. Performance of the modern scheme is verified in MATLAB/SIMULINK software.

1   INTRODUCTION

The power system not only refers to the grid operations but also includes a number of dynamically varying parameters and states. The modern distance relays are popularly used for the protection of long distance EHV and UHV transmission lines (>250 km) (Yingyu (2019), Kasztenny (2008), Roberts (1993)). But, the readily available numerical relays do not serve the purpose of protecting the lines against all types of faults under all the possible circumstances (Roberts, 1993). The researchers have proved mathematically and experimentally that the impedance measured by the relay during different faulty conditions depends on a number of factors, such as type of fault, fault location, value of fault resistance, power angle, type of reactive power compensation, etc. (Das (2005), Phadke (1977), Eissa (2006)).

Another variation seen in the grid includes a compensating device connected in shunt or series with the transmission lines. FCTCR is a shunt FACTS device connected to provide capacitive as well as inductive compensation at the point of connection. For better performance, i.e., to reduce the lower order harmonics, 12-pulse FCTCR is used instead of 6- pulse FCTCR. It not only reduces the harmonics but also nullifies the effect of zero-sequence current.

Yingyu (2019) has proposed a new method to calculate fault impedance and the location of the fault. This method doesn't change the impedance measured by the relaying scheme against the variations in fault resistance, fault location and power angle. Its accuracy is better in this method and response time of the method is faster than the other scheme considered by the authors for comparison. Kasztenny (2008) has proved the effect of non-homogeneity in the system on the impedance seen at the relaying point. This literature provides a strong base to the study of distance protection.

Roberts (1993) has proved the effect of zero sequence current and mutual impedance on the impedance measured at the relaying point. Albasri (2007) has presented the deleterious effects of shunt compensation on the performance of protection algorithm. The authors have proved that the presence of compensating devices leads to over-reaching or under-reaching of the conventional distance relay. It even leads to delay in response time and incorrect phase selection.

The co-author of this paper had designed a distance protection scheme which was tested for a L-G fault and simultaneous open conductor and ground fault including high fault resistance. This paper provides an extension of the same scheme for FCTCR compensated transmission line. A 12-pulse FCTCR is connected to the transmission line at the mid-point and a number of variations in system and fault parameters are considered for the study. It includes variation in fault location, power angle, and fault resistance. The authors have also covered the effect of capacitive and inductive mode of compensation on the impedance seen at relaying point.

2 ANALYSIS OF THE SYSTEM

2.1 System configuration

Figure 1 represents a three-phase double infeed transmission line model. A 12-pulse FCTCR is connected in shunt at the mid-point for improvement in power factor and reactive power management. The transmission line is connected between buses A and B. The compensator is connected at the point C (at the mid-point). An L-G fault is present at point P, which is at m percentage line length from bus A.

2.2 Analysis of L-G fault

The denotations used to make the analysis simple are: V_R and I_R are the voltage and current, respectively, measured at bus A, Z_L is the line impedance, a is the percentage line length where fault is assumed, R_F is the fault resistance, I_F is the fault current, U_{120} represent Positive, Negative and Zero sequence component of any of the above quantities, assumed as U, k_0 is the zero sequence compensation factor, δ is the power angle and ω is the deviation between I_R and I_F.

2.2.1 Impedance measured by distance relay according to conventional scheme
The impedance measured by the distance relay depends upon the type of fault. The impedance seen by the relay for L-L and L-G fault is given as (Roberts (1993))

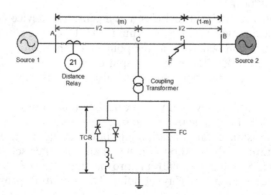

Figure 1. Three-phase double infeed transmission line.

$$m.Z_{L1} = \frac{V_R}{I_R + k_o I_{RO}} \tag{1}$$

The conventional scheme measures the impedance with a huge error as shown in results and discussion in the next section. The advanced scheme proposed by (Makwana, 2012) minimizes this error and reduces it to almost negligible value.

2.2.2 *Impedance measured by distance relay according to advanced scheme*
According to the advanced scheme given by (Makwana, 2012), the impedance of the faulted portion of transmission line, seen by the relay for L-G fault is given as

$$Z'_{L1} = \frac{X_R + R_R \sec\omega.\sin\omega}{(X_L/R_L) + \sec\omega.\sin\omega}\left(1 + j\frac{X_L}{R_L}\right) \tag{2}$$

Here, R_L is the resistance of the transmission line per kilometre and X_L is the reactance of the transmission line per kilometer. X_R and R_R are the reactance (ohm) and resistance (ohm) of the line, respectively, measured by conventional scheme.

3 RESULTS AND DISCUSSION

In this section, the implementation of conventional and advanced scheme is presented. The transmission line is made up of ACSR twin bundle moose conductor which is used for 400 kV, 50 Hz system. The line length is 300 km. The entire analysis is done on L-G fault. The software used for analysis is MATLAB/SIMULINK. The variations are considered in δ (-25° and +25°), RF (5 Ω, 10 Ω, 25 Ω, 50 Ω), m (0% to 80%, in steps of 10%) and FCTCR operation mode (Capacitive and Inductive). The result shown in tabular format gives the values of resistance and reactance in ohm, denoted by R_{conv} and X_{conv}, respectively, for conventional scheme and R_{adv}, X_{adv}, respectively, for advanced scheme. R_{true} and X_{true} represent the actual resistance and reactance (in ohm) of the faulted portion of the transmission line.

3.1 *Variation in 'm' and 'δ'*

Table 1 shows the comparison in error in the reactance measured by conventional scheme and advanced scheme considering variations in the fault locations from relay point (0%) to remote end (80%) in steps of 10%. The fault resistance is 50 Ω. δ is +25° and -25°. The same table can be understood by the graph as well, as shown in Figure 2.

Table 1. Comparison between conventional scheme and advanced scheme (δ = 25° and -25°).

m	R_{true} (Ω)	X_{true} (Ω)	R_{conv} (Ω)	X_{conv} (Ω)	Error in X_{conv} (%)	R_{adv} (Ω)	X_{adv} (Ω)	Error in X_{adv} (%)	R_{adv} (Ω)	X_{adv} (Ω)	Error in X_{adv} (%)
	True Value		Conventional Scheme			Advanced Scheme (δ = 25°)			Advanced Scheme (δ = -25°)		
0	0	0	25.45	0	NA	0	0	NA	0	0	NA
10	0.89	9.96	28.52	9.01	-9.54	0.89	9.96	0	0.89	9.95	-0.1
20	1.79	19.92	32.08	17.88	-10.24	1.79	19.93	0.05	1.79	19.91	-0.05
30	2.68	29.88	36.31	26.55	-11.14	2.69	29.93	0.17	2.68	29.88	0
40	3.58	39.84	41.47	34.91	-12.37	3.59	39.97	0.33	3.58	39.87	0.08
50	4.47	49.8	47.94	42.8	-14.06	4.49	50.06	0.52	4.48	49.88	0.16
60	5.36	59.76	55.93	49.9	-16.5	5.39	60.06	0.50	5.34	59.53	-0.38
70	6.25	69.72	67.07	55.79	-19.98	6.29	70.1	0.55	6.194	69.03	-0.99
80	7.15	79.68	83.81	59.44	-25.40	7.2	80.24	0.70	7.14	79.45	-0.29

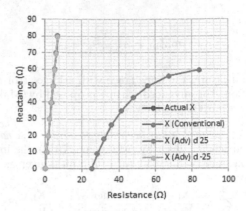

Figure 2.   Effect of variation in m seen in conventional and advanced scheme.

3.2   *Variation in 'R$_F$'*

Another variation considered is variation in the fault resistance, shown in the Table 2. δ = -25°, and m is varied from relay point (0%) to remote end (80%). The results are obtained for R$_F$ = 25 Ω and R$_F$ = 50 Ω. Figure 3 shows the graphical representation of the results obtained in Table 2. It can be seen that the value of reactance according to conventional scheme is higher than the respective actual value when R$_F$ = 25 Ω, whereas, it is lower than the respective actual value when R$_F$ = 50 Ω.

3.3   *Variation in operating mode of FCTCR*

The firing angle of the TCR can be varied in order to change the effect of capacitor in FCTCR. The FCTCR operates in capacitive mode when firing angle, α is between 90° to 145°, and in inductive mode when α is between 165° to 180°. However, the FCTCR cannot be operated from 145° to 165° as the circuit faces resonance effect within this range.

The results obtained in the Tables 1 and 2 are for α = 90° and similar results are obtained for α = 145° and α = 170°. However, the results are not shown here due to space limitation. Table 3 shows the variation in reactance for α = 145° and α = 170°. δ is -25°, R$_F$ is 50 Ω, and m is varied from 0% to 80%. Figure 4 shows the graphical representation of the results obtained. Comparing the results from Table 2 for α = 90°, the variation in reactance is higher for α = 145° and α = 170°.

Table 2.   Variation in the fault resistance R$_F$ = 25 Ω.

	True Value		Conventional Scheme			Advanced Scheme (RF = 25 Ω)			Advanced Scheme (RF = 50 Ω)		
m	R_{true} (Ω)	X_{true} (Ω)	R_{conv} (Ω)	X_{conv} (Ω)	Error in X_{conv} (%)	R_{adv} (Ω)	X_{adv} (Ω)	Error in X_{adv} (%)	R_{adv} (Ω)	X_{adv} (Ω)	Error in X_{adv} (%)
0	0	0	14.5	0.9	NA	0	0	NA	0	0	NA
10	0.89	9.96	16.82	11.44	14.86	0.89	9.93	-0.30	0.89	9.95	-0.10
20	1.79	19.92	19.52	22.16	11.24	1.78	19.88	-0.20	1.79	19.91	-0.05
30	2.68	29.88	22.73	33.13	10.88	2.68	29.83	-0.17	2.68	29.88	0.00
40	3.58	39.84	26.7	44.46	11.60	3.57	39.8	-0.10	3.58	39.87	0.08
50	4.47	49.8	31.86	56.38	13.21	4.47	49.79	-0.02	4.48	49.88	0.16
60	5.36	59.76	38.63	68.84	15.19	5.34	59.46	-0.50	5.34	59.53	-0.38
70	6.25	69.72	48.91	82.97	19.00	6.19	68.99	-1.05	6.194	69.03	-0.99
80	7.15	79.68	66.95	101.5	27.38	7.11	79.35	-0.41	7.14	79.45	-0.29

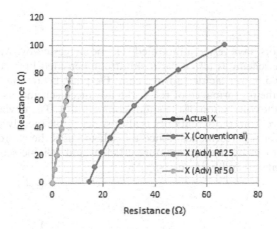

Figure 3.   Effect of variation in R_F seen in conventional and advanced scheme.

Table 3.   Variation in reactance for α = 145° and α = 170°.

	True Value		Conventional Scheme			Advanced Scheme (α = 145°)			Advanced Scheme (α = 170°)		
m	R_{true} (Ω)	X_{true} (Ω)	R_{conv} (Ω)	X_{conv} (Ω)	Error in X_{conv} (%)	R_{adv} (Ω)	X_{adv} (Ω)	Error in X_{adv} (%)	R_{adv} (Ω)	X_{adv} (Ω)	Error in X_{adv} (%)
0	0	0	30.53	1.972	NA	0	0	NA	0	0	NA
10	0.89	9.96	34.62	13.2	32.53	0.89	9.96	0.00	0.89	9.94	-0.20
20	1.79	19.92	39.58	24.85	24.75	1.79	19.92	0.00	1.79	19.92	0.00
30	2.68	29.88	45.8	37.06	24.03	2.68	29.89	0.03	2.68	29.89	0.03
40	3.58	39.84	53.88	50.19	25.98	3.58	39.88	0.10	3.58	39.88	0.10
50	4.47	49.8	64.86	64.88	30.28	4.48	49.89	0.18	4.48	49.89	0.18
60	5.36	59.76	81.52	81.99	37.20	5.39	60.03	0.45	5.4	60.22	0.77
70	6.25	69.72	109.3	104.7	50.17	6.3	70.1	0.55	6.29	69.7	-0.03
80	7.15	79.68	165.6	144.3	81.10	7.22	80.4	0.90	7.1	79.4	-0.35

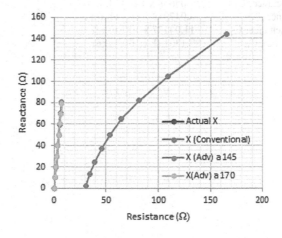

Figure 4.   Effect of variation in operating mode seen in conventional and advanced scheme.

4 CONCLUSION

In this paper the authors have tested a modified scheme suggested by (Makwana, 2012) in his early publication. The scheme is tested for a shunt compensated line, connected for reactive power management and power factor improvement. Despite the fact that the presence of compensator deteriorates the performance of relay, the advanced scheme, as tested on the circuit, works well even with the connection of the shunt compensator. The deviation of the measured values of the impedance from the actual ones is almost negligible (nearly zero), which indicates high accuracy in the relay performance. Since the scheme uses the source side parameters to find the impedance, it can be marked as more efficient and reliable scheme as compared to those which require remote- end parameters.

REFERENCES

Albasri, Fadhel A., et. al. 2007. Performance Comparison of Distance Protection Schemes for Shunt-FACTS Compensated Transmission Lines. IEEE Transactions on Power delivery. 22(4): 2116–2125.

Das, Biswarup, and J. Vittal Reddy. 2005. Fuzzy-logic-based Fault Classification Scheme for Digital Distance Protection. IEEE transactions on Power Delivery. 20(2): 609–616.

Kasztenny, et al. 2008. Fundamentals of distance protection. Protective Relay Engineers. 61st IEEE Annual Conference: 1–34.

M. M. Eissa. 2006. Ground Distance Relay Compensation based on Fault Resistance Calculation. IEEE Transactions on Power Delivery. 4(21): 1830–1835.

Makwana, V.H. and Bhalja, B.R. 2012. A new digital distance relaying scheme for compensation of high- resistance faults on transmission line. IEEE Transactions on Power Delivery. 27(4): 2133–2140.

Phadke, Arun G., et. al. 1977. Fundamental Basis for Distance Relaying with Symmetrical Components. IEEE Transactions on Power Apparatus and Systems. 96(2): 635–646.

Roberts, et al. 1993. Z= V/I does not make a distance relay. Schweitzer Engineering Laboratories. inc. Pullman. Washington. 20th Annual Western Protective Conference: 19–21.

Yingyu Liang et al., 2019. A Novel Fault Impedance Calculation Method for Distance Protection Against Fault Resistance. IEEE Transactions on Power Delivery. DOI 10.1109/ TPWRD.2019.2920690.

Zhizhe, Zhang, and Chen Deshu. 1991. An Adaptive Approach in Digital Distance Protection. IEEE Transactions on Power Delivery. 6(1): 135–142.

ANNEX
Transmission line data.

Positive sequence resistance, R_1	0.029792 Ω/km
Zero sequence resistance, R_0	0.16192 Ω/km
Positive sequence inductance, L_1	1.05678 x 10^{-3} H/km
Zero sequence inductance, L_0	3.947042 x 10^{-3} H/km
Positive sequence capacitance, C_1	11.04137 x 10^{-9} F/km
Zero sequence capacitance, C_0	7.130141 x 10^{-9} F/km

Technologies for Sustainable Development – Mahajan, Patel & Sharma (eds)
© 2020 Taylor & Francis Group, London, ISBN 978-0-367-33737-7

Solving of economical load dispatch using efficient group leader optimization technique

Kathan Shah, Kuntal Bhattacharjee & Shanker Godwal
Electrical Engineering Department, Institute of Technology, School of Engineering, Nirma University, Ahmedabad, Gujarat, India

ABSTRACT: Economical Load Dispatch (ELD) is an imperative aspect in power system operations, controls and scheduling. Convex and Non-Convex economical load dispatch can be resolved by using classified based techniques and various soft computing techniques. This paper presents the Group Leader Optimization technique algorithm to solve ELD for minimizing the fuel cost of power generation with valve point loading used as constraint. Group Leader Optimization Algorithm (GLOA) is recent and efficient technique in optimization domain. Mathematical formulation-based simulation proves the efficiency, quality of solution. The mathematical formulation can successfully optimize ELD problems. Finally, from the simulation results it has been concluded that the proposed technique has given improved results than existing optimization techniques. Numerical results are provided in the paper to verify implementation of GLOA on ELD.

1 INTRODUCTION

ELD is inevitable requirement for power utility to provide reliable cost-effective power to consumer. Economical Load Dispatch (ELD) is a method which decides the generating unit output according to definite load demand economically so the power system can be operated economically. It is a method to manage the generation of electrical units and maintaining transmission line capability so that unenterable power is available to consumer at reliable cost with satisfying the various different constraints. To cope up with above requirements the fuel cost curve can be highly non-linear but several optimization techniques are developed to obtain the nearest solution. There are several methods like Linear Programming (LP) methods and calculus method for solving ELD. Problem with calculus method is that it requires smooth and differentiable objective function for solving ELD. Linear Programming (LP) Fanshel & Lyne (1964) have some problems related with piecewise linear cost estimation. In recent years some various modern intelligent techniques are used to solve complex ELD problems. Techniques which are used are Particle Swarm Bacterial Foraging Algorithm (BFA) Panigrahi & Pandi (2008), Evolutionary Programming (EP) Sinha, Chakrabarti & Chattopadhyay (2003), sine cosine algorithm K. Bhattacharjee, N. Patel (2019), Search Group Optimization K. Bhattacharjee, N. Patel (2019) etc. All these various methods confirmed closet optimal global solution. Computational speed of methods is also superior. In the case of speed and accuracy modern techniques are superior. Some hybrid and modified techniques are also efficiently applied in ELD problem like Directional search genetic algorithm and IGA-MU. Some other population - based bio inspired algorithm techniques have common disadvantages as these methods have intricate computation and they take many restraints for computation and they are very difficult to understand for beginners. Recently a new effectual optimization technique has developed called GLOA in computational techniques. This method is developed from general idea of leaders in social groups and cooperative co-evolutionary algorithm.

2 MATHEMATICAL FORMULATION OF THE ELD PROBLEM

The conventional composition of ELD problem is to reduce the fuel costs of individual generators with consideration of power balance with respect to the real power and the demand power and also is subject to limits on generator output. The ELD may be composited as nonlinear constrained optimization problem. Here one of the complex types of test system of ELD problems have been composited and solved by GLOA method.

2.1 ELD - quadratic cost function, different constraints

Here, Purpose of economic dispatch problem is to reduce total cost of fuel (C_F) at power plant with consideration of operating constraints of system. As shown below

$$C_F = min\left(\sum_{i=1}^{n} C_F(E_i)\right) = min\left(\sum_{i=1}^{n} X_i + Y_i E_i + Z_i P_i^2\right) \tag{1}$$

E_i is power generation unit i and $C_F(E_i)$ are cost function and expressed as quadratic polynomial. $X_i, Y_i, \& Z_i$ = cost coefficient of ith generator. n = number of generators.

2.1.1 Constraint of balancing real power

$$\sum_{i=1}^{n} E_i - E_d = 0 \tag{2}$$

E_d is power demand for lossless transmission consideration,

2.1.2 Constraint of generation capacity

There are some maximum and minimum power generation limits for each unit should not violated otherwise it may cause imbalance in the system. It is necessary that the generation should be between max and min limit for stable operation.

$$E_i^{maximum} \le E_i \le E_i^{minimum} \tag{3}$$

Here $E_i^{maximum}$ and $E_i^{minimum}$ are maximum & minimum power generation from generator i^{th} unit.

2.1.3 Constraints for power balance

Once equality constraint is satisfied than power balance is easy to achieve. Summation of total demand and total load should equal to total generation.

$$\sum_{i=1}^{n} E_i = E_d \tag{4}$$

2.2 Quadratic cost function for ELD

For quadratic cost function for ELD, the objective function is the same as (1). Here it can be observed that the objective function having equal as equation (1) objective cost function is required to be minimized concerning (2), (3). losses during transmission are ignored. $E_l=0$.

2.3 Valve-point effect for ELD

$$C_F = min\left(\sum_{i=1}^{n} C_F(E_i)\right)$$
$$= min\left(\sum_{i=1}^{n} X_i + Y_i E_i + Z_i P_i^2 + |k_i * sin\{C_i * (E_i^{min} - E_i)\}|\right) \tag{5}$$

k_i and C_i are cost coefficient.

182

3 GROUP LEADER OPTIMIZATION TECHNIQUE

Group leader optimization is newly developed technique Daskin & Kais (2011). Here group members are influenced by the group leader. The leader represents the nature of the group but for the leader should have potential and abilities better than the other members. The quality of members of a group might lead to changes like blight or enhancement in new etiquette and peculiarities under the influence of the leader. The formation of the group is totally based on random selection and not based on the similar nature of the members. Different groups are created and each group has their individual group leader. In GLOA method every group tries over and over to obtain global solutions under the influence of each group leader who are the nearest and closest members of the group to local or global minima. The fitness value of each leader is the best among the group. After some iterations, if any other member from the group achieves the best fitness value from the group then that member is the new leader for that group. Some part of algorithm is creating new members. Hence the leader affects the other members of the group and revolution occurs at every iteration and group members come closer to each other. This way solution space is created between the leader and the members and the group is able to find search area for an optimum (Global or Local) solution quickly. After some iteration it may be possible that there is no much variance among fitness value of the leader and members so transfer of such variable is done between groups randomly to maintain diversity of the group. This crossover benefits the group to emerge from local minima solutions and encourages to search for new solution spaces.

Sequential Steps of GLOA.

Five steps are for GLOA algorithm can be classified as Initialization, Fitness evaluation, New member creation, Transfer(crossover).

1) Create (N) number of total populations in every group randomly within Max and Min boundary range and addressing their different constrains. So, total population is G*N where G = number of groups. Formation of groups and members are with highest randomness. Specify number of unknown variables and their Maximum and Minimum range.
2) Calculation of Fitness function for entire group - members.
3) Select leader for each group according to best fitness function.
4) Produce new members using process of mutation and recombination by using old members.

$$NEW = R_1 * old + R_2 * leader + R_3 * ran \qquad (6)$$

ran is random number between 0 to 1. Here R_1, R_2, and R_3 are the values defining the parts of old member, leader, and random values of group members while generating the new population. Summation of values of R_1, R_2, R_3 is 1. Proper values of R_1, R_2, R_3 is must require for accurate and optimum solution. Created new member or other member from group have better fitness value than leader than it will replace the leader.

5) Now, In the Crossover process parameter or variables or members are transferred from one group to another group. Transferred members are chosen randomly from the group.
Here transfer rate is H times. At one time one parameter is transferred.

$$1 \leq C_R(var/2) + 1 \qquad (7)$$

'var' is variable and its value is same as population. C_R is random number between 0 to 1.
6) Repeat with step-3 to step-5. Go with the same process until number of iteration or neck and neck accuracy achieved than dismiss the process or go to step-4 for continue process.

4 SIMULATION AND RESULT

For getting simulation results 40 generator units have taken with an assumption of no transmission losses. Data required for input taken from Ghorbani & Babaei (2016). 10500 MW is total demand. Here, the output result of EMA- QPSO Ghorbani & Babaei (2016) and IPSO Mohammadi-ivatloo, Rabiee, Soroudi & Ehsan (2012) are compared with GLOA. In Table 1, minimum fuel cost for 40 generator units is 121412.5354 \$/hr. obtained by the GLO algorithm, Better than EMA - QPSO and IPSO. The minimum, maximum, average fuel cost obtained from 50 trials are shown in Table 2. From Table 2 it is seen that GLOA is the fastest as well

Table 1. Optimum power output and fuel cost for GLOA and other techniques comparison for 40-unit system.

	Power Output (MW)			
Unit	GLOA	EMA	QPSO	IPSO
01	110.7989	110.7998	111.2000	110.8000
02	110.7989	110.7998	111.7000	110.8000
03	97.3995	97.3999	97.4000	97.4000
04	179.7333	179.7331	179.7300	179.7330
05	87.79844	87.7999	90.1400	87.8000
06	139.9998	140.0000	140.0000	140.0000
07	259.6014	259.5996	259.6000	259.6000
08	284.5974	284.5996	284.8000	284.6000
09	284.6008	284.5996	284.8400	284.6000
10	130.0001	130.0000	130.0000	130.0000
11	94.00008	94.0000	168.8000	94.0000
12	94.00170	94.0000	168.8000	94.0000
13	214.7593	214.7598	214.7600	214.7600
14	394.2783	394.2793	304.5300	394.2790
15	394.2793	394.2793	394.2800	394.2790
16	394.2764	394.2793	394.2800	394.2790
17	489.2794	489.2793	489.2800	489.2790
18	489.2794	489.2793	489.2800	489.2790
19	511.2787	511.2793	511.2800	511.2790
20	511.2784	511.2793	511.2800	511.2790
21	523.2802	523.2793	523.2800	523.2790
22	523.2797	523.2793	523.2800	523.2790
23	523.2785	523.2793	523.2900	523.2790
24	523.2797	523.2793	523.2800	523.2790
25	523.2794	523.2793	523.2900	523.2790
26	523.2806	523.2793	523.2800	523.2790
27	10.0000	10.0000	10.0100	10.0000
28	10.0000	10.0000	10.0100	10.0000
29	10.0008	10.0000	10.0000	10.0000
30	87.8012	87.7999	88.4700	87.8000
31	189.9999	190.0000	190.0000	190.0000
32	189.9975	190.0000	190.0000	190.0000
33	189.9985	190.0000	190.0000	190.0000
34	164.8005	164.7998	164.9100	164.8000
35	199.9990	200.0000	165.3600	194.4000
36	194.4057	194.3977	167.1900	199.9999
37	109.9998	110.0000	110.0000	110.0000
38	110.0000	110.0000	107.0100	110.0000
39	110.0000	110.0000	110.0000	110.0000
40	511.2802	511.2793	511.3600	511.2790
Fuel Cost(\$/hr)	121412.5354	121412.5355	121448.2100	121412.5455

Table 2. Generating cost($/hr), Time & No. of hits to minimum solution obtained by GLOA, EMA & QPSO optimization techniques for 40 generator units (50 trials).

Methods	Generation cost ($/hr.)			Time/Iteration(S)	No. of hits to minimum solution
	Maximum	Minimum	Average		
GLOA	121414.2353	121412.5355	121412.7060	0.24	45
EMA	121416.2031	121412.5355	121414.6617	0.29	21
QPSO	121455.9510	121448.2100	121453.6287	0.65	15

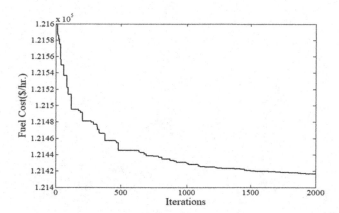

Figure 1. Convergence characteristic of GLOA for 40 generator units.

as it gives most optimum solution. The convergence characteristic of GLOA is displayed in Figure 1. The net power delivered to the system comes out to be 10500 MW. So, the level of the result accuracy is 100.00% based on (4) with transmission losses ignored.

5 CONCLUSION

An efficient GLOA population-based algorithms is proposed to tackle ELD problem. From the paper, it is clear that proposed technique is flexible, efficient and comfortable in global minima and rarely gets trapped in local minima. In this method no computationally, expensive derivatives are significant so it is quite easy. The numerical results output shows that GLOA is capable of finding extraordinary ELD solution as compared to well-regarded optimizers. Obtained output numerical results ensure the excellent capability of GLOA in convergence characteristics, solution quality and heftiness compared to other optimizers. Hence GLOA techniques results revealed that it can tackle complex ELD problems.

REFERENCES

Fanshel, S., Lynes, E. S.1964. Economic Power Generation Using Linear Programming. In IEEE Transactions on Power Apparatus and Systems 83(4): 347–356.
Panigrahi,B. K.& Pandi, V. R.2008. Bacterial foraging optimization: Nelder-Mead hybrid algorithm for economic load dispatch In IET Generation, Transmission, Distribution 2(4): 556–565.
Sinha, N., Chakrabarti, R., Chattopadhyay, P.K.2003. Evolutionary programming techniques for economic load dispatch. In IEEE Transaction Evolutionary Computing 7 (1): 83–94.

Bhattacharjee, K & Patel, N.2019. A comparative study of Economic Load Dispatch using Sine Cosine Algorithm. In International Journal of Science and Technology, Scientia Iranica, Article in press, DOI: 10.24200/sci.2018.50635.1796,

Bhattacharjee, K & Patel, N.2019., An experimental study regarding Economic Load Dispatch using Search Group Optimization In International Journal of Science and Technology, Scientia Iranica Article in press, DOI: 10.24200/sci.2019.51798.2367

Daskin, A., & Kais, S. 2011. Group leaders optimization algorithm. In Molecular Physics: An International Journal at the Interface Between Chemistry and Physics 109(5): 761–772.

Ghorbani,N & Babaei,E.2016.Exchange market algorithm for economic load dispatch. In International Journal of Electrical Power & Energy Systems 75: 19–27.

Mohammadi-ivatloo, B., Rabiee, A., Soroudi, A., & Ehsan, M. 2012. Iteration PSO with time varying acceleration coefficients for solving non-convex economic dispatch problems.In Electrical Power and Energy Systems.42: 508–516.

Artificial neural network based space vector PWM controller applied to three-level NPC shunt active power filter

Siddharthsingh K. Chauhan, Nisarg Mankodi, Malhar Patel & P.N. Tckwani
Department of Electrical Engineering, Institute of Technology, Nirma University, India

ABSTRACT: Increased penetration of non-linear loads in the power system has led to power quality deterioration. Shunt active power filters help in mitigating harmonics and thereby improving the power quality. Artificial Neural Network (ANN) based Three-level Shunt Active Power Filter (SAPF) for mitigating harmonics is presented in the paper. Neural Synchronous method based reference current generation technique is used for the control of SAPF. ANN based Space Vector Pulse Width Modulation controller is developed for the effective control of SAPF. Simulation analysis of Diode clamped based three-level SAPF is presented. Both steady-state as well as transient analysis of the proposed ANN controlled three-level SAPF is presented. Experimental investigation of the proposed SAPF using dSPACE is also presented.

1 INTRODUCTION

Extensive usage of non-linear loads has deteriorated power quality as they inject harmonics in the power system. In order to compensate for the harmonics and reactive power produced by non-linear loads, active power filters (APF) – shunt, series and hybrid are extensively used [Abdeslam 2007], [Esfahani 2015], [Pandya 2018]. The shunt active power filter (SAPF), compensates load current harmonics by injecting equal but opposite harmonic compensating currents [Bhattacharya 2011].

Artificial Intelligence (AI) techniques, like Artificial Neural Network (ANN), Fuzzy Logic (FL), Particle Swarm Optimization (PSO), etc., are finding usage in applications like, drives, industrial process control, image processing, diagnostics, medicine, space technology, power system optimization and stability, etc. [Kumar 2009].

Applications of ANN techniques is now also extended to power quality [Bose 2007]. Compared to conventionally used techniques of reference compensating current generation and current controllers for the control of active power filter, usage of ANN techniques, provides faster dynamic response due to reduced complexity and computational time requirement.

Artificial Neural Network (ANN) resembles human brain and comprises of number of interconnected processing elements termed as neurons [Wilamowski 2010]. Artificial Neural Network (ANN) techniques are reported for reference compensating current generation of SAPF [Nguyen 2011; Qasim 2014; Tekwani]. Various controllers for application to SAPF, such as, proportional-integral (PI) controller, hysteresis current controller (HCC), deadbeat controller, and predictive controller have been reported [Holtz 1992] of which HCC is the most popularly used controller due to its simplicity in implementation. However, it suffers from limitations like random switching of voltage vectors, variation of switching frequencies, limit cycle oscillations, shoot-up of current errors, etc. [Chauhan 2013, 2014; Tekwani 2015, 2017; Chudasama 2017]. Space vector pulse width modulation (SVPWM) based current control overcomes the limitations of HCC by providing optimal switching of voltage vectors [Sharma 2017, Bhandankar 2016, Arora 2018].

This paper focuses on simulation analysis as well as dSPACE based digital implementation of ANN controlled three-level SAPF. In this paper, three-level SAPF using ANN based technique – Neural Synchronous method for reference compensating current generation and ANN based space vector Pulse Width Modulation controller for effective tracking of actual compensating current with the generated reference current is developed and analyzed. The proposed ANN based SAPF provides benefits of effective compensation as well as reduced computations.

2   SHUNT ACTIVE POWER FILTER

Shunt active power filter, as shown in Figure 1, helps in mitigation of harmonic components from the source side by providing compensating current at the point of common coupling (PCC) and in this process improves the power quality of the supply. SAPF injects this 180-degree phase shifted current into at PCC, hence improves power quality. Here, three-level Neutral point clamped converter is used as SAPF.

3   ARTIFICIAL INTELLIGENCE BASED CONTROL OF SAPF

Neural network as shown in Figure 2, is a system of functions inter connected with each other where each neuron's (an activation function like sigmoid), output is dependent on the weighted sum of the output of layer before it. This system of interconnected function can be used to predict, cluster or classify data, conventional algorithms like feed forward or back

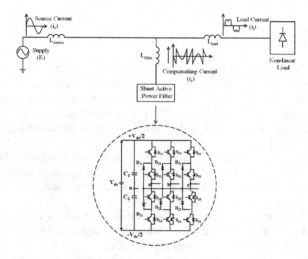

Figure 1.   Shunt active power filter.

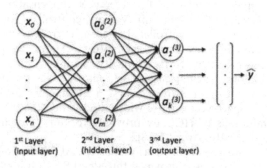

Figure 2.   Conventional neural network.

188

propagation are used for fitting. Basically, an ANN estimates a relationship without any mathematical model and only with representative data. Data are measured inputs and outputs in the case of a real system. Furthermore, by iteratively adjusting their parameters, ANNs are able to take into account complex systems with time-varying behavior. They have therefore been applied successfully for the identification and control of dynamical complex systems.

3.1 *Reference current generation technique*

Adaline network based neural synchronous technique is used for generation of reference currents. By performing a Fourier analysis we can expand the periodic waveform as a Fourier analysis is basically the sum of all the cos and sin frequency components in the system. The signal model to be estimated is given by:

$$F(t) = \sum\nolimits_{n=1,2....n} [X_n \cos(nwt) + Y_n \sin(nwt)] \tag{1}$$

Here Xn and Yn show the amplitude of the cos components and sin components of the order-n^{th} harmonic. In vector form -

$$F(t) = W^T . x(t) \tag{2}$$

$$\text{here } W^T = [X_1 Y_1 \ldots \ldots X_N Y_N] \tag{3}$$

We sample the signals at a uniform discrete sampling rate. An Adaline neuron carries out the dot product. Again the W^T term refers to the weight vector of the network. Post the first iteration the network updates the weights using an adaptive algorithm in order to converge to the actual result. As shown in Figure 3, the fundamental current I_{Lf} is extracted from the weights of Adaline networks. This current is then synchronized with the source voltage V_{sabc} through the Phase Locked Loop (PLL) system. The synchronized reference compensating current is then obtained as:

$$I_{ref} = I_L - I_{Lfsync} \tag{4}$$

3.2 *ANN based SVPWM controller*

Space Vector Pulse width Modulation (SVPWM) is a fixed switching frequency based controller. SVPWM being a voltage controlled approach, a reference output voltage (V_o^*) is generated form the reference compensating currents. Voltage space phasor structure of three-level SAPF is shown in Figure 4. Identification of the sector of SAPF voltage space phasor structure in which V_o^* lies is based on the comparative relationship of instantaneous PCC voltages.

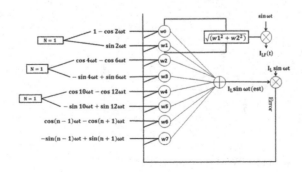

Figure 3.   Neural synchronous method.

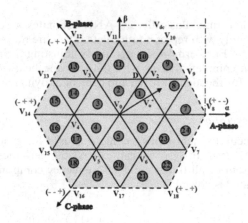

Figure 4. Voltage space phasor structure of three-level SAPF.

Once sector is identified, further determination of switching time duration in a sector is carried out. The proposed ANN based SVPWM Controller utilizes MATLAB nftool for training purpose of gate pulses generated by SVPWM. The Neural Network is trained using the Levenberg-Marquardt training function.

4 SIMULATION RESULTS

Performance analysis of ANN based three-level SAPF using neural synchronous method for reference compensating current generation and ANN based SVPWM is carried out using MATLAB.

The system parameters are: Source Voltage (V_s) = 325 V, DC-link Voltage of SAPF (V_{dc}) = 600 V, and SAPF Inductor (L_{SAPF}) = 12 mH. Non-linear load used for injection of current harmonics is a diode bridge rectifier with inductance (L_{load}) of 1.5 mH. Switching frequency of the controller is 5kHz. Behavior of the proposed ANN based SAPF is presented in Figure 6. It is clearly evident from Figure 6 that the proposed ANN based SAPF mitigates load current harmonics effectively. The source current, after being compensated using a neural network based shunt active filter, closely resembles a sine wave. This clearly indicates that the harmonics have been eliminated and the THD has improved. The proposed SAPF reduces the Total Harmonic Distortion (THD) in the source current from 30.68 % (without compensation) to 3.63 %. Dynamic behavior of proposed SAPF for load changes is shown in Figure 7. It is evident from the presented results that the developed ANN based SAPF effectively mitigates the harmonics even during load changes.

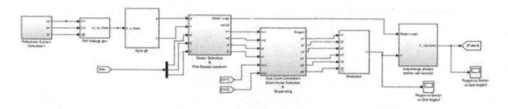

Figure 5. Block schematic of SVPWM for proposed SAPF.

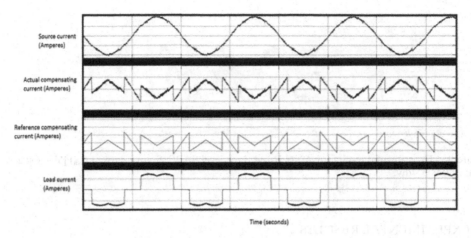

Figure 6.   Behavior of ANN based SAPF: A-phase source current (i_{sa}), actual compensating current (i_{ca}), reference compensating current (i_{ca}^*), and load current (i_{la}) [Y-Axis: 5 A/div.; X-Axis: 0.1 s/div].

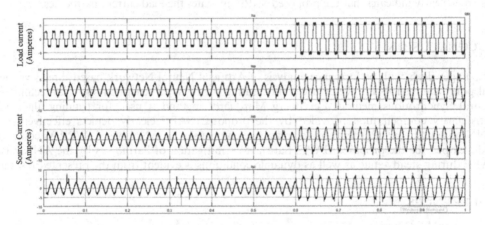

Figure 7.   Dynamic behavior of ANN based SAPF for load changes: A-phase load current (i_{la}) and source currents of all phases (i_{sa}, i_{sb}, i_{sc}) [Y-Axis: 5 A/div.; X-Axis: 0.1 s/div].

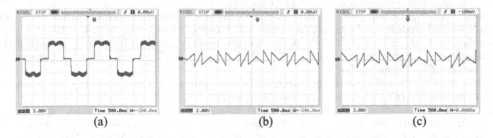

| (a) | (b) | (c) |

Figure 8.   Experimental results for ANN based three-level SAPF: (a) Load current (i_{la}), (b) actual compensating current (i_{ca}), (c) reference compensating current (i_{ca}^*) [Y-Axis: 2 A/div.; X-Axis: 0.005 s/di.].

191

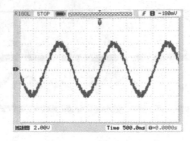

Figure 9. Experimental results for ANN based three-level SAPF: Source current (i_{sa}) [Y-Axis: 2 A/div.; X-Axis: 0.005 s/div.

5 EXPERIMENTAL RESULTS

Real-time implementation of the developed ANN based three-level SAPF is carried out using DS 1104. Satisfactory performance of the ANN based three-level SAPF is presented in Figure 8 and Figure 9 clearly indicates that the proposed SAPF mitigates the load current harmonics.

6 CONCLUSION

This paper deals with performance analysis of Artificial Neural Network based three-level neutral point clamped SAPF. Reference compensating currents are generated by ANN technique – Neural Synchronous method. Levenberg Marquardt method is used for training algorithm. Satisfactory compensation provided by the proposed SAPF clearly depicts effectiveness of ANN based scheme of generation for reference compensating. ANN based SVPWM controller is developed for the proposed SAPF. Satisfactory mitigation of harmonics by the ANN based SAPF during steady-state as well as dynamic conditions is evident from the presented results.

REFERENCES

Abdeslam, D. O., Wira, P., Mercklé, J., Flieller, D., & Chapuis, Y. 2007. A Unified Artificial Neural Network Architecture for Active Power Filters. *IEEE Trans. Ind. Electron.* 54(1): 61–76.

Arora, K., Bhatt, H., Tekwani, P. N., Chauhan, S. K., & Shah, M. T. 2018. Performance Investigations of Adaptive ANN-SVPWM Controlled Shunt Active Power Filter. *in Proc. IEEE Int. Conf. on Power Electronics, Drives and Energy Systems (PEDES 2018)*: 1–6.

Bhattacharya, A., & Chakraborty, C. 2011. A shunt active power filter with enhanced performance using ANN-based predictive and adaptive controllers. *IEEE Trans. Ind. Electron.* 58(2): 421–428.

Bhandankar, V. V., & Naik, A. J. 2016. Artificial neural network based implementation of space vector modulation for three phase VSI. in *Proc. 2016 IEEE International Conference on Electrical Energy Systems (ICEES)*, Chennai:145-150.

Bose, B. K. 2007. Neural network applications in power electronics and motor drives - an introduction and perspective. *IEEE Trans. Ind. Electron.* 54(1): 14–33.

Chauhan, S. K., & Tekwani, P. N. 2013. Current error space phasor based hysteresis controller for two-level and three-level converters used in shunt active power filters. in *Proc. 2013 39th Annual Conference of the IEEE Industrial Electronics Society (IECON 2013)*, Vienna, Austria: 8514-8519.

Chauhan, S. K., Shah, M. C., Tiwari, R., & Tekwani, P. N. 2014. Analysis, Design and Digital Implementation of Shunt Active Power Filter with Different Schemes of Reference Current Generation. *IET Pow. Electron.* 7(3): 627–639.

Chudasama, M. J., Tekwani, P. N., Chauhan, S. K., & Patel, V. 2017. Simulation, design, and analysis of three-level shunt active harmonic filter using t-typed npc (tnpc) topology (ICRISET 2017, Feb. 2017), *Kalpa Publications in Engineering* 1: 9–18.

Esfahani, M. T., Hosseinian, S. H., & Vahidi, B. 2015. A new optimal approach for improvement of active power filter using FPSO for enhancing power quality. *Elsevier International Journal of Electrical Power and Energy Systems* 69: 188–199.

Holtz, J. 1992. Pulsewidth modulation-a survey. *IEEE Trans. Ind. Electron.* 39(5): 410–420.

Kumar, P., & Mahajan, A. Soft computing techniques for the control of an active power filter. 2009. *IEEE Trans. Power Del.* 24(1): 452–461.

Nguyen, N.K., Flieller, D., & Wira, P., Abdeslam, D. O., & Merckle, J. 2011. Harmonic Identification with artificial network: application to active power filtering. *International Journal of Emerging Electric Power Systems.* 12(5): 1–27.

Pandya, N., Tekwani, P. N., & Patel, V. 2018. Analysis, design and simulation of three-phase active power filter with series capacitor topology for current harmonic compensation. *Inderscience International Journal of Power Electronics*, Genève, Switzerland 9(4): 426–446.

Qasim, M., & Khadkikar, V. 2014. Application of Artificial Neural Networks for Shunt Active Power Filter Control. *IEEE Trans. Ind. Info.* 10(3): 1765–1774.

Sharma, D., Bhat, A. A., & Ahmad, A. 2017. ANN based SVPWM for three phase improved power quality converter under distributed AC mains. in *Proc. 2017 IEEE International Conference on Computer Application in Electrical Engineering- Recent* Advances *(CERA)*: 533–538.

Tekwani, P. N., & Chauhan, S. K. 2015. Implementation of sector change detection schemes for current error space phasor hysteresis controller based Shunt Active Power Filters," in *Proc. 2015 11th IEEE International Conference on Power Electronics and Drive Systems (PEDS)*, Sydney, Australia: 1029–1034.

Tekwani, P. N., Chudasama, M. J., Chauhan, S. K., & Patel, V. 2017. Simulation as well as experimental investigations on T-type NPC topology based three-level shunt active power filter. in *Proc. 2017 IEEE Biennial International Conference on Technological Advancements in Power & Energy (TAP Energy 2017)*, Amrita Vishwa Vidyapeetham University, Kollam, India: 161–166.

Tekwani, P. N., Chandwani, A., Sankar, S., Gandhi, N., & Chauhan, S. K. Artificial neural network-based power quality compensator. *Inderscience International Journal of Power Electronics*, Genève, Switzerland, forthcoming article.

Wilamowski, B., & Yu, H. 2010. Improved computation for Levenberg-Marquardt Training. *IEEE Trans. Neu. Net.* 21(6): 930–937.

Technologies for Sustainable Development – Mahajan, Patel & Sharma (eds)
© 2020 Taylor & Francis Group, London, ISBN 978-0-367-33737-7

Generalized sector detection scheme for space vector modulation based modular multi-level inverter

M.T. Shah
Assistant Professor, Electrical Engineering Department, Institute of Technology, Nirma University, Ahmedabad, India

Ekansh Dadheech & Abhishek Singh
B. Tech Student, Electrical Engineering Department, Institute of Technology, Nirma University, Ahmedabad, India

ABSTRACT: This paper presents a generalized sector detection technique for space vector modulation applied to any N-level modular inverter topology. The proposed sector detection scheme only requires the three-phase voltage vectors in order to find the triangular sector without d-q transformation. Proposed scheme is able to detect the appropriate sector even if there is variation in line side or load side conditions without involving any additional circuitry or logic. The simulation results are presented for various steady-state and dynamic conditions, which makes five-level converter to operate from two-level mode of operation to fivtese-level including all intermediate level of operation. In all the conditions, proposed scheme is able to select the appropriate sector without involving any complex mathematical calculation or additional circuitry, which makes it generalized for any N-level modular inverter topology.

1   INTRODUCTION

Multi-level converters are used in many high-power applications such as HVDC transmission, motor drive applications as discussed in Gupta & Khambadkone (2006). Many topologies like Neutral-Point Clamped (NPC), Flying capacitor, Cascaded H-Bridge are proposed and discussed in Nabae, Takahashi & Akagi (1981), Saeedifard, Iravani & Pou (2007), Lai & Peng (1996), Amini (2014), Malinowski, Gopakumar, Rodriguez & Pérez (2010). A new multi-level inverter scheme was introduced in the early 2000's namely Modular Multi-Level Converters (MMC) by Lesnicar & Marquardt (2003). MMC consists of number of identical sub-modules (SM) connected in series in each phase to achieve desire level of inverter. Due to the possibility of modularity and scalability of MMC, this scheme is more explored in recent research. Carrier based modulation PWM techniques are used to control modular multi-level inverters as discussed in Hagiwara & Akagi (2009). Both, the phase shifted carrier based modulation technique as mentioned in Mohammadi & Bina (2011), Tu, Xu & Xu (2011), or level shifted carrier based modulation technique as explained in Solas & Abad (2013), Montesinos-Miracle (2013) are explored to control the MMC. Carrier based PWM technique gives uniform harmonic spectrum but dc-link utilization is less as compare to Space Vector Modulation (SVM) techniques as mentioned in Van der Broeck, Skudelny & Stanke (1988). As the level of inverter increases, voltage space phasor structure becomes more complex. The voltage space phasor structure is divided into many triangular regions called sectors. To select the appropriate voltage vector to control the inverter, information of sector is required. Sector selection is become more difficult as level of inverter increase. SVM based PWM technique is more explored up to three-level and five-level inverters due to complexity involved in finding appropriate sector. In literature, generalized methods were proposed by researchers for sector selection for multi-level inverters like sector mapping techniques and reverse sector mapping i.e., shifting the origin to smaller two-level hexagon which is a sub-part of the bigger multi-level space vector diagram as proposed in Seo, Choi & Hyun (2001), Jacob & Baiju (2013), Deng, Teo & Harley (2014), Deng, Teo & Harley (2013). These techniques involve complex mathematical calculations as number of level increases. Hence, straight forward generalized method is proposed in this paper to find the appropriate sector by locating reference voltage vector for any N-level multi-level inverters.

2 MULTI-LEVEL SPACE VECTOR PULSE WIDTH MODULATION

The voltage space phasor structure of five-level inverter is shown in Figure 1. It consist of 96 triangular sectors, 65 voltage vector and 125 switching states. In Figure 2, numeric value 0, 1, 2, 3 and 4 indicates the voltage level 0 V, $V_{dc}/4$, $2V_{dc}/4$, $3V_{dc}/4$ and $4V_{dc}/4$ respectively for phase-A, B and C. A reference voltage vector (V_{ref}), is continuously rotating in the voltage space phasor structure of five-level MMC inverter. The tip of the reference vector would be enclosed by a triangle composed of three adjacent voltage vectors as shown in Figure 1. Reference voltage vector can take any position in voltage space phasor structure due change in magnitude. By locating the tip of supply voltage, three adjacent voltage vectors are to be switched to control the MMC. Hence, for the selection of appropriate voltage vector, the first step is to have information of present sector in which tip of V_{ref} is lying.

2.1 Sector calculation

As observed from the space vector diagram in Figure 2, in any horizontal layer, as one moves from left to right the difference between the voltage level of B and C stage remains same, only A-phase vector changes. From vector algebra, the direction of vector "B-C" lies in the vertically upwards direction as shown in Figure 2. Therefore, when moving from bottom to upwards the difference between B and C increases, having the value "zero" for base layer (if the point lies along A phase), value "one" for first value in upward direction and so on, similarly when moving in lower direction the value of vector "B-C" decreases. For unified representation, the greatest integer function of the vector gives the same value between layer "zero" and "one" i.e., "1", value "2" between layer "one" and "two" and so on, as shown in Figure 3 (a). In similar manner as vector "B-C", the vector "A-C" and "A-B" would also show the similar characteristics. Therefore, by using the value of "B-C" and "A-C" together, a unique combination is formed, which may identify a particular quadrilateral as shown in Figure 3 (b). (Different colours in diagram represents that every quadrilateral has a different value and can be differentiated for sector identification). In order to identify the modulation triangle, there

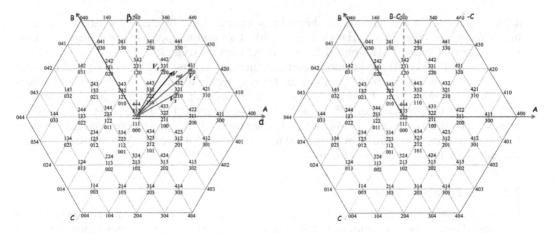

Figure 1. SVM with V_{ref} & modulation triangle.

Figure 2. SVM with indication of the direction if B-C.

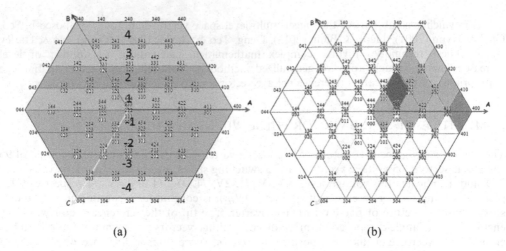

Figure 3.　(a) Layer identification for sector calculation, (b) quadrilateral identification technique.

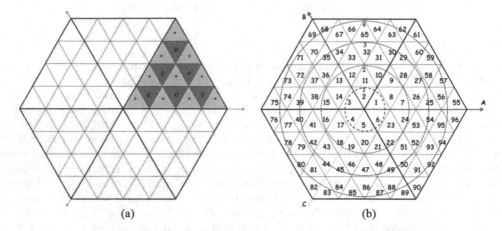

Figure 4.　(a) Determination of upper facing or lower facing triangle, (b) Sector numbering.

are two triangles in a quadrilateral, therefore, a separate logic to identify whether V_{ref} lies in upper facing or lower facing triangle is necessary. To find the upper or lower facing triangle, the inclusion of vector "A-B" may differentiate between the triangles.

If p = [B-C] (where f(x)=[x] determines the greatest integer function), q=[A-C], and r=[A-B], then the difference between (p+r) and q, i.e., (p+r)-q, would differentiate between the triangles as shown in Figure 4 (a). The above logic holds for only for the section between angle 0° to 60°. By using the symmetry of 60°, the logic could be extended for the complete voltage space phasor structure any N-level MMC. Thus, in the proposed technique sector detection is done without involving d-q transformation. Various trajectories followed by V_{ref} in order to satisfy the load demand is as shown in Figure 4 (b).

2.2 *Duty cycle calculation*

For a very small time period T_s, the position of rotating vector, V_{ref} could be considered as constant. Hence, the semiconductor switch can be switched numerous times in that period in order to obtain the value same as that of the reference period. The time period T_s is called as sampling time and the frequency is called as sampling frequency. The time for which a given switching state is turned on is called duty cycle. The expression is used to find the duty cycle is given by (1), (2) and (3).

From volt-second balance along α-β axis (with reference to Figure 2):

$$V_1\cos(\theta_1)T_1 + V_2\cos(\theta_2)T_2 + V_3\cos(\theta_3)T_3 = V_s\cos(\theta_s)T_s \qquad (1)$$

$$V_1\sin(\theta_1)T_1 + V_2\sin(\theta_2)T_2 + V_3\sin(\theta_3)T_3 = V_s\sin(\theta_s)T_s \qquad (2)$$

$$T_1 + T_2 + T_3 = T_s \qquad (3)$$

Where V_1, V_2, & V_3 are the voltage magnitude of the voltage points from which the sector is formed. $\theta1$, $\theta2$ and $\theta3$ are the inclinations of V_1, V_2 and V_3 with respect to x-axis, T_1, T_2, and T_3, are the duty cycle of voltage states V_1, V_2, and V_3.

3 SIMULATION AND RESULTS

The proposed sector detection technique for space vector modulation is simulated in MATLAB for a modular five-level converter. To verify the dynamic behavior of proposed scheme, V_{ref} is given different value considering per unit system. When V_{ref} is kept at 0.5 pu and 1 pu respectively,

V_{ref} follows the trajectory of two-level and three-level mode of operation (Circle 1 and 2 respectively as shown in Figure 4 (b)) as validate in simulation result of Figure 5 (a) and (b). Sudden change in V_{ref} from 1 pu to 1.5 pu and then to 2.11 pu as shown in Figure 6 (a) and (b). Sector detection pattern changed from three-level mode of operation to four-level and then to five-level as indicated in Figure 6 (a) and (b). This proves the self-adaptive nature of

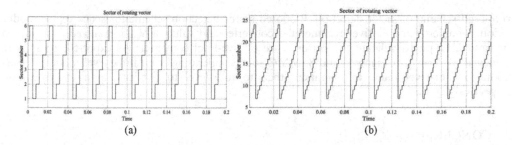

Figure 5. Sector detection for (a) two-level operation when V_{ref} is kept at 0.5 pu, (b) three-level operation. when V_{ref} is kept at 1 pu.

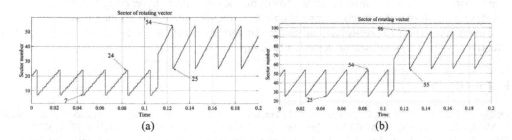

Figure 6. Sudden change in reference vector magnitude from (a) 1pu to 1.5pu (Three to four-level operation) (b) 1.5pu to 2.11pu (Four to five-level operation).

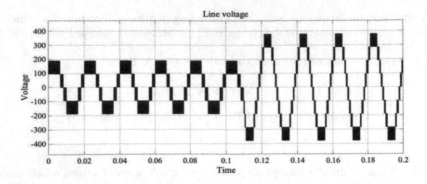

Figure 7.   Five-level MMC operated three-level to four-level mode of operation.

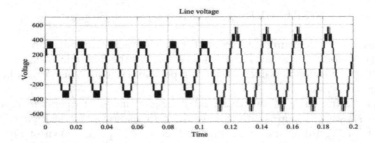

Figure 8.   Five-level MMC operated four-level to five-level mode of operation.

proposed scheme of sector detection. Proposed sector detection scheme for space vector modulation is implemented on five-level modular converter and value of V_{ref} is changed to 1 pu, 1.5 pu, and 2.11 pu to validate the performance of five-level converter. As expected, results are achieved as shown in Figure 7 and 8, MMC started operating from three-level mode of operation to four-level mode of operation, then to four-level to five-level mode of operation respectively.

4   CONCLUSION

Generalized sector detection technique without involving d-q transformation is proposed for space vector modulation scheme applied any modular multi-level converter. Furthermore, this scheme does not require any additional circuitry or fast real time controller for implementation. Proposed scheme is also validated for dynamic conditions by varying the per unit value of reference voltage. Proposed scheme is able to detect the sectors in all the range of five-level converter (two-level to five-level) and make converter operate accordingly which proves the effectiveness of proposed sector detection technique.

REFERENCES

Deng, Y. Teo, K.H., Duan, C., Habetler, T. G Harley, R.G. 2014. A Fast and Generalized Space Vector Modulation Scheme for Multi-level Inverters in *IEEE Trans. Power Electron.* 29 (10): 5204–5217.

Deng, Y. Teo, K.H. Harley, R.G. 2013 Generalized DC-Link Voltage Balancing Control Method for Multi-level Inverters. *in Proc. IEEE Applied Power Electronics Conference and Exposition (APEC):* 1219–1225.

Gupta, A. Khambadkone, A. 2006. A space vector PWM scheme for multi-level inverters based on two-level space vector PWM. *IEEE Trans. Ind. Electron.* 53 (5): 1631–1639.

Hagiwara, M. Akagi, H. 2009 Control and experiment of pulse width modulated modular multi-level converters. *IEEE Trans. Power Electron.* 24 (7): 1737–1746.

Jacob, B. Baiju, M. R. 2013 Vector-Quantized Space-Vector-Based Spread Spectrum Modulation Scheme for Multi-level Inverters Using the Principle of Oversampling ADC. *IEEE Trans. Ind. Electron.* 60 (8): 2969–2977.

Lesnicar, A. Marquardt, R. 2003 An innovative modular multi-level converter topology suitable for a wide power range. *in proc. IEEE Power Tech Conf. Proc.*: 1–6.

Malinowski, M. Gopakumar, K. Rodriguez, J. Pérez, M. A. 2010. A Survey on Cascaded Multi-level Inverters. *IEEE Trans. Ind. Electron.* 57(7): 2197–2206.

Mei, J. Xiao, B. et al. 2013 Modular Multi-level Inverter with New Modulation Method and its Application to Photovoltaic Grid-Connected Generator. *IEEE Trans. Power Electron.* 28(11): 5063–5073.

Mohammadi, H. Bina, M. T. 2011 A Transformer less Medium-Voltage STATCOM Topology Based on Extended Modular Multi-level Converters. *IEEE Trans. Power Electron.* 26(5): 1534–1545.

Montesinos-Miracle, D. Massot-Campos, M. Bergas-Jane, J. Galceran-Arellano, S. Rufer, A. 2013 Design and Control of a Modular Multi-level DC/DC Converter for Regenerative Applications. *IEEE Trans. Power Electron.*, 28(8): 3970–3979.

Nabae, A., Takahashi, I. Akagi H. 1981 A new neutral-point-clamped PWM inverter. *IEEE Trans. Ind. Appl.*, IA-17 (5):518–523.

Saeedifard, M. Iravani, R. Pou J. 2007 Analysis and Control of DC- Capacitor-Voltage-Drift Phenomenon of a Passive Front-End Five-Level Converter. *IEEE Trans. Ind. Electron.* 54(6): 3255–3266.

Seo, J. Choi, C. Hyun, D. 2011 A new simplified space-vector PWM method for three-level inverters. *IEEE Trans. Power Electron*, 16(4): 545–550.

Solas, E. Abad, G. et al. 2013 Modular Multi-level Converter with Different Submodule Concepts—Part I: Capacitor Voltage Balancing Method. *IEEE Trans. Ind. Electron.*, 60(10): 4525–4535.

Tu, Q., Xu, Z. Xu, L. 2011 Reduced Switching-Frequency Modulation and Circulating Current Sup - pression for Modular Multi-level Converters. *IEEE Trans. Power Del.* 26(3): 2009–2017.

Van der Broeck, H.W., Skudelny, H. Stanke, G.V. 1988 Analysis and realization of a pulsewidthmodulator based on voltage space vectors. *IEEE Trans. Ind. Appli.* 24(1): 142–150.

Effect of synthetic inertia controller on frequency response in a multi-machine power system with high penetration of renewable energy sources

Chintan R. Mehta, Prasad D. Deshpande, Bhavik D. Nathani & Santosh C. Vora

Department of Electrical Engineering, Institute of Technology, Nirma University, Ahmedabad, India

ABSTRACT: This paper addresses the impacts of high penetration of renewable energy (RE) sources on frequency stability of multi-machine power system. Renewable energy sources are intermittent in nature and they are generally decoupled from the grid at the time of system disturbances which will reduce the overall inertia of the system. This results in reduction of the frequency support available to the grid. Thus, to address the frequency stability issue and to extract the hidden inertia available in the rotor of wind turbine generators, a synthetic inertia control scheme proposed in literature is implemented in this paper. The impact of synthetic inertia control scheme on multi-machine power system consisting of doubly fed induction generators, conventional synchronous generators and solar PV farm is shown by the authors considering modified WSCC 9 bus system. The effectiveness of frequency regulation has been demonstrated by using dynamic simulations in PSCAD/EMTDC software.

Keywords: Doubly Fed Induction Generator (DFIG), frequency stability, inertia response, PV solar farm, Synthetic Inertia Control (SIC), WSCC 9 bus system

1 INTRODUCTION

In recent times, the need of cleaner energy and minimizing the dependency on fossil fuel has led to integration of large-scale renewable generations in power grid. The bulk power generation is achieved through wind energy and solar energy. With the emerging structural changes in power systems particularly due to the rapid penetration of wind resources sets notable concerns regarding reliable and secure operation of power system, amongst which one of the major concerns regarding secure operation is frequency instability.

As the penetration of wind generation is increased, the characteristics of power systems is also changing. Unlike the conventional synchronous generators (SG), power converter-based wind power generation behaves differently in terms of the prominent features it provides to the grid among which one feature is the lack of inertia [He *et al.* 2017]. For keeping the stability in the grid, the inertia of a power system is an important property. However, the existing control schemes for wind power generation contributes an inappreciable amount of inertia to the grid. Thus, the total inertia of the power system will decrease with the increase in penetration of renewable generation. Based on the type of conventional generator used, the inertia constant typically lies in the range of 2-9 s [John J. Grainger and Stevenson, 1994] which can improve the system when there is a frequency disturbance or mismatch in generation and demand. On the other hand the inertia constants of wind turbine (WT) lies typically between 4-6 s, but are unable to participate in frequency control because of the power converter based interface with the grid [Ackermann 2005]. In case of a doubly fed-induction generator (DFIG), the back-to-back converter is connected to the rotor winding of the DFIG and is sized accordingly to carry 30% of the generated power by the machine. The stator winding on the other hand is directly connected to the grid and operates at grid frequency. Although the stator winding of DFIG is connected directly with the grid, the current control loops of the converter strictly controls the generated electrical power by DFIG. Such a configuration of the converter helps to restore the power output in case of any perturbation, thus decoupling the inertial response of DFIG [Nguyen *et al.* 2017]. A novel probabilistic scheme for estimating the inertial response from WTs at a particular location under constantly changing wind scenarios which is achieved by implementing a Gaussian probability distribution function for representing the variations in wind speed w.r.t. time is proposed by [Wu and Infield 2013]. A novel method to provide inertia support by directly adjusting dynamic response of phase locked loop (PLL) has been proposed in [Wang *et al.* 2015]. The proposed technique accounts the impacts of the different controller parameters of WTs, its operating points, and specifically the mechanical power variations which are caused by the rotational speed or pitch angle changes during the inertial response period. Authors of [Ochoa and Martinez 2017] have presented a method of analyzing the dynamics of frequency in a large power system with high share of REs and have prepared a simplified model of Type-3 based WT. Additionally, [Ochoa and Martinez 2017] have also proposed an optimized version of conventional control i.e. MPPT and have implemented it for Type-3 based WT. The prepared model by [Ochoa and Martinez 2017] incorporates an additional control system allowing the WT for providing fast-frequency response to the grid in case of any perturbation. For providing synthetic inertia an alternative strategy of droop control for supplying inertial response is proposed by the authors of [Van de Vyver *et al.* 2016]. They have focused upon the droop control without accounting for the inertial gain. [Wang-Hansen *et al.* 2013] have investigated various control strategies for modelling of wind turbine frequency control, however the work is generally concentrated on Type-4 based wind farms. [Zhang *et al.* 2013] has carried out comparison of various methods for inertia control for WFs, but have not mentioned the detailed control methods. A novel scheme for Type-3 based WTGs is proposed by [Zhu *et al.* 2018], enabling the converter to exhibit an inertial response which is similar to that of a synchronous machine in response to system frequency disturbances. Also, the synthetic inertia control (SIC) scheme proposed by [Zhu *et al.* 2018] avoids the dependency on df/dt measurement, which may lead to the introduction of instability in the control system. Dynamic model which contains the control strategy and simulation of a PV farm and WF to a smart grid power system is proposed in [Natsheh *et al.* 2011].

This paper focuses on inertial response from WT and frequency stability for short time with wind and solar penetration in the power system. Wind, solar and conventional synchronous generators are working together within PSCAD software environment. Frequency stability is studied in the modified WSCC 9-bus system when wind speed and solar insolation is assumed to be constant. The authors of this paper considered the following two cases for analysis:

- DFIG based wind farms with Synthetic Inertia Controller (SIC) connected to a modified WSCC 9 bus system having SGs.
- DFIG based wind farms with Synthetic Inertia Controller (SIC) connected to a modified WSCC 9 bus system having SG and Solar farm.

2 SYNTHETIC INERTIA CONTROL

In steady state operation of a power system, the total power generated (P_{gen}) is always equal to the sum of total power demand (P_D) and line losses (P_L) ensuring the frequency being stable in a tolerable range. In case, if disturbance occurs, the kinetic energy stored in the rotating masses of the machines remaining connected to the system, will react to the change. The frequency stability is achieved if the active power balance is maintained. If there occurs a sudden load increase, the output power of the generators cannot increase instantaneously, so the initial frequency variations will be dependent upon the released kinetic energy from the rotating masses J. This phenomenon is termed as "inertial response" and has an important impact on the system frequency stability. However, as the penetration of WFs is increasing in the present power system, the total system inertia is decreasing because of the converter based

WTs. Although, it is possible to extract the hidden inertia from the rotating masses of WT by implementing an appropriate control strategy in the power converter to provide inertial response in case of frequency events.

2.1 Controller design for SIC

Figure 1 shows the Synthetic Inertia Controller (SIC) implemented in the rotor side converter (RSC) of WT, which mimics the inertial response behavior as that of the SGs. The method is proposed by [Nguyen *et al.* 2017]. The SIC either absorbs or releases the hidden energy providing support for the system during power imbalances. When the system is operating in the steady state condition, the active power delivered to the system is controlled by the maximum power point tracking algorithm (MPPT) and it does not react to the variations in system frequency. In case of a frequency event, there is an inertial response signal P_{extra} added to the pre-defined reference active power. P_{extra} originates from two control loops: the frequency derivative df/dt loop and the frequency deviation Δ f loop as shown in Figure 2.

2.2 Activation scheme

The activation scheme required to activate the synthetic inertia controller plays a vital role and needs a proper concern. Also the implementation of SIC for WTGs is an important perquisite for a low inertia system. The proper co-ordination of SIC and activation scheme enables a better inertial response for WT. [Nguyen *et al.* 2017] have presented two activation strategies namely frequency gradient trigger scheme and over/under frequency trigger scheme for the SIC controller. However, in this paper the over/under frequency trigger scheme has been implemented.

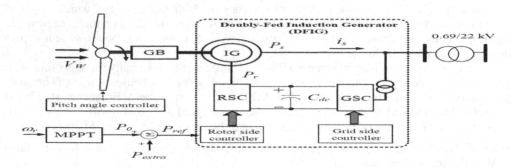

Figure 1.   Block diagram of DFIG connected with SIC scheme [Nguyen *et al.* 2017].

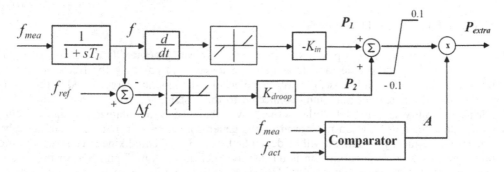

Figure 2.   Synthetic inertia controller [Nguyen *et al.* 2017].

202

3 SIMULATION AND RESULTS

3.1 *Conventional WSCC 9 bus system*

A WSCC 9 bus system is described in [Anderson *et al.* 1979]. It represents a simple approximation of the WSCC to an equivalent system. Three generators are connected at bus number 1, 2 and 3 respectively. Bus 1 is considered as slack bus. Bus 2 and 3 are considered as PV bus. The base voltage levels are 16.5 kV, 18 kV and 13.8 kV at bus 1, 2 and 3 respectively. Base voltage at other buses are 230 kV. Base apparent power in the system is 100 MVA. Bus 5, bus 6 and bus 8 are load buses. Total load connected to the system is 315 MW and 105 MVAR [Anderson *et al.* 1979]. The generator data for WSCC 9 bus system is given in [Anderson *et al.* 1979]. To see the effect of change in active power demand on the frequency response, active power demand of the load connected at Bus 5 and Bus 8 will be changed by introducing an additional load to create a frequency event. All the other parameters of the system are kept unaltered. This modified WSCC 9 bus system is simulated in PSCAD/EMTDC software.

3.2 *WSCC 9 bus modified test system with DFIG WF*

A modified WSCC 9 bus system is used in this paper. As the work is concentrated on the DFIG based WFs, the conventional WSCC 9 bus system is modified by replacing the synchronous generator based power plant at Bus 2 by an equivalent DFIG based wind farm. The wind farm is considered to supply the active power and reactive power similar to the conventional system in the steady state condition at the same voltage. The generation from the other two generators are also kept unaltered. To see the effect of change in active power demand on the frequency response, active power demand of the load connected at Bus 5 and Bus 8 is altered by introducing an additional load of 50 MW at 10 second for 5 seconds for creating a frequency event. All the other parameters of the system are kept unaltered. The gain K_p and K_i have been modified. This modified WSCC 9 bus system is simulated in PSCAD/EMTDC software [Deshpande 2019]. It is assumed that the wind speed is constant at 11 m/s throughout the simulation period. Each DFIG in the wind farm is rated for 5 MVA and 0.69 kV. As can be seen from the Figure 3, as with the conventional control technique DFIG fails to provide the inertial response thereby increasing the burden of extra load to be supplied by the remaining two generators. However, when the DFIG is being equipped with SIC, it exhibits the property of inertial response during perturbation and provides the extra demanded power to the system which is quite evident from Figure 4.

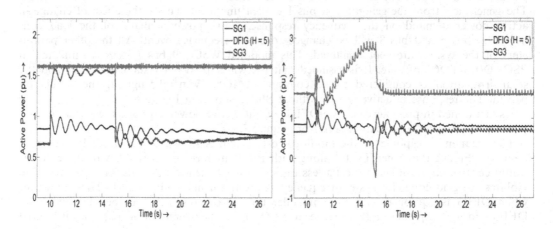

Figure 3. Active power output response of WSCC 9 bus system with DFIG WF at bus 2.

Figure 4. Active power output response with SIC of WSCC 9 bus system with DFIG WF a Bus 2.

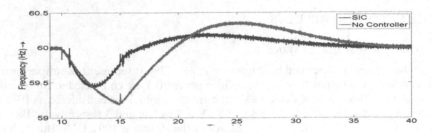

Figure 5.   Frequency response of WSCC 9 bus system with DFIG WF at bus 2.

Table 1.   Magnitude of frequency during a load change in WSCC
9 bus test system.

Frequency Response	Conventional Control	SIC
Min. (Hz)	59.18	59.43
Max. (Hz)	60.37	60.20

The frequency response of the test system is shown in Figure 5. It is evident from the Figure that under conventional control the DFIGs frequency response for a perturbation violates the grid codes for frequency as the frequency nadir is 59.18 Hz/s and also after the removal of perturbation, the system takes a huge span of time to get restored to the steady state. On the other hand when DFIG is equipped with the controller the frequency nadir increases to 59.43 Hz/s which is within permissible limits. Also the overshoot of frequency is limited with SIC scheme. The magnitude of the frequency response for a perturbation is given in Table 1

3.3   WSCC 9 bus modified test system with DFIG WF & PV solar farm

A modified WSCC 9 bus system is used for the simulation purpose. In this case, the conventional WSCC 9 bus system is modified by replacing the synchronous generators at Bus 2 and 3 by equivalent capacity DFIG based wind farm and PV solar farm respectively. The wind farm and solar farm are considered to supply the active power and reactive power similar to the conventional system in the steady state condition at the same voltage. The generation from the generator at bus 1 is kept unaltered. To see the effect of change in active power demand on the frequency response, active power demand of the load connected at bus 5 and bus 8 will be changed to create frequency event. All the other parameters of the system are kept unaltered. This modified WSCC 9 bus system is simulated in PSCAD/EMTDC software. It is assumed that the wind speed is constant at 11 m/s and the solar irradiation is considered to be constant at 1000 W/m^2 throughout the simulation period. The response of active power without SIC is shown in Figure 6.

As mentioned in previous cases the DFIGs output active power supports the system in case of any perturbation which can be seen from Figure 7.

The frequency response of the modified test system is shown in Figure 8. It is evident from the Figure that when DFIG along with PV solar farm is operated with the conventional control, as both being inertia-less the systems frequency response for a perturbation violates the grid codes for system frequency as the frequency nadir is 59.15 Hz/s. However when DFIG is equipped with the controller the inertial response can be obtained from DFIG and the frequency nadir increases to 59.46 Hz/s. The overshoot of frequency is limited

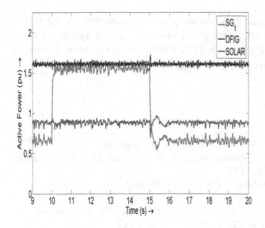

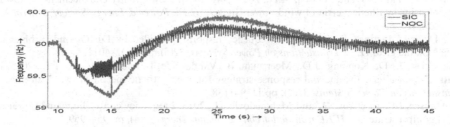

Figure 6. Active power output response of WSCC 9 bus system with DFIG & PV under conventional control.

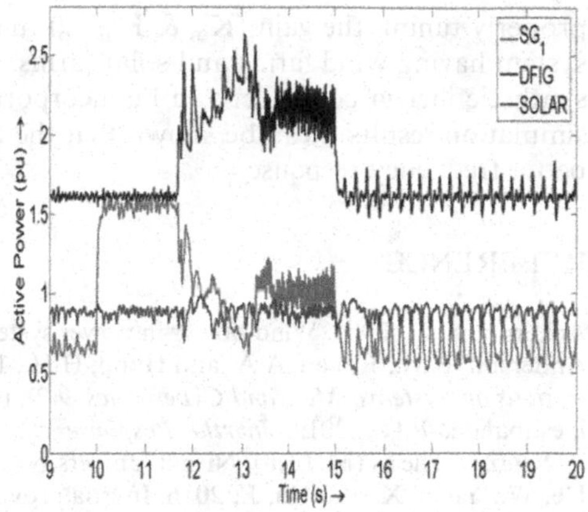

Figure 7. Active power output response of WSCC 9 bus system with DFIG & PV under synthetic inertia control.

Figure 8. Frequency response with SIC of WSCC 9 bus system with solar PV farm at bus 3.

Table 2. Magnitude of frequency during a load change in modified WSCC 9 bus test system.

Frequency Response	Conventional Control	SIC
Min. (Hz)	59.15	59.46
Max. (Hz)	60.45	60.32

with SIC scheme enabled. The magnitude of the frequency response for a perturbation is given in the Table 2.

4 CONCLUSION

The drastic increase of penetration of REs in the power system, the requirements of grid codes for enhancement of frequency stability especially for integrating large WFs and solar farms are necessary for the stable operation of the power system. The inertial response of DFIG based wind turbines is almost unavailable for frequency support because of the decoupling between the system frequency and rotor speed. Solar farms do not having any rotating part so the inertia is negligible. However, by implementing a SIC into the rotor of DFIG and by properly tuning the gain (K_{in} & K_{droop}) the frequency stability can be achieved in the power system having wind farms and solar farms. It can be seen from the simulation results that the synthetic inertia controller can be incorporated into DFIG for frequency support. From the simulation results it can be shown that the more kinetic energy can be extracted with SIC for better frequency response.

REFERENCES

Ackermann, T., 2005. Wind power in power systems (Vol. 140): Wiley Online Library.

Anderson, P.M., Fouad, A.A. and Happ, H.H., 1979. Power system control and stability. *IEEE Transactions on Systems, Man, and Cybernetics*, 9(2), pp.103–103.

Deshpande, P.D., 2019. *Inertia Response and Frequency Control Techniques for Renewable Energy Sources*. Thesis (M.Tech). Nirma University.

He, W., Yuan, X. and Hu, J., 2016. Inertia provision and estimation of PLL-based DFIG wind turbines. *IEEE Transactions on Power Systems*, 32(1), pp.510–521.

John J. Grainger and Stevenson, W.D., 1994. *Power system analysis* (Vol. 67). New York: McGraw-Hill.

Natsheh, E.M., Albarbar, A. and Yazdani, J., 2011, December. Modeling and control for smart grid integration of solar/wind energy conversion system. In *2011 2nd IEEE PES International Conference and Exhibition on Innovative Smart Grid Technologies* (pp. 1–8). IEEE.

Nguyen, H.T., Yang, G. and Nielsen, A.H., 2017, July. Frequency stability enhancement for low inertia systems using synthetic inertia of wind power. In *2017 IEEE Power & Energy Society General Meeting* (pp. 1–5). IEEE.

Ochoa, D. and Martinez, S., 2016. Fast-frequency response provided by DFIG-wind turbines and its impact on the grid. *IEEE Transactions on Power Systems*, 32(5), pp.4002–4011.

Van de Vyver, J., De Kooning, J.D., Meersman, B., Vandevelde, L. and Vandoorn, T.L., 2015. Droop control as an alternative inertial response strategy for the synthetic inertia on wind turbines. *IEEE Transactions on Power Systems*, 31(2), pp.1129–1138.

Wang-Hansen, M., Josefsson, R. and Mehmedovic, H., 2013. Frequency controlling wind power modeling of control strategies. *IEEE transactions on sustainable energy*, 4(4), pp.954–959.

Wang, S., Hu, J., Yuan, X. and Sun, L., 2015. On inertial dynamics of virtual-synchronous-controlled DFIG-based wind turbines. *IEEE Transactions on Energy Conversion*, 30(4), pp.1691–1702.

Wu, L. and Infield, D.G., 2013. Towards an assessment of power system frequency support from wind plant—Modeling aggregate inertial response. *IEEE Transactions on Power Systems*, 28(3), pp.2283–2291.

Zhang, Z., Wang, Y., Li, H. and Su, X., 2013, June. Comparison of inertia control methods for DFIG-based wind turbines. In *2013 IEEE ECCE Asia Down under* (pp. 960–964). IEEE.

Zhu, J., Hu, J., Hung, W., Wang, C., Zhang, X., Bu, S., Li, Q., Urdal, H. and Booth, C.D., 2017. Synthetic inertia control strategy for doubly fed induction generator wind turbine generators using lithium-ion supercapacitors. *IEEE Transactions on Energy Conversion*, 33(2), pp.773–783.

Technologies for Sustainable Development – Mahajan, Patel & Sharma (eds)
© 2020 Taylor & Francis Group, London, ISBN 978-0-367-33737-7

Analyzing the effect of circuit parasitic on performance of DC-DC buck converter using circuit averaging technique

Annima Gupta, Akhilesh Panwar & Akhilesh Nimje
Electrical Engineering Department, Nirma University, Ahmedabad, India

ABSTRACT: This paper presents a detailed state-space average model for the step down buck regulator including circuit parasitic. Presented work focuses on analyzing the effect of circuit non-idealities such as ESR of inductor and capacitor, switches on time resistant and series voltage drop on circuit performance. The averaging method is used to evaluate the system matrix of the time varying circuit model. Developed model is further used to analyze the effect of the circuit non-idealities on selection of duty cycle, stability margin for different loading condition and the open loop frequency response. Proposed model can be used to design robust controller that can satisfy the design requirement.

1 INTRODUCTION

In modern days, most of power delivery system are operated with some of the electronic converter. These converters are provided for numerous purposes such as DC-DC conversion, AC regulation, rectification, inversion and many more. Out of all available converters, DC-DC converter is widely used. This includes circuit topologies such as Buck, Boost, Buck-Boost and other complex configuration. These regulators are specialized in much needed application such as charge control of battery, voltage isolation etc. DC voltage regulator maintains constant output voltage irrespective of the perturbation in the load demand or supply voltage. Extensive research has been done for voltage regulator with a wide variety of control techniques. However, with improved technical demand more sophisticated control technique is needed for accurate and fast regulation.

The topology for DC-DC converter includes linear (resistor, inductor and capacitor) and non-linear (switch, diode) element. Due to difference in the characteristics of these elements the behavior is largely time varying. As these converters exhibits time varying non-linear response, a linearized model is needed to design a linear control. *Cuke et. al* first developed the state space model of buck regulator without including all circuit non-idealities. Since then most of the work carried so far is mainly concerned about the designing of the controller. The control design technique without inclusion of non-idealities (switch on time voltage, conduction resistance, ESR of inductor and capacitor, forward voltage drop and on time diode resistance) does not guarantee the satisfactory operation in real time.

In the recent articles about the designing of controller for buck regulator, various non-idealities are not considered. *Cristri et. al* has done computer based simulation to propose the efficacy of the designed controller. These simulation fails to elucidate physical interpretation. *Sarikhani et.al* has demonstrated the application of DC converter for photovoltaic application. *Bayram et. al* has done analysis of multilevel DC regulator topology. The effect non-ideality is prominent for multilevel DC regulator.

The proposed work is focused on developing the state space model of the DC buck regulator. This state space model for time varying buck regulator with non-idealities is developed using averaging technique. Evaluated expression is used to determine the accurate value of duty cycle. In the presence on non-linearity, the plant is further linearized around steady state point. This helps in evaluating the plant performance for ideal and non-ideal circuit using frequency domain analysis. At last, the comparison is done between ideal and non-ideal converter using frequency plots and by determining the stability margin.

2 BUCK CONVERTER MODELING

Buck converter is a class of DC voltage regulator which enables the reduction in the voltage level from input to output. This converter uses high frequency switch to control the time at which input voltage is supplied to load. Apart from this, it uses diode, load resistor and a filter circuit. In this work, various associated non-liberalities are included. These non-idealities mainly are switch on time resistance (r_s), switch on time voltage drop (V_s), diode on time resistance (r_d), diode on time voltage drop (V_d), inductor ESR (r_l) and capacitor (r_c). The typical layout of the buck regulator in shown in the Figure 1

The buck regulator topology is modelled using state space modelling approach. The detailed modelling is given in the following sub section.

2.1 *On state operation and state space modeling*

The controlled switch (MOSFET) of converter remains ON during dTs. During ON period the equivalent circuit is shown in Figure 2.

During dT_s, *eq. 1-4* governs system dynamics during ON state. The capacitor voltage (v_c) and inductor current (i_l) are chosen as state variable of the buck regulator. During ON state v_c and i_l are defined in *eq. 1* and *2* respectively.

$$\frac{di_l}{dt} = \frac{1}{L}\left[V_{in} - \frac{i_L}{(R+r_c)}(r_sR + r_LR + r_cr_s + r_Lr_c + r_cR) - \frac{RV_c}{R+r_c} - V_s + \frac{Rr_c}{(R+rc)}i_z \right] \quad (1)$$

$$\frac{dV_c}{dt} = \frac{1}{C}\left[\frac{Ri_L}{R+r_c} - \frac{V_c}{R+r_c} + \frac{r_ci_z}{R+r_c} \right] \quad (2)$$

$$\begin{bmatrix} \frac{di_l}{dt} \\ \frac{dV_c}{dt} \end{bmatrix} = \begin{bmatrix} \frac{-K_1}{L(R+r_c)} & \frac{-R}{L(R+r_c)} \\ \frac{R}{C(R+r_c)} & \frac{-1}{C(R+r_c)} \end{bmatrix} \begin{bmatrix} i_L \\ v_c \end{bmatrix} + \begin{bmatrix} \frac{1}{L} & \frac{Rr_c}{(R+r_c)L} & \frac{-1}{L} & 0 \\ 0 & \frac{r_c}{(R+r_c)C} & 0 & 0 \end{bmatrix} \begin{bmatrix} V_{in} \\ i_z \\ V_s \\ V_d \end{bmatrix} \quad (3)$$

Where

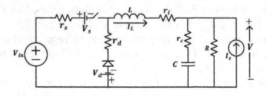

Figure 1. Buck regulator.

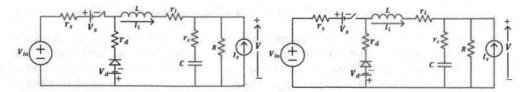

Figure 2. Equivalent circuit during on state.   Figure 3. Equivalent circuit during off state.

$$K_1 = r_s R + r_L R + r_c r_s + r_L r_c + r_c R$$

$$
\begin{bmatrix} V \\ i_0 \\ i_L \end{bmatrix} = \begin{bmatrix} \frac{Rr_c}{(R+r_c)} & \frac{R}{(R+r_c)} \\ \frac{R}{(R+r_c)} & \frac{-1}{(R+r_c)} \\ 1 & 0 \end{bmatrix} \begin{bmatrix} i_L \\ v_c \end{bmatrix} + \begin{bmatrix} 0 & \frac{-Rr_c}{(R+r_c)} & 0 & 0 \\ 0 & \frac{r_c}{(R+r_c)} & 0 & 0 \end{bmatrix} \begin{bmatrix} V_{in} \\ i_z \\ V_s \\ V_d \end{bmatrix}
\tag{4}
$$

Here, for ON time the state space equation are given as

$$
\dot{x} = A_1 x + B_1 u; y = C_1 x + D_1 u; x = \begin{bmatrix} i_L \\ v_c \end{bmatrix}; u = \begin{bmatrix} V_{in} \\ i_z \\ V_s \\ V_d \end{bmatrix}; y = \begin{bmatrix} V \\ i_0 \\ i_L \end{bmatrix}
$$

Where

$$
A_1 = \begin{bmatrix} \frac{-K_1}{L(R+r_c)} & \frac{-R}{L(R+r_c)} \\ \frac{R}{C(R+r_c)} & \frac{-1}{C(R+r_c)} \end{bmatrix} B_1 = \begin{bmatrix} \frac{1}{L} & \frac{Rr_c}{(R+r_c)L} & \frac{-1}{L} & 0 \\ 0 & \frac{r_c}{(R+r_c)C} & 0 & 0 \end{bmatrix} C_1 = \begin{bmatrix} \frac{Rr_c}{(R+r_c)} & \frac{R}{(R+r_c)} \\ \frac{R}{(R+r_c)} & \frac{-1}{(R+r_c)} \\ 1 & 0 \end{bmatrix} D_1 = \begin{bmatrix} 0 & \frac{-Rr_c}{(R+r_c)} & 0 & 0 \\ 0 & \frac{r_c}{(R+r_c)} & 0 & 0 \end{bmatrix}
$$

2.2 Off state operation and state space modeling

The controlled switch (MOSFET) of converter remains OFF during $(1-d)Ts$. During OFF period the equivalent circuit is shown in Figure 2.

For $(1-d)T_s$, equ. 5-8 governs system dynamics during OFF state. During OFF state v_c and i_l are given in equ. 5 and 6 respectively.

$$
\frac{di_L}{dt} = \frac{1}{L}\left[-\frac{i_L}{(R+r_c)}(r_d R + r_L R + r_c r_d + r_L r_c + r_c R) - \frac{Rv_c}{R+r_c} - V_s + \frac{Rr_c}{(R+rc)}i_z - V_d \right]
\tag{5}
$$

$$
\frac{dV_c}{dt} = \frac{1}{C}\left[\frac{Ri_L}{R+r_c} - \frac{V_c}{R+r_c} + \frac{r_c i_z}{R+r_c} \right]
\tag{6}
$$

$$
\begin{bmatrix} \frac{di_l}{dt} \\ \frac{dV_c}{dt} \end{bmatrix} = \begin{bmatrix} \frac{-K_2}{L(R+r_c)} & \frac{-R}{L(R+r_c)} \\ \frac{R}{C(R+r_c)} & \frac{-1}{C(R+r_c)} \end{bmatrix} \begin{bmatrix} i_L \\ v_c \end{bmatrix} + \begin{bmatrix} 0 & \frac{Rr_c}{(R+r_c)L} & 0 & \frac{-1}{L} \\ 0 & \frac{r_c}{(R+r_c)C} & 0 & 0 \end{bmatrix} \begin{bmatrix} V_{in} \\ i_z \\ V_s \\ V_d \end{bmatrix}
\tag{7}
$$

Where

$$K_2 = r_d R + r_L R + r_c r_d + r_L r_c + r_c R$$

$$
\begin{bmatrix} V \\ i_0 \\ i_L \end{bmatrix} = \begin{bmatrix} \frac{Rr_c}{(R+r_c)} & \frac{R}{(R+r_c)} \\ \frac{R}{(R+r_c)} & \frac{-1}{(R+r_c)} \\ 1 & 0 \end{bmatrix} \begin{bmatrix} i_L \\ v_c \end{bmatrix} + \begin{bmatrix} 0 & \frac{-Rr_c}{(R+r_c)} & 0 & 0 \\ 0 & \frac{r_c}{(R+r_c)} & 0 & 0 \end{bmatrix} \begin{bmatrix} V_{in} \\ i_z \\ V_s \\ V_d \end{bmatrix}
\tag{8}
$$

Here, for OFF state the state space equation are given as

$$\dot{x} = A_1 x + B_1 u; y = C_1 x + D_1 u$$

$$A_1 = \begin{bmatrix} \frac{-K_1}{L(R+r_c)} & \frac{-R}{L(R+r_c)} \\ \frac{R}{C(R+r_c)} & \frac{-1}{C(R+r_c)} \end{bmatrix} B_1 = \begin{bmatrix} \frac{1}{L} & \frac{Rr_c}{(R+r_c)L} & \frac{-1}{L} & 0 \\ 0 & \frac{r_c}{(R+r_c)C} & 0 & 0 \end{bmatrix} C_1 = \begin{bmatrix} \frac{Rr_c}{(R+r_c)} & \frac{R}{(R+r_c)} \\ \frac{R}{(R+r_c)} & \frac{-1}{(R+r_c)} \\ 1 & 0 \end{bmatrix}$$

$$D_1 = \begin{bmatrix} 0 & \frac{-Rr_c}{(R+r_c)} & 0 & 0 \\ 0 & \frac{r_c}{(R+r_c)} & 0 & 0 \end{bmatrix}$$

In this modelling *eq. (3), (4), (7)* and *(8)* shown the system dynamics during ON and OFF state. The accurate duty cycle is calculated using these equation. The output voltage after inclusion of circuit non-idealities is given in *eq. (9)*.

$$V = \frac{DV_{in} - DV_s - (1-D)V_d}{1 + \left(\frac{Dr_s + r_L + (1-D)r_d}{R}\right)} \tag{9}$$

Iterative process is selected to calculate the accurate duty cycle as per design requirement. These equations are combined by circuit averaging technique over T_s period. The average model of the time varying buck regulator is given by

$$\dot{x} = A_{avg} x + B_{avg} u; y = C_{avg1} x + D_{avg} u \tag{10}$$

Where

$$A_{avg} = A_1 D + A_2(1-D); B_{avg} = B_1 D + B_2(1-D)$$

$$C_{avg} = C_1 D + C_2(1 - {}'D); D_{avg} = D_1 D + D_2(1-D)$$

2.3 Linearization of system around operating point

The state equation given in the *equ. (10)* exhibits linear response for fixed duty cycle. However, duty cycle is continuously regulated to maintain the constant output voltage in accordance with change in load demand or change in supply voltage. In order to design the controller, the system must be linearized around operating point. This linearization helps in designing the controller for small signal variation around operating point. The small signal variation for the proposed modeling is modeled for system states (x), input (u), and duty cycle (d). The small change in these quantities is given by

$$x(t) = X + \hat{x}; u(t) = U + \hat{u}; d(t) = D + \hat{d} \tag{11}$$

The small signal model of the system is given by eq. (11)

$$\frac{\widehat{dx}}{dt} = A_{avg}\hat{x} + B_{avg}\hat{u} + G\hat{d}; y = C_{avg}X + D_{avg}\hat{u} + H\hat{d} \tag{12}$$

Where

$$G = [(A_1 - A_2)X + (B_1 - B_2)U]\hat{d}; H = [(C_1 - C_2)X + (D_1 - D_2)U]\hat{d}$$

210

3 EFFECT OF PARASITICS ON BUCK REGULATOR

In order to investigate the effect of circuit non-idealities, the transfer functions of output voltage to duty cycle, output voltage to load current and output voltage to input voltage are derived using small signal state space equation. Extensive analysis is done for

$$V_{in} = 24V, V = 12V, R_L = 2.6\Omega, 10\Omega, 11\Omega, , L = 335\mu H, C = 10\mu F, V_s = 0.5V, V_d = 0.8V,$$
$$r_c = r_d = r_s = 0.1\Omega, r_L = 0.2\Omega$$

.

3.1 *Effect of the non-idealities and load variation on duty cycle*

The duty cycle variation with respect of circuit non-idealities as well as load variation is shown in the Figure 4. It is observed that due to circuit parasitic, larger value of the duty cycle is needed to maintain the same output voltage. During the analysis, it is evident from the Figure 4 that selection of duty cycle further depends on load parameters. The amount of losses are significant in the presence of heavy load current. Hence, larger value of duty cycle is needed.

3.2 *Effect of load current on system stability*

Based on the transfer function developed using state space approach, the pole zero graph is plotted for voltage to duty cycle transfer function. It is evident from Figure 5 that with increase in the load demand the pole tends to cross imaginary axis. This signify that for heavier load, the system has less stability margin.

$$G_{vdideal}(S) = \frac{7.164 \times 10^9}{s^2 + 9091s + 2.985 \times 10^8}; G_{vdnonideal}(S) = \frac{7188s + 7.188 \times 10^9}{s^2 + 1.02 \times 10^4 s + 3.039 \times 10^8}$$

3.3 *Effect of load current on system stability*

The frequency response analysis is done using bode plot and shown in Figure 6. The comparison for ideal and non-ideal converter is done by evaluating the value of gain margin, phase margin, gain cross-over frequency and phase cross-over frequency. The result is shown in the Table 1. It is evident from Figure 6, that the converter modeled after considering non-idealities of the system has larger phase margin. This indicates that the controller designed through approximate design may leads to the false performance in real time. On the other hand, open loop transfer function of inductor current to load current variation has marginal difference in the phase margin.

Table 1. Stability margin parameter.

Output voltage to duty cycle

G_{vd}	G_m	PM	W_{gc}	W_{pc}
Ideal	Inf	6.275158409	Inf	86138.47697
Non-Ideal	Inf	11.95412955	Inf	86401.60121

Inductor current to duty cycle

G_{vd}	G_m	PM	W_{gc}	W_{pc}
Ideal	inf	90.37495	inf	75535.72766
Non-Ideal	inf	91.29328	inf	76350.26313

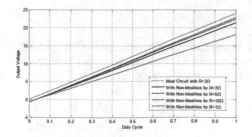

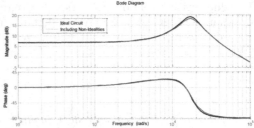

Figure 4. Variation of duty cycle with non-idealities and load variation.

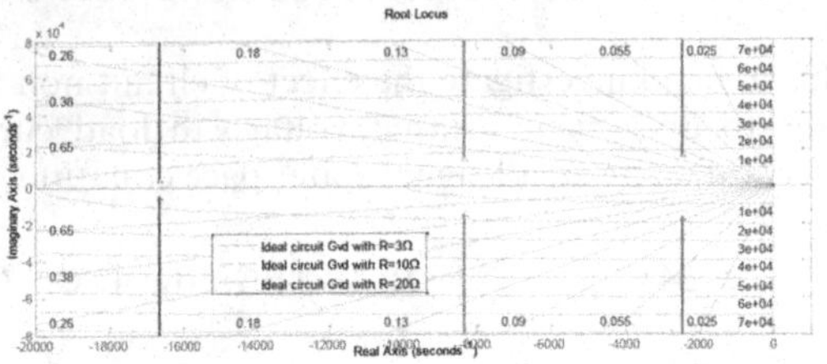

Figure 5. Change in pole location with load variation.

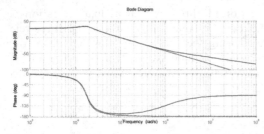

Figure 6. Bode plot for output voltage to duty cycle transfer function.

Figure 7. Bode plot for inductor current to duty cycle transfer function.

4 CONCLUSION

There are several uncertainties associated with DC voltage regulator. While designing controller for such sophisticated DC converter topology, it is important to include non-idealities such as switch on time voltage, conduction resistance, ESR of inductor and capacitor, forward voltage drop and on time diode resistance. In order to avoid complexity in the design, most of the uncertainty is ignored. However, this leads to approximate designing of the controller. Such controller does not provide the fast and accurate control over duty cycle variation. In the proposed work, extensive modelling of the buck regulator is done using state space averaging technique. Developed model is used to determine the open loop frequency response under ideal and non-ideal case. The evaluated results shows the efficacy of the developed model for designing the controller.

REFRENCES

Bayram M. Baha, Sefa Ibrahim, Balci Selami, 2017 A static exciter with interleaved buck converter for synchronous generators, International Journal of Hydrogen Energy, Volume 42, Issue 28, Pages 17760–17770.

Cristri A.W., Iskandar R.F., 2017 Analysis and Design of Dynamic Buck Converter with Change in Value of Load Impedance, Procedia Engineering, Volume 170, Pages 398–403.

Erickson, R. W.,2007. DC–DC Power Converters. In Wiley Encyclopedia of Electrical and Electronics Engineering, J. G. Webster (Ed.).

Middlebrook R. D. and Cuk, S. 1976 A general unified approach to modelling switching-converter power stages, *IEEE Power Electronics Specialists Conference*, Cleveland, OH, pp. 18–34.

Sarikhani Ali, Allahverdinejad Babak, Hamzeh Mohsen, Afjei Ebrahim, 2019, A continuous input and output current quadratic buck-boost converter with positive output voltage for photovoltaic applications, Solar Energy Volume 188, Pages 19–27.

Technologies for Sustainable Development – Mahajan, Patel & Sharma (eds)
© 2020 Taylor & Francis Group, London, ISBN 978-0-367-33737-7

Hybrid technique for DC-link capacitor voltage balancing in entire operating range of five-level NPC inverter using redundancy of switching states and two quadrant DC-DC converter

Siddharthsingh K. Chauhan, Selarka Viraj, Shah Prem & P.N. Tekwani

Department of Electrical Engineering, Institute of Technology, Nirma University, India

ABSTRACT: Neutral point clamped (NPC) inverter faces the problem of fluctuating neutral-point potential (NPP), which leads to severe problems like increase in total harmonics distortion (THD) of output voltage and increase in voltage stress of the power semiconductor switches. This paper is focused on hybrid techniques used for reducing fluctuations in NPP by means of balancing the voltage across dc-link capacitors. A technique of providing offset in carrier waves using phase disposition pulse width modulation (PDPWM) is proposed to ensure the voltage balancing. The drawback of this scheme is it does not work satisfactorily for four-level and five-level mode of operations of the inverter. It works as intended up to three-level operation. Other approach to achieve the task is based on carrier less PWM technique. In this technique, the redundancy of switching states in voltage state phasor diagram is used. The limitation of this technique is that it provides dc-link balancing up to modulation index 0.5 due to its inherent characteristics. To provide dc-link voltage balancing above 0.5 modulation index, external hardware is required. Using hybrid combination of bidirectional buck-boost chopper and SVPWM scheme, the dc-link voltage balancing for the entire range of operation of inverter is obtained. MATLAB simulation of this technique is carried out on three-phase, 50 Hz, 4 pole, 400 V squirrel cage induction motor. For experimental verification of the proposed technique, three-phase five-level NPC inverter is used with three-phase star connected 100 Ω resistors.

1 INTRODUCTION

The necessity of high-power inverters for induction motor drives in industries has increased. For high-power medium-voltage drives, voltage source inverter (VSI) operated on high switching frequency PWM technique is not preferred because of increased switching losses, high dv/dt, and EMI issues. Multi-level inverters are widely used in motor drives, active and hybrid power filters, high voltage dc transmission (HVDC), static VAR compensators, railway traction, hydro pumped storage, wind energy conversion etc. [Rodriguez 2002, 2007; Kouro 2010; Nabae 1981]. From proposed multilevel converter topologies, the most popular topologies are NPC or diode clamped, cascaded H-bridge (CHB), and flying capacitor (FC), [Rodriguez 2002, 2010; Kouro 2010; Nabae 1981; Tallam 2005]. NPC inverters have several advantages. The quality of output voltage and current waveforms is better, bearing voltages, shaft current and stress on insulation winding can be significantly reduced if proper PWM technique is introduced in NPC inverter [Tallam 2005]. NPC inverter suffers from some disadvantages, such as imbalance in dc-ink capacitor voltages due to fluctuation in NPP [Tallam 2005], main structural drawbacks include complex power bus structure, difference in loss distribution amongst switches and the resulting unsymmetrical semiconductor-junction temperature [Rodriguez 2010].

Fluctuation of NPP happens due to following reasons:

- Non-uniform capacitance leakage currents
- Asymmetrical tracking of current, non-uniform delays in semiconductor devices
- Existence of dc components in the neutral current etc. [Srikanthan 2009]

Main reason for fluctuation of NPP is uneven flow of dc-link neutral current. The NPP fluctuation leads to inherent imbalance in the dc-link capacitor voltages, which causes deteriorated quality of voltage waveform, presence of lower-order harmonics at the inverter output voltage, torque pulsation and increased voltage stress on the dc-link capacitors and power semiconductor devices [Tekwani 2005].

There are various PWM schemes available to switch the inverter such as carrier based PWM (CBPWM), selective harmonic elimination (SHE), carrier-less or space vector PWM (SVPWM). [Rodriguez 2002, 2007, 2010; Kouro 2010; Nabae 1981, Tallam 2005; Srikanthan 2009; Tekwani 2005; Tamasas 2014]. From above all schemes; the CBPWM is having less computational requirement and hence have inherent simplicity. It is further classified into phase-shifted PWM (PSPWM) and level-shifted PWM (LSPWM). LSPWM is further classified into phase-disposition PWM (PDPWM), phase-opposition-disposition PWM (PODPWM), and alternate phase-opposition-disposition PWM (APODPWM) [Tamasas 2014; McGrath 2002; Bharatiraja 2012; Mohan 2012]. Capacitor voltage balancing technique reported in literatures are mainly about balancing capacitor voltage by adding extra hardware circuitry to the inverter which requires addition of power semiconductor switches, inductance and capacitance to control the charging and discharging current of dc-link capacitors. The other one is modification in modulation technique of the inverter [Chaturvedi 2014]. In [McGrath 2002] conventional CBPWM technique with offset voltage injection for a single-phase three-level NPC inverter is used to get NPP control. Two modulation techniques, carrier-based PWM-offset voltage injection (CBPWM-OVI) and carrier-based PWM-maximum offset voltage injection (CBPWM-MOVI) are discussed. CBPWM-MOVI has better dynamic response compared to previous method [McGrath 2002]. The SVPWM allows a particular output phase or line-to-line voltage to be generated by more than one combination of switches. Effective utilization of these redundant switching states helps in reducing capacitor voltage imbalance. However, this technique performs poorly at high modulation indices since few redundant states can be used in the outer vector space hexagons.

2   DC-LINK CAPACITOR VOLTAGE IMBANCING IN FIVE-LEVEL NPC

One leg of a five-level NPC inverter is shown in Figure 1. Its switching states are listed in Table 1. Problem of capacitor voltage imbalance is quite evident from the results presented. Figure 2 shows the inverter output voltage during five-level operation when each capacitor is supplied by

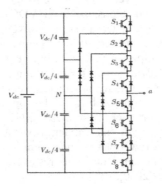

Figure 1.   One leg of a five-level NPC inverter.

Table 1. Switching logic for five-level NPC inverter.

V_{aN}	S_1	S_2	S_3	S_4	S_5	S_6	S_7	S_8
$+V_{dc}/2$	1	1	1	1	0	0	0	0
$+V_{dc}/4$	0	1	1	1	1	0	0	0
0	0	0	1	1	1	1	0	0
$-V_{dc}/4$	0	0	0	1	1	1	1	0
$-V_{dc}/2$	0	0	0	0	1	1	1	1

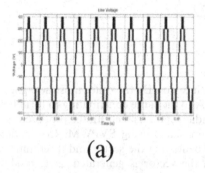

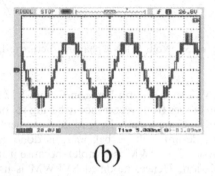

(a)   (b)

Figure 2. (a). Line voltage of five-level NPC, (b) Line voltage V_{AB}.

a separate dc source. Figure 2(a) shows simulation results and Figure 2(b) shows real-time results. In order to obtain the gating pulses for five-level NPC inverter, dSPACE controller board is used. Dead-band of 1.2 µs is provided externally between gating signals of complementary switches. Results are taken on 100 Ω star connected load for modulation index 0.9. However, when a single dc source is used (after t = 1 s) to supply all the capacitors and no specific measures are taken for dc-link capacitor voltage balancing, then a worst case divergence in capacitor voltages and distortion in output voltage is observed. This situation demands a dc-link capacitor voltage balancing scheme.

3 DC-LINK CAPACITOR VOLTAGE BALANCING TECHNIQUE

In this paper, a hybrid technique is proposed for achieving balancing of dc-link capacitor voltages in the entire operating range of the five-level NPC inverter. Balancing of capacitor voltage is achieved in two-level and three-level modes of operations by effectively exploiting the redundancy of switching states of NPC inverter. Whereas, the balancing of voltages in four-level and five-level modes of operation is achieved by using a two-quadrant chopper as buck-boost converter [Sipai 2016].

3.1 DC-link voltage balancing using redundant switching states of voltage space phasor structure

SVPWM is the most attractive modulation strategy for multilevel inverters because it provides significant flexibility for optimizing switching waveforms, and because SVPWM is well suitable for digital signal processor implementation. The voltage space phasor structure of conventional five-level NPC inverter comprises a total of 125 switching states, 61 vectors and 96 triangular sectors as shown in Figure 3. These 61 vectors combined can be assumed to make four hexagons: A, B, C and D. Vectors V1 to V6 contribute to form hexagon A; V7 to V18 form hexagon

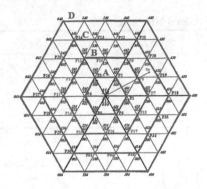

Figure 3. Voltage space phasor structure (5-L NPC).

B; V19 to V36 form hexagon C, and V37 to V60 form hexagon D (the outer most hexagon). If one moves from vectors contributing to form hexagon A to the same contributing to form hexagon B, C and finally D; gradual decrement in the redundant states is seen. In the proposed strategy, dc-link voltage balancing is done using redundant states using SVPWM. Conventional five-level SVPWM is complex because it incorporates identifying the sector and dwell time calculation, Hence, modified SVPWM is used in which the vector is generated using modified SPWM. Capacitor voltages Vc1, Vc2, Vc3 and Vc4 are sensed. To keep voltages within the check, the logic of Schmitt trigger is applied. So, if Vc(i)≥ 101 V, it needs to discharge the capacitor; for that variable L(i) is defined which indicates the action for the corresponding capacitor, here as Vc(i)≥101, L(i) = -1 i.e. to discharge C(i). Similarly, if Vci≤ 99 V: Li=1; in this condition it needs to charge the capacitor, where i = 1,2,3,4. Four index variables are selected: n1, n2, n3, n4. They are pre-set to 0 at the start of the algorithm. Taking the inner most hexagon, each vector has four redundant states and effect of each state on capacitor voltage is shown in Table 2. The n1 corresponds to the first redundant vector, n2, n3, and n4 correspond to second, third, and fourth redundant vectors, respectively. Now if L(1)=1,then that condition is satisfied in n1, n2, n3, then it will be incremented by one. And if L(1)=-1 then n4 is incremented by one. Similar process is done for L(i), where i=2, 3, 4. Now the maximum of n(i) is selected which represents the optimum switching state. Four multiplication factors are selected as a, b, c, d which dictate about redundant state to be selected. If n(1) is max, then a=1, i.e. first redundant vector is to be selected, and so on. The study of effect of each switching state on dc-link capacitors is carried out.

When Vdc1 crosses the upper limit, so L1= -1 (discharge C1 needed). Vdc2, Vdc3, Vdc4 cross the lower band, so L2=1, L3=1, L4=1. This condition is satisfied in fourth redundant vector which can be justified from table. So a=0, b=0, c=0, d=1.

S = S[a, b, c] = a(0 0 1) + b(1 1 2) + c(2 2 3) + d(3 3 4)

S = S[a, b, c] = (3 3 4)

To justify the control technique, simulation results were taken on R load. The modulation index is increased linearly and finally set to 1. Up to modulation index 0.6, the dc-link voltages are seen balanced (in Figure 4); beyond which it is not the case (Figure 5 and 6).

Table 2. Effects of redundant states on capacitor voltage for vector no. 5.

Vector No.	Sa	Sb	Sc	Vc1	Vc2	Vc3	Vc4
	0	0	1	C	C	C	D
	1	1	2	C	C	D	C
5	2	2	3	C	D	C	C
	3	3	4	D	C	C	C

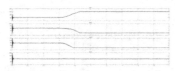

Figure 4.   DC-link capacitor voltages.

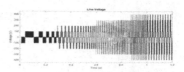

Figure 5.   Line voltage using redundant states.

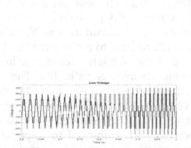

Figure 6.   Distorted line voltage for ma = 0.6 to 1.

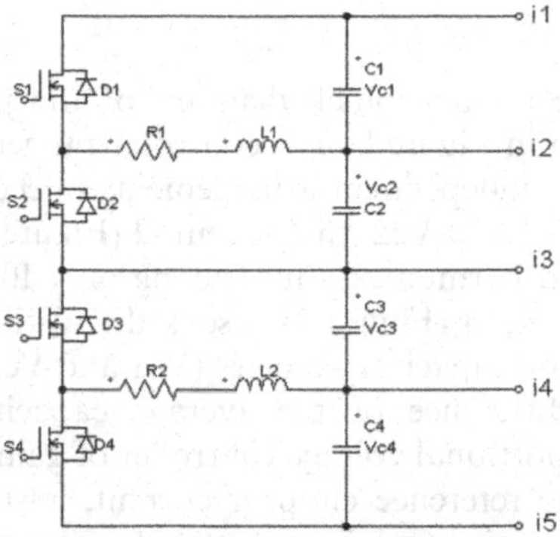

Figure 7.   Two-quadrant chopper.

3.2   *Capacitor voltage balancing by bidirectional buck-boost converter and hybrid technique*

With more than three-level operation, dc-link voltage imbalance is a major technical draw-back of NPC. Buck-boost chopper can be connected with the dc-link of NPC for voltage balancing. It is a reliable and robust solution to this problem. Capacitor voltage equalization can be achieved by various control technique of chopper. Such as single-pulse, multi-pulse and hysteresis band current control schemes. Single-pulse scheme involves switches with high power rating and the switching frequency is also at higher magnitude. So, overall switching losses increase. The current rating of chopper can be kept within limit by multi-pulse scheme, but it involves faster switching actions and offers slower response. The hysteresis band current control scheme offers faster dynamics, lower current rating of the chopper devices and can nullify the initial voltage imbalance as well, however, it involves much faster switching actions which may not be feasible for some of its applications. Therefore, depending on the system requirements and ratings, one of these schemes may be used. Here, the hysteresis band current control method is adopted for dc voltage balancing of five-level NPC inverter [Tekwani 2005].

To prevent the imbalance; bidirectional buck-boost chopper circuit is used on dc-link capacitors of the inverter, which is shown in Figure 7. The resistances R1 and R2 represent internal resistances of the inductors L1 and L2, respectively. This circuit is helping to balance the capacitor voltages by transferring energy from overcharged to undercharged capacitor. L1 is used to exchange the energy between Cd1 and Cd2 using SC1, SC2, DC1, DC2. Whereas L2 exchanges the energy between Cd3 and Cd4 using SC3, SC4, DC3 and DC4. If there are operating conditions in which the charge is to be transferred from Cd2 (Cd3) to Cd1 (Cd4), especially under transients, then the single-quadrant chopper circuit is not sufficient. However, using the two-quadrant chopper circuit of Figure 7, it is possible to transfer energy both from Cd2 (Cd3) to Cd1 (Cd4) or from Cd1 (Cd4) to Cd2 (Cd3). Therefore, this circuit is more useful to restore the voltage balancing under transient conditions. Furthermore, most of the power semiconductor switches have built in anti-parallel diodes and not much extra cost is involved in this.

217

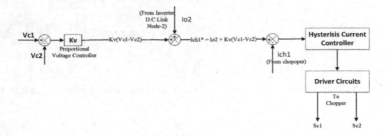

Figure 8.   Switching logic for controlling S1 and S2 of chopper.

The control block diagram of the proposed hysteresis band chopper current control is shown in Figure 8. The control arrangement only for the upper chopper circuit is shown, and similar independent arrangement is needed for the lower chopper circuit as well. For controlling Vc1 and Vc2, the current i2 (Figure 7) is sensed and filtered out to remove the fundamental and harmonics, thus leaving only I02 (average value of i2). Similarly, for controlling Vc3 and Vc4, i4 (Figure 7) is sensed and filtered out to obtain its average value I04. The corresponding capacitor voltages (Vc1 and Vc2) are measured and filtered out to remove the ripples. The difference of the average capacitor voltages ($\Delta Vc = Vc1 - Vc2$) is passed through a proportional voltage controller of gain Kv to output $Kv\Delta Vc$, which is added to I02 to generate the reference chopper current, $ich1^*$. The chopper circuit is employed to track $ich1^*$ in a hysteresis band. This control loop ensures that the average chopper current is equal to $ich1^*$. Generation of gating pulses of the chopper is carried out by hysteresis current control technique. When the error ($ich1^*-ich1$) hits the upper limit of the band; Vc1 is overcharged so, S1 switch is turned "ON" and S2 switch is turned "OFF" and energy transfer takes place from C1 to C2. If it hits the lower band; reverse control action from previous case is taken. The capacitor voltage balancing is maintained throughout the operating range of the inverter as evident from the results shown in Figure 9, 10, and 11.

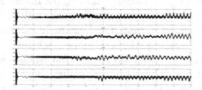

Figure 9.   Capacitor voltages.

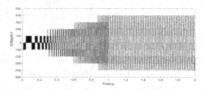

Figure 10.   Balanced line voltages.

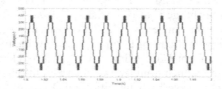

Figure 11.   Line Voltage (five-level).

4 CONCLUSION

Proposed Control technique leads to minimum discrepancy in dc-link capacitor voltages. For modulation index below 0.6, there is no need of external hardware. It just needs to sense the capacitor voltages and that information is provided to space vector algorithm. Depending upon the values of voltage of dc-link capacitors, the suited redundant state is selected and the voltages are kept in limit provided. The main advantage of this technique is it does not require any external hardware. So, for most of the time the motors in the industries are not always operated on full-speed and hence the inverter is not needed to operate on higher modulation index. In such cases this is very good solution. For operation nearer to full-speed (at higher modulation index), the bottleneck of SVM algorithm is that it can provide voltage balancing up to 0.5 modulation index due to the inherent space phasor structure of NPC. So, above 0.5 modulation index bidirectional buck-boost chopper is installed and it provides the voltage balancing by energy transfer amongst capacitors.

REFERENCES

Bharatiraja, C., Latha, S., and Dash, S. 2012. A Space Vector Pulse Width Modulation Approach for DC Link Voltage Balancing in Diode- Clamped Multilevel Inverter. In Proc. AASRI Conf. on Modeling & Control.

Chaturvedi, P., Jain, S., and Agarwal, P. 2014. Carrier-Based Neutral Point Potential Regulator With Reduced Switching Losses for Three Level Diode-Clamped Inverter. *IEEE Trans. Ind. Electron.* 61(2): 613–624.

Kouro, S., Malinowski, M., Gopakumar, K., Pou, J., Franquelo, L., Wu, B., Rodriguez, J., Prez, M., and Leon, J. 2010. Recent Advances and Industrial Applications of Multilevel Converters. *IEEE Trans. Ind. Electron.* 57(8): 2553–2580.

McGrath, B., and Holmes, D. 2002. Multicarrier PWM Strategies for Multilevel Inverters. *IEEE Trans. Ind. Electron.* 49(4): 858–867.

Mohan, D. and Kurub, S. 2012. A Comparative Analysis of Multi Carrier SPWM Control Strategies using Fifteen Level Cascaded H-bridge Multilevel Inverter. Int. Journal of Computer Applications. 41 (21): 7–11.

Nabae, A., Takahashi, I., and Akagi, H. 1981. A New Neutral-Point- Clamped PWM Inverter. *IEEE Trans. Ind. Appl.* Ia-17(5): 518–523.

Rodriguez, J., Lai, J., and Peng, F. 2002. Multilevel Inverters: A Survey of Topologies, Controls, and Applications. *IEEE Trans. Ind. Electron.* 49(4): 724–738.

Rodriguez, J., Bernet, S., Wu, B., Pontt, J., and Kouro, S. 2007. Multilevel Voltage-Source-Converter Topologies for Industrial Medium-Voltage Drives. *IEEE Trans. Ind. Electron.* 54(6): 2930–2945.

Rodriguez, J., Bernet, S., Steimer, P., and Lizama, I. 2010. A Survey on Neutral-Point-Clamped Inverters. *IEEE Trans. Ind. Electron.* 57(7): 2219–2230.

Shukla, A., Ghosh, A., and Joshi, A. 2012. Control and DC capacitor voltages in diode-clamped multilevel inverter using bidirectional buck-boost choppers. *IET Power Electron.* 5(9): 1723–1732.

Sipai U. 2016. Implementation of Close-Loop Control of Five-Level Neutral-Point Clamped Inverter for DC-link Capacitor Voltage Balancing. Major Project Report for M. Tech.

Srikanthan S., Mishra, M., KalyanKumar, B., and Bhaskar, K. 2009. Capacitor Voltage Balancing in Neutral Clamped Inverters for DSTATCOM Application, *IEEE POWERENG*.

Tallam, R., Naik, R., and Nondahl, T. 2005. A Carrier- Based PWM Scheme for Neutral-Point Voltage Balancing in Three-Level Inverters. IEEE Trans. Ind. Appl. 41(6): 1734–1743.

Tamasas, M., Saleh, M, Shaker, M., and Hammoda, A. 2014. Evaluation of Modulation Techniques for 5-Level Inverter Based on Multicarrier Level Shift PWM. In proc. 17th IEEE Mediterranean Electrotechnical Conf.

Tekwani, P., Kanchan, R., and Gopakumar, K. 2005. Five-Level Inverter Scheme for an Induction Motor Drive with Simultaneous Elimination of Common-Mode Voltage and DC-Link Capacitor Voltage Imbalance. *IEE Proceedings – EPA.* 152(6): 1350–2352.0

Electronics and communication engineering track

Technologies for Sustainable Development – Mahajan, Patel & Sharma (eds)
© *2020 Taylor & Francis Group, London, ISBN 978-0-367-33737-7*

Performance evaluation of MIMO under time-varying and spatially correlated channels with feedback delay

Dhaval J. Upadhyay*
Space Applications Centre, Indian Space Research Organization, Ahmedabad, India

Y.N. Trivedi
Institute of Technology, Nirma University, Ahmedabad, India

S.C. Bera
Space Applications Centre, Indian Space Research Organization, Ahmedabad, India

ABSTRACT: The extent of advantages of coherent detection and antenna selection schemes depend on accuracy of available Channel State Information (CSI). In this paper, imperfect CSI at receiver (CSIR) is assumed in MIMO with M_t transmit antenna and M_r receive antenna for detection of symbols using receive diversity technique and single transmit antenna selection (TAS). Imperfect CSIR is assumed to be available with error of fixed variance (σ^2_e) due to presence of channel estimation error. Channels are considered as time selective with correlation coefficient a, frequency non-selective, flat fading and identically distributed using Rayleigh distributions. Therefore, channels change during the feedback of selected transmit antenna index from receiver to transmitter due to finite transmission delay. Time varying channel is characterized by first order autoregressive (AR1) process. We also consider equal spatial correlation (ρ) at receiver and transmitter ends and between any two channels. Performance of M_txM_r MIMO with single TAS and receive diversity technique is evaluated by BER simulations for different antenna configurations under the compound effect of a, ρ, (σ^2_e) and feedback delay for BPSK.

Keywords: AR1 process, spatial correlation, time varying channel, imperfect CSIR, MIMO

1 INTRODUCTION

Perfect knowledge of Channel State Information (CSI) is required at Multiple Input Multiple Output (MIMO) receiver end for coherent detection and transmit and/or receive antenna selection schemes. Perfect knowledge of CSI at transmitter (CSIT) is required for spatial multiplexing and transmit beamforming schemes (David T. et. al. 2005). Whereas, perfect knowledge of CSIR is required for the coherent detection, transmit and receive antenna selection schemes (David et. al. 2005, Simon et. al. 2000, Proakis 2001). In MIMO, CSI is required for each symbol and each channel at transmitter and receiver ends. However, it is practically difficult to have the perfect CSIR and CSIT due to limited transmitted pilot symbol energy and presence of error of fixed variance in CSI estimation at receiver end which leads to imperfection in CSI. In (Chen et. al. 2008, Mi et. al. 2017, Yu et. al. 2018, Anvar et. al. 2015), performance of MIMO system is analysed using imperfect CSI. The imperfection in CSIR leads to error floor in SER performance of MIMO at higher SNR. Therefore, performance of above schemes depends upon the accuracy of the CSI availability at receiver and transmitter ends. Channels are also assumed to be time varying in nature due to presence of doppler effects. Therefore, channel

---

* Corresponding author: djupadhyay@sac.isro.gov.in

changes during the feedback of selected transmit antenna indices from receiver to transmitter end due to presence of large transmit overheads and transmission delay. In (Chen et. al. 2008, Yu et. al. 2018, Trivedi et. al. 2011), performance of the MIMO system is analysed using delayed CSIT under time varying channels. In transmit antenna selection scheme, best antenna is not selected due to delayed CSIT which degrades the SER performance of the system. Further, due to size constraints of receiver and transmitter design, multiple antennas are placed closely with each other. Therefore, channels are spatially correlated with each other in MIMO. In (Chiani et. al. 2003, Jacobs et. al. 2013, Hajjaj et. al. 2016, Kulkarni et. al. 2014, Xu et. al. 2009), performance of the MIMO system is analysed under spatially correlated channels. SER performance of MIMO under spatially correlated channels degrades due to degradation in diversity order of the system. Therefore, it becomes necessary to analyse the performance of MIMO under time varying & spatially correlated channels and feedback delay using imperfect CSI.

In this paper, performance of $M_t x M_r$ MIMO system with single TAS and receive diversity technique is analyzed by simulations under channels with temporal correlation & spatial correlation and feedback delay using imperfect CSI. We considered time varying channel with temporal correlation coefficient a, flat fading and identically distributed using Rayleigh distributions. The time varying channel is approximated by AR1 model. We consider equal spatial correlation at receiver and transmitter ends and also equal spatial correlation (ρ) between any two channels. It is assumed that receiver tracks the CSI for each channel and each symbol. It is also assumed that CSIR is available with error of fixed variance (σ_e^2). Imperfect CSIR is used at receiver end for detection of symbols and selection of single transmit antenna. Maximum channel frobenius norm criteria is used to select transmit antenna index. Channel is also assumed to be changing during the feedback of selected transmit antenna index from receiver to transmitter end for antenna selection. BER simulations are presented for BPSK under the compound effect of ρ, a, σ_e^2 and feedback delay to find out the impact on MIMO performance.

2 SYSTEM MODEL

$M_t x M_r$ MIMO system is considered in the proposed system model with single transmit antenna selection out of M_t antennas and receive diversity technique. The proposed system model is used in the uplink and downlink scenarios of user equipment (UE) and base station (BS) of 4G and 5G wireless systems where UE and BS has more than one antenna which is shown in Figure 1.

Time selective, frequency non-selective and identically distributed Rayleigh fading channels are assumed. Here, two dimensional CSI matrix (H) is defined as follows

$$H = \begin{bmatrix} h_{1,1} & \cdots & h_{1,M_r} \\ \vdots & \ddots & \vdots \\ h_{M_t,1} & \cdots & h_{M_t,M_r} \end{bmatrix},$$

Where, $h_{i,j}$ is channel coefficient of ith transmit antenna and j^{th} receive antenna, $i = 1,2,...,M_t$ and $j = 1,2,...,M_r$. $h_{i,j} \sim CN(0,1)$. The low pass equivalent complex received symbol is denoted by

$$y = hx + n \tag{1}$$

Here, $y = [y_{I,1}\ y_{I,2}\ \cdots\ y_{I,Mr}]^T$ and $y_{I,j}$ is the received symbol at j^{th} receive antenna from I^{th} selected transmit antenna. Where, $I=1,2,...,M_t$, $j=1,2,...,M_r$. x is M-PSK modulation symbol with average power per symbol E_s. $n = [n_1\ n_2\ ...\ n_{Mr}]^T$, where n_j is circularly symmetric complex Gaussian random variable with zero mean and variance N_0. $h = [h_{I,1}\ h_{I,2}\ \cdots\ h_{I,Mr}]^T$, where $h_{I,j}$ is channel coefficient of Ith selected transmit antenna and j^{th} receive

224

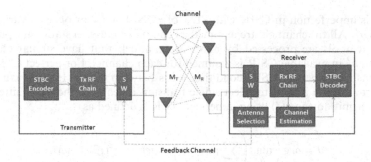

Figure 1. System model of M_txM_r MIMO with single TAS and receive diversity technique.

antenna. Time varying channel is characterized by the first order autoregressive (AR1) model as

$$H = a\hat{\text{H}} + \sqrt{1 - a^2}\,W \tag{2}$$

where $a=E[\hat{h}_{i,j}{}^* \, \hat{h}_{i,j}]$ is temporal correlation coefficient, $0\leq a\leq 1$. Here, $\hat{H}$ is defined as two dimensional CSI matrix at previous symbol position. W is the time varying component of the channel which is an independent and identically distributed (i.i.d.) Gaussian random variable with probability density function (PDF) of $\sim CN(0, 1)$. For static channel condition $a = 1$, whereas for time varying channel condition $a<1$. Relation of temporal correlation a with Doppler frequency and symbol duration is assumed as (Jakes et. al. 1994, Jarinová 2013, Gomadam et. al. 2007),

$$a = I_0(2\pi f_d T_s),$$

where f_d is doppler frequency in Hz, and T_s is the symbol duration in seconds. Further, we considered spatial correlation in all M_txM_r channels due to close placement of antennas at transmitter and receiver. Here, equal spatial correlation is assumed at receiver and transmitter (Kulkarni et. al. 2014). Further, for lower number of transmit and receive antennas and closely placed antennas, equal spatial correlation (ρ) between a pair of any two channels is considered (Simon et. al. 2000, Kulkarni et. al. 2014). For large number of M_t and M_r, our assumption of equal spatial correlation between channels provides an upper-bound on MIMO performance compared to any other spatial correlation models such as exponential correlation model, Kronecker correlation model. Thus, spatial correlation coefficient (ρ) is defined as

$$\rho = E\big[h_{i_1 j_1} h_{i_2 j_2}\big], where, i_1, i_2 = \{1, 2, \ldots, M_t\}, j_1, j_2 = \{1, 2, \ldots, M_r\}, i_1 \neq i_2, \ j_1 \neq j_2.$$

We assume that imperfect CSIR is available with error of fixed variance (σ_e^2). Imperfect CSIR is denoted as $\tilde{h}$. Here, imperfect CSIR is used at receiver end for detection of symbols and selection of single transmit antenna index. Due to presence of channel estimation error, the imperfect CSIR is considered (Chen et. al. 2008, Mi et. al. 2017, Yu et. al. 2018, Anvar et. al. 2015). The imperfection in CSIR is considered to be additive independent complex Gaussian random variable. The imperfect CSIR ($\tilde{h}_{i,j}$) for i^{th} transmit antenna and j^{th} receive antenna is defined as

$$\tilde{\text{h}}_{i,j} = \sqrt{1 - \sigma_e^2}\,\text{h}_{i,j} + \sigma_e h_{i,j}^e,$$

where, $h_{i,j}^e$ is imperfection in CSIR with PDF of ~$CN(0,1)$. Error of fixed variance in CSIR is denoted as σ_e^2. All the channels are assumed as complex Gaussian stationary process and all the received symbols are processed by hardware channels that have similar characteristics. Thus, uniform σ_e^2 in available CSIR is assumed for each channel. For perfect CSIR condition, σ_e^2 is 0 whereas for imperfect CSIR condition, σ_e^2 is between 0 to 1. Here, maximum channel frobenius norm criteria is used to select single transmit antenna at the transmitter end. Index of selected transmit antenna (I) using imperfect CSIR is defined as

$$I = arg\left(\max_i \left(\sum_{j=1}^{M_r} \left| \widetilde{h}_{i,j} \right|^2 \right) \right), where\ i = 1, 2, \ldots, M_t \tag{3}$$

We assume feedback delay of one symbol duration in transmitting selected antenna index from receiver to transmitter end. The detection variable for the symbol with receive diversity technique is expressed as

$$d = \widetilde{\mathbf{h}}^* y$$

3 PERFORMANCE ANALYSIS

The expression of average SNR for the considered MIMO ($M_t, 1; M_r$) system is derived in this section. Detection variable can be expanded as

$$d = \widetilde{h}_{I,1}^* y_{I,1} + \widetilde{h}_{I,2}^* y_{I,2} + \ldots + \widetilde{h}_{I,M_r}^* y_{I,M_r} = \|h\| \sqrt{1 - \sigma_e^2} x + \delta, \quad where, \quad z_j \quad is \quad defined \quad as$$

$$z_j = \sqrt{1 - \sigma_e^2} \widetilde{h}_{I,j}^* n_j + \sigma_e h_{I,j} h_{I,j}^{e*} x + \sigma_e h_{I,j}^{e*} n_j$$

δ is the complex Gaussian random variable with variance of $\sigma_e^2 E_s + N_0$. We can express instantaneous SNR (γ_i) at receiver with i^{th} selected transmit antenna as

$$\gamma_i = \frac{\|h^2\| \left(1 - \sigma_e^2\right) E_s}{\sigma_e^2 E_s + N_0}.$$

We can express average SNR ($\bar{\gamma}_i$) at receiver with i^{th} selected transmit antenna as

$$\gamma_i = \frac{\left(1 - \sigma_e^2\right) E_s}{\sigma_e^2 E_s + N_0}. \tag{4}$$

Average probability of error, P_e can be defined as (Kulkarni et. al. 2014)

$$P_e = \int_0^\infty P_e(x) P_\gamma(x) dx = -\int_0^\infty P_e^{'}(x) F_\gamma(x) dx \tag{5}$$

where $P_e^{'}(x)$ is the derivative of $P_e(x)$ which is the conditional error probability (CEP) as a function of instantaneous SNR. For BPSK, $P_e(x) = Q(\sqrt{2x})$ and hence we obtain

$$P_e = \frac{1}{2\pi} \int_0^\infty \frac{e^{-x}}{\sqrt{x}} F_\gamma(x) dx$$

Cumulative distribution function (CDF) of γ can be defined as

$$F_\gamma(x) = P\left(\max_i\left(\frac{\sqrt{1-\sigma_e^2}E_s}{\sigma_e^2 E_s + N_0}\sum_{j=1}^{M_r}\left|\widetilde{h}_{i,j}\right|^2\right) < x\right) = P\left(\gamma_1 < x, \gamma_2 < x, \ldots, \gamma_{M_t} < x\right) \qquad (6)$$

where, $\gamma = \left[\gamma_1\gamma_2\ldots\gamma_{M_t}\right]^T$.

In this work, BER simulations are presented for BPSK under spatially correlated & temporally correlated channels, imperfect CSIR and feedback delay to find out the impact on performance.

4   SIMULATION RESULTS

BER simulation results are presented in Figure 2 and Figure 3 for MIMO (2,1;2) TAS & receive diversity technique for $M = 2$, $a = 0.999, 0.99$, $\rho = 0.5, 0.9$, and $\sigma_e^2 = 0.01, 0.1$ with feedback delay. It can be seen from both the results that impact of imperfect CSIR is significant compared to spatial correlation on BER performance of MIMO with feedback delay.

It can also be seen from the results that degradation in SNR is <1dB for $a = 0.99$ as compared to $a = 0.999$. Hence, it can be concluded that impact of large scale fast fading channel is

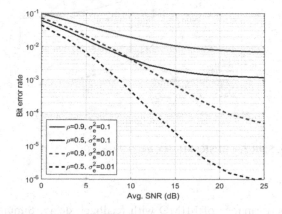

Figure 2.   BER vs Avg. SNR for BPSK, MIMO (2,1;2), a=0.999.

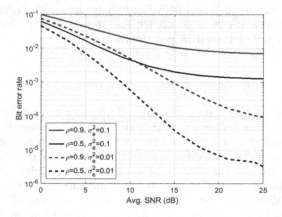

Figure 3.   BER vs Avg. SNR for BPSK, MIMO (2,1;2), a = 0.99.

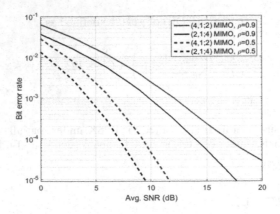

Figure 4.   BER vs Avg. SNR for BPSK, MIMO, $a=0.999$, $\sigma_e^2=0.01$.

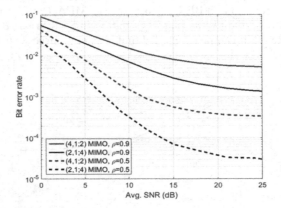

Figure 5.   BER vs Avg. SNR for BPSK, MIMO, $a=0.999$, $\sigma_e^2=0.1$.

negligible on BER performance of MIMO with feedback delay. Simulation results are also presented in Figure 4 and Figure 5 for MIMO $(4,1;2)$ and $(2,1;4)$ TAS & receive diversity technique configurations for $M = 2$, $a = 0.999$, $\rho = 0.5,0.9$ and $\sigma_e^2 = 0.01,0.1$. The error floor in BER performance is seen for different MIMO configurations with higher error of fixed variance (σ_e^2) in CSIR. Hence, it can be concluded that impact of imperfect CSIR is significant compared to spatial correlation on BER performance of MIMO.

5   CONCLUSION

We considered $M_t x M_r$ MIMO system with single TAS and receive diversity technique. Channels are assumed to be flat fading and identically distributed using Rayleigh distributions. Time varying and spatially equi-correlated channels are considered in analysis. We also considered that CSIR is available with error of fixed variance. Imperfect CSIR is used at receiver end for detection of symbols and selection of single transmit antenna. Channel is assumed to be changing during the feedback of selected transmit antenna index from receiver to transmitter. Performance of MIMO is analysed for proposed scheme using BER simulations. Impact of imperfect CSIR is significant compared to spatial correlation on BER performance of

MIMO in presence of feedback delay. Impact of large scale fast fading channel is negligible on BER performance of MIMO in presence of feedback delay.

REFERENCES

David T. and Pramod V. 2005. Fundamentals of Wireless Communication. Cambridge University Press, New York, NY, USA.

Simon, M. K. and Alouini, M.S. 2000. Digital Communication over Fading Channels-A Unified Approach to Performance Analysis. Hoboken: Wiley.

Proakis, J.G. 2001. Digital Communications (4th ed.). New York: McGraw Hill.

Chen, C.Y., Sezgin, A., Cioffi, J.M., Paulraj, A. 2008. Antenna Selection in Space-Time Block Coded Systems: Performance Analysis and Low-Complexity Algorithm. IEEE Trans. Signal Process., vol. 56 (7), 3303–3314.

Mi, D., Dianati, M., Zhang, L., Muhaidat, S., Tafazoli, R. 2017. Massive MIMO Performance with Imperfect Channel Reciprocity and Channel Estimation Error. IEEE Trans. Commun., vol. 65(9), 3734–3749.

Yu, X., Xu, W., Leung, S.H., Wang, J. 2018. Unified Performance Analysis of Transmit Antenna Selection with OSTBC and Imperfect CSI Over Nakagami-m Fading Channels. IEEE Trans. Veh. Technol., vol. 67(1), 494–508.

Anvar, S.M.M., Khanmohammadi, S., Museviniya., Javad 2015. Game theoretic power allocation for fading MIMO multiple access channels with imperfect CSIR. Telecommun. Syst., vol. 61(4), 875–886.

Trivedi, Y.N., Chaturvedi, A.K. 2011. Performance analysis of multiple input single output systems using transmit beamforming and antenna selection with delayed channel state information at the transmitter. IET Commun., vol. 5(6), 827–834.

Chiani, M., Win, M.Z., Zanella, A. 2003. On the Capacity of Spatially Correlated MIMO Rayleigh-Fading Channels. IEEE Trans. Inf. Theory, vol. 49(10), 2363–2371.

Jacobs, L., Moeneclaey, M. 2013. Accurate Closed-Form Approximation of BER for OSTBCs with Estimated CSI on Spatially Correlated Rayleigh Fading MIMO Channels. IEEE Commun. Lett., vol. 17 (3), 533–536.

Hajjaj, M., Chainbi, W., Bouallegue, R. 2016. Low-Rank Channel Estimation for MIMO MB-OFDM UWB System Over Spatially Correlated Channel. IEEE Commun. Lett., vol. 5(1), 48–51.

Kulkarni, M., Choudhary, L., Kumbhani, B., Kshetrimayum, R.S. 2014. Performance Analysis Comparison of Transmit Antenna Selection with Maximum Ratio Combining and Orthogonal Space Time Block Codes in Equicorrelated Rayleigh Fading Multiple Input Multiple Output Channels. IET Commun., vol. 8(10), 1850–1858.

Xu, Z., Sfar, S., Blum, R.S. 2009. Analysis of MIMO Systems with Receive Antenna Selection in Spatially Correlated Rayleigh Fading Channels. IEEE Trans. Veh. Tech., vol. 58(1), 251–262.

Jakes, W.C., Cox, D.C., (eds.). 1994. Microwave Mobile Communications. New York: Wiley-IEEE Press.

Jarinová, D. 2013. On autoregressive model order for long-range prediction of fast fading wireless channel. Telecommun. Syst., 52(3), 1533–1539.

Gomadam, K.S., Jafar. S.A., 2007. Modulation and Detection for Simple Receivers in Rapidly Time Varying Channels. IEEE Trans. Commun., vol. 55, pp. 529–539.

TAP (Time, Area, and Power) trade off at lower geometries in data center ASIC: A deep dive into executional view and results

Ruchita Shah*
M.Tech EC (VLSI & EMBEDDED Systems), U.V.Patel College of Engineering, Ganpat University, Mehsana, India

Nilesh Ranpura*
Delivery Manager ASIC, Einfochips an Arrow Company, Ahmedabad, India

Bhavesh Soni*
Assistant Professor (EC Department), U.V.Patel College of Engineering, Ganpat University, Mehsana, India

ABSTRACT: Industry has come up with new strategies and techniques to cope up with the lower geometry challenges. These challenges are design scale, complexity, manufacturing requirements for lower node technology challenges like double patterning, FinFET statistical analysis and optimization capabilities to achieve Time, Area, and Power (TAP) on vigorous project timelines. The most important one is taping out ASIC on schedule is a milestone. Partitioning, geometry usage, routing/resource distribution, block execution has its own challenges and there is huge dependability on each block's quality physical verification closure. As the transistor density increases, the designer is responsible for choosing the suitable or best trade off between Time, Area, and Power for the fast growing and improving technology at lower geometry. In this paper, we will see several Timing optimization techniques which can be used at different PNR stages, how the power, area like parameters are affected and how the best trade off should be selected which suits our design, at 28nm technology node of a complex networking ASIC chip.

1 INTRODUCTION

There are different types of optimization techniques proposed to cope up with Timing, Area and Power tradeoff. (W.Luk, 2016) The new synthesis tools will not only perform the place-and-route stages of an ASIC, but also can create a design-specific standard cell layout and can characterize them during placement for optimal design performance and power. (K. Golshan, 2006) Each ASIC design physical implementation has well thought-out stages from synthesis, DFT, PnR (Place and Route) to sign-off. At every stage to perform timing analysis and to meet all the violations leads to an ASIC design with higher performance along with optimum power and area.

The main objective, early analysis of design including static timing analysis, enables to identify and resolve technical challenges even before the design is completed because same approach may be insufficient for newer technologies. As a result, variation imposed performance alteration can overcome, ultimately high yield.

To understand the importance of meeting timing goals, one has to perform each step of ASIC flow, starting from Synthesis to sign-off. To improve Time, Area and Power various optimization techniques need to be analyzed. Designer has to select the best strategy to reduce

* Corresponding author: ruchita.shah@einfochips.com; nilesh.ranpura@einfochips.com; bhavesh.soni@ganpatuniversity.ac.in

time slack, to decrease power consumption and to reduce area as well. It is an iterative process to find out the best suitable technique which stands out with the Timing, Area, and Power tradeoff that is described well in research paper titled Timing, Power and Area correlation at lower geometries of complex networking ASIC. (R. Shah, C. Panchal, B. Soni, 2019)

2 MOTIVATION

Generally, the issue in VLSI design is Time, Area and Power trade-off, that will arise while fixing for either area improvement or timing met or power optimization with respect to all performance. Because techniques to improve time slack, cell interchange or sizing will affect the area and furthermore if low power design and area is the concern then timing will affect. (N. Kulkarni, A. Dengi, S. Vrudhula, 2017) There is a very complex correlation between Timing, Area, and Power which needs to be balanced in order to get the best ASIC chip.

As shown in (Figure 1), the Network is growing very fast, and nowadays world is going fast in speed of any device, at the same time the devices are getting compact which reduced the area and the power consumption should be low, so in this paper various timing optimization techniques, its results and affecting parameters at 28nm node of a complex networking Asic are described.

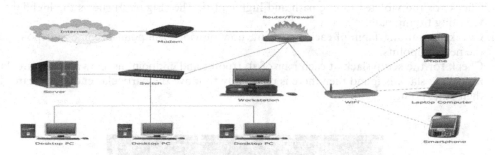

Figure 1. Network device.

3 SELECTION OF TECHNIQUES

It is very difficult to choose optimization techniques for power, area and time as techniques which are best at one technology node, may not cause good optimization at lower than that technology node which is well described in research paper titled Timing, Power and Area correlation at lower geometries of complex networking ASIC. (R. Shah, C. Panchal, B. Soni, 2019), with comparison of various parameters at different technology node. For example, the power optimization technique like power gating gives best result at 90nm node but it may not give good optimization at 28nm node, maybe DVFS (A. B. Chong, 2014) can be used for better result. Also, the techniques used for timing optimization may increase or decrease the power and area of the design. For power optimization (1) Replace normal flop into multibit flops which also cause area reduction due to shared transistor. (J. Shah, P. Gandhi, D. Parikh, 2017) (2) Clock skew scheduling technique can be used to reduce the glitch power. (A. Vijayakumar, S. Kundu, 2014) One more technique proves effective to reduce power, area and leakage of STD cell by replacing it with threshold logic flipflops at prior stage. (N. Kulkarni, J. Yang, et al., 2016).

3.1 *Various technique for optimization of timing at different stages of PNR*

There are different techniques like cell swapping, cell up sizing or down sizing, buffer insertion (O. Coudert, 2002) for solving the time violation such as setup and hold violation but these techniques are generally used in sign off stage. The hierarchical design planning method can be used before PNR stages to improve time which partition the hierarchical design into flat design. (S. Xin Xu, L. Min Dong, X. Hong Peng, 2011) In this paper some new techniques which are used during placement and routing stages are discussed in following sub sections.

3.2 *Fanin/Fanout*

Using Fanin and Fanout technique all the connected cells of particular macro can be checked, which will help in macro placement during floor planning stage. Sometimes there may happen in the design that one macro is not connected at very first level to another macro, so level by level connection also can be checked. This technique also gives the idea of spacing between macros, e.g. suppose there are10000 cells connected to macro, more spacing can be kept between two macros so that later on it will not convert into congestion violation. The (Figure 2) shows the Fanin and Fanout of the macro highlighted with orange color. As shown in (Figure 2) not only standard cells or macros but I/O ports are also highlighted so macro can be placed near to connected port.

3.3 *Data path optimization technique - bound creation*

Bound of cell can be made and force them to place near to its start point which can solve the time violation. Command: create-bounds -name b1 -coordinate -color green [cell name]

  i. Grep all unique start points and Endpoints with their slack from timing report.
 ii. Check for the violated timing path and highlight it. The (Figure 3) shows the highlighted violating timing path.
iii. Check fanout and fanin of each Endpoint, now previous Endpoint become Startpoint for the new endpoints.
 iv. Check for the setup slack at each pair of startpoint and endpoint as shown in (Figure 4a), if positive slack is found then there is possibility to move the particular endpoint nearer to their start point.

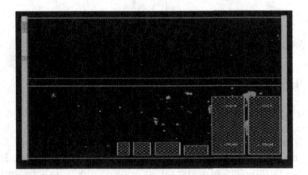

Figure 2.   Fanin and Fanout of macro.

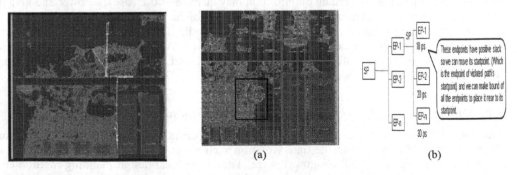

(a)   (b)

Figure 3.   Highlighted timing path.   Figure 4.   (a) Bound created, (b) Checks for setup slack.

232

v. The combinational cell between the violated paths is not recommended to include in the bound. bound all violated paths end point nearer to its start point as shown in (Figure 4b).

3.4 *Clock path optimization technique*

In clock path optimization technique, optimization in clock path can be done by using early clocking or late clocking. Where latency can be set at violated path (at starpoint or at endpoint) for particular clock source. *Command: set-clock-latency *value -source -early/late [get-clocks clk-name]*

   i. Check for the Data arrival time, clock network delay and data required time.

   ii. Check for the local skew and latency of violated path's source clock.

   iii. On the bases of that one can decide what amount of latency is to be set to source clock.

   iv. The latency should be set before the CTS run, by using either early or late clocking technique. It is not advisable to use both techniques like early and late clocking at the same time.

   v. If there is single violated start point and multiple endpoints than it is better to change latency at start point rather than changing at multiple endpoints.

   vi. In this paper early clocking is used at source clock which is source clock of violated startpoint and endpoint. While using early clocking at start point to remove setup violation, also needs to check for hold violation at endpoint. Because sometimes it may be possible that while removing setup violation it turns in to hold violation in your design.

   vii. For example, as shown in (Figure 5) below if there is setup violation of 1 ns at Startpoint, combinational n/w and Endpoint, either clock can be made to reach at Startpoint 2ns early to make it 0 or late source clock 2ns late to make 0 at Endpoint.

Figure 5. Early/late clocking example.

3.5 *Useful skew technique*

Generally, data path optimization techniques are used to solve most of the timing violations. With useful skew technique, the timing violations like setup and hold can be fixed by accommodating clock arrival times at registers or latches. (N. Kulkarni, J. Yang, et al., 2016) Suppose there is negative slack in one path and positive slack in other path, so by increasing clock arrival time at sink of the negative path slack and decreasing the clock arrival time at the sink of the positive path slack can fix the timing violation. The useful skew technique can be used in two different manner, 1) PreCTS 2) PostCTS or Post Route. In PreCTS the larger latency to adjust the target can be achieved which means there is more useful skew. Before CTS the clock is ideal so time derate does not come into picture on the clock path. After CTS or Route stage useful skew can be apply incrementally, skew can be more effective as after routing the timing is accurate. Useful skew techniques have some limitations like it cannot improve self looping path (reg to itself) and feedthrough path (in to out port). (Solvnet, 2019).

4 RESULTS

In this section, how the parameter like Time, Area and Power get affected when different techniques are used, is discussed. Design Statistics- In this paper the complex networking design is used for tool realization. The CNC design is partitioned at block level. Author has performed the PNR flow on one of the blocks of the design at 28nm technology node. The tool used for design is IC compiler from Synopsys. The block contains 6 macros, around 2.5

Mn of standard cells and 928 input/output ports. There are M1-M9, AP total ten metal layers. The operating frequency and the voltage of the block is 75 MHz and 0.81V respectively.

4.1   Floorplan stage

Table 1. Shows the results of four iteration for the floorplan stage. There is no technique used at floorplan or any other stage in Iteration 1, it is simple PNR iteration. In 2nd iteration Fanin and Fanout technique is used for floorplanning. 3rd and 4th iteration have the same floorplanning as in 2nd iteration. Using Fanin and Fanout technique the worst negative slack (WNS), total negative slack (TNS) and violating path for setup is reduced. The result for 3rd and 4th iteration is same as 2nd iteration due to the techniques are used in later stage like placement, CTS.

Table 1.   Timing report of floorplan stage.

Parameters	Iter.1	Iter.2	Iter.3	Iter.4 (with bounds)
Setup WNS (ns)	-0.55	-0.43	-0.43	-0.43
Setup TNS (ns)	-2552.39	-1970	-1970	-1970
Setup NVP	8228	5992	5992	5992
Std. cell count	234830	234830	234830	234830
buff/Inv count	14420	14420	14420	14420

4.2   Placement stage

Table 2. Shows the timing report after placement stage. There is no technique used in 1st and 2nd iteration for placement stage. In 3rd iteration bound creation technique mentioned in section 4.2 is used. As it can be seen, the setup WNS is reduced from -150 ps to -340 ps, No. of violating paths are also reduced so that TNS. DRVs are also reduced to 34 to 5. Congestion noted is 0 overflow for both horizontally as well as vertically. Iteration 4 is same as the 3rd iteration as same technique is used for placement stage which will differ in CTS stage. Hold violation is not checked at Placement stage as there is ideal clock, it will come into picture after CTS stage. Congestion also noted 0 for this block, so the main focus is to solve the setup violation at placement stage.

Table 2.   Timing report of placement stage.

Parameters	Iter.1	Iter.2	Iter.3(With Bound)	Iter.4(With Bound)
Setup WNS	-0.39	-0.34	-0.15	-0.15
Setup TNS	-1997.52	-1434.65	-588.51	-588.51
Setup NVP	8497	7259	6102	6102
DRV	16	34	5	5
Std.cell	245071	246181	244892	244892
Buff/Inv	23762	24933	23533	23533
Congestion	0 GRC with 0 Overflow	0 GRC with 0 Overflow	0 GRC with 0 Overflow	0 GRC with 0 Overflow

4.3   CTS stage

In iteration 4 clock path optimization technique early clocking is used as mention in section 4.3, As the block needs to clean in terms of setup violation and successfully clean the Setup

Table 3. Timing report of CTS stage.

Parameters	Iter .1	Iter .2	Iter .3	Iter .4
Setup WNS	-0.56	-0.29	-0.10	0.02
Setup TNS	-1948.70	-166.32	-22.52	0.00
Setup NVP	6014	1950	716	0.00
Hold WNS	-0.13	-0.17	-0.06	-0.07
Hold TNS	-471.52	-2639	-131.04	-130.12
Hold NVP	34575	84875	7979	7764
DRV	2584	17442	937	717
Core Utilization	63.20%	63.43%	60.83%	69.61%
Std.cell	250577	257526	254634	297013
Buff/Inv	29151	36148	33188	75690

Table 4. Timing report of routing stage.

Parameters	Iter .1	Iter .2	Iter .3(With Bound)	Iter .4(Early Clocking)
Setup WNS (ns)	-0.70	-0.27	-0.12	0.02
Setup TNS (ns)	-2809	-142.53	-37.52	0.00
Setup NVP	8674	1339	617	0.00
Hold WNS	-0.12	-0.06	-0.06	-0.09
Hold TNS	-241.42	-414.76	-467.6	-444.5
Hold NVP	17893	23701	26340	24731
DRV	7	87	74	48
Std.cell	250577	257526	254634	254400
Buff/Inv	29151	36148	33188	33077

Violation which is +20 ps as shown in Table 3. The result also shows that the TNS and Violating paths for Setup are now zero, but hold is increased by 30 ps. Number of DRVs are reduced from 74 to 48. Here one more thing can be noticed that is by using this technique number of std. cell and buff/inv count also reduced which indicate the overall area is also improved.

4.4 *Routing stage*

Table 4 shows there is no change in setup violation, but the hold is reduced by 20 ps in iteration 4, there is change in iteration 4 in setup violation but no change in hold. So, there is always challenging job to improve timing violation because setup will affect while solving hold and vice versa. In routing there is no other techniques used in this stage.

5 COMPARISON OF POWER, AREA AND TIMING

As shown in Table 5 the three parameters Power, Area and Timing, are compared for each and every stage. Different techniques have been used for different experiments. In Ite-1 Power and Area are very less as compared to Ite-2, 3, and 4 but at the cost of Timing (highest setup violation -560 ps). In Ite-3 where Data path optimization technique is used the Timing improves drastically from -560 ps to -100 ps and also Power reduced by some amount, but Area is increased. In Ite-4 with the Clock path optimization technique Timing which was -100 ps in Ite-3 is turned in to positive 20 ps which means the block is clear in terms of setup violation but at the same time Power and Area are increased. (A. B. Chong, 2014) The graphical representation of time, area and power is shown in (Figure 6). So there is always tradeoff

Table 5. Comparison of power, area and timing.

| Experiment | Power (mW) | Area (um^2) | Timing (ns) | |
			SETUP WNS	HOLD WNS
Iter.1	171.62	706953	-0.56	-0.13
Iter.2	179.62	707837	-0.29	-0.17
Iter.3 [Data Path Opt.]	167.88	744367	-0.10	-0.06
Iter.4 [Clk Path Opt.]	173.60	743526	0.02	-0.07

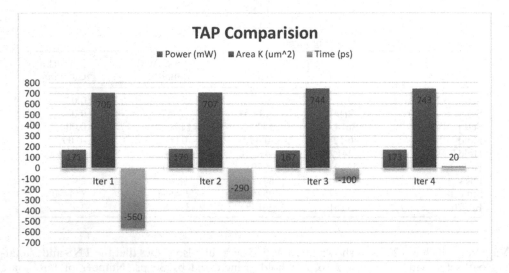

Figure 6. Graphical representation of TAP comparison.

between these three parameters. Therefore, designer needs to check which parameter is more important for the design and according to that best results for the design has been chosen.

6 CONCLUSION

The back end design flow including physical design, the ideas driving every progression of the procedure have been concentrated on. At that point the limitations, reasons and the prerequisites bringing about the current implementation of the back end design have been considered. Further, different analysis has been performed to enhance the design in different angles like power, area and timing. The trade-offs required and the advantages of each of the trials have been differentiated and investigated.

So, in this paper the block of 28nm on which different techniques have been performed at the back end flow is discussed. So, that one can realize what types of hurdles are to be faced and how they can be solved? Can same techniques work at lower geometries like 16nm or 7nm that is applied in 28nm of the complex networking chip? If yes, then what other parameters affect and if no, then what techniques can be used for those technology nodes to cope up with Time, Area and Power trade off.

For example, networking ASICs are power hungry and power is critical. So, area can be compromised in networking application chip to certain limit, but not the power. Vice versa, mobile/consumer Asic is area critical where area is more important as well as power.

Needless to mention that time is most critical at GHZ applications. Where application time and Testing time is immense critical. So, it is needed to balance AREA, TIME and POWER such way that it meets functional, electrical and physical specification. All mentioned techniques results may vary in different technological node - FINFET Vs CMOS transistor and application level.

ACKNOWLEDGMENT

This research was fully supported by einfochips. I thank my colleague who provided perception and proficiency that greatly assisted the research work. I would like to thank three "anonymous" reviewers for their perception.

REFERENCES

A. B. Chong. (2014) ASIC Leakage, Performance and Area Trade Off at IEEE International Conference on Circuits and Systems (ICCAS), p. 6.

A. Vijayakumar, S. Kundu. (2014) Glitch Power Reduction via Clock Skew Scheduling at IEEE Symp. On VLSI, pages 504–509.

J. Shah, P. Gandhi, D. Parikh. (2017) Power Optimization using Multi BIT flops and MIMCAPs in 16nm technology and below at Design and Reuse, p. 5.

K. Golshan. (2006) PHYSICAL DESIGN ESSENTIALS (Book) Newport Beach, CA: Springer.

N. Kulkarni, A. Dengi, S. Vrudhula. (2017) A Clock Skewing Strategy to Reduce Power and Area Of ASIC Circuits at IEEE, p. 6.

N. Kulkarni, J. Yang, J.S. Seo, and S. Vrudhula. (2016) Reducing Power, Leakage and Area of Standard Cell ASICs Using Threshold Logic Flipflops at IEEE Transactions on VLSI, p.6.

O. Coudert. (2002) Timing and Design Closure in Physical Design Flows at IEEE Symposium On quality Electronics design, p. 5.

R. Ewetz, C. Kok Koh. (2016) MCMM clock tree optimization based on slack redistribution using A reduced slack graph at IEEE ASP–DAC, p 6.

R. Shah, C. Panchal, B. Soni. (2019) Timing, Power and Area correlation at lower geometries of Complex networking ASIC at STEM, p.5.

Solvnet, useful skew technique [online] Available on: www.synopsys.com (accessed March 12, 2019).

S. Xin Xu, L. Min Dong, X. Hong Peng, (2010) Timing optimization for deep sub-micron hierar- chical design at 10th IEEE International Conference on Solid-State and Integrated Circuit Technology, p. 6.

W.Luk, (2016) Chip Basics: Time, Area, Power, custom Computing for Advanced Digital Systems, 3rd addition, p. 23.

Technologies for Sustainable Development – Mahajan, Patel & Sharma (eds)
© 2020 Taylor & Francis Group, London, ISBN 978-0-367-33737-7

Design, development and testing of SmaTra: A smart traffic light control system

Nitesh Kalwani, Punit Malpani & Sachin Gajjar*

Department of Electronics & Communication Engineering, Institute of Technology, Nirma University, Ahmedabad, Gujarat, India

ABSTRACT: Increasing traffic due to the increase in population and number of vehicles in metro cities calls for a smart traffic management system. Vehicle queues at traffic lights leads to an increase in petrol consumption and pollution. This paper proposes SmaTra, a smart traffic management system using hardware as well as a software approach. The hardware approach uses an array of Infrared Radiation (IR) sensors placed at the roadside which could be tripped by moving vehicles to indicate the density of traffic on each side of the road. This input from the IR sensors is used to change the vehicle waiting timing at the traffic junction. An alternate approach for smart traffic light using neural networks computation technique running on a neural network hardware engine is also developed. The hardware does image classification to identify the density of traffic and modifies the vehicle waiting timing at the junction. Out of the two approaches, the software approach gives more accurate and precise results up to an accuracy of 97% of identifying the traffic density.

Keywords: Smart traffic light control, hardware design, neural network computing

1   INTRODUCTION

According to Ministry of Roads and Highways, Government of India there is an increase of 10.1 % CAGR (Compounded Annual Growth Rate) in the number of vehicles in 2016 (Highways, 2019). Under such circumstances, implementation of smart traffic light system would help in traffic management. Several Smart traffic control systems using hardware as well as software models have been proposed in literature. (Kriesel, 2019) computes traffic density based on image detection and classification. Google Maps Navigation System provides traffic density information through web mapping service. It uses Google Earth as source for satellite imaging and traveler's GPS location for identifying traffic density (Schutzberg, 2019). The approach cannot be used in real time smart traffic control systems because the latency in providing traffic information varies between 10 seconds to 5 minutes. SmaTra, the proposed smart traffic light control system is accurate and can control traffic in real time. SmaTra's hardware design approach is described in Section II. Section III presents the software approach. Section IV discusses the comparison of both the approaches. The paper is concluded in Section V.

2   SMATRA DESIGN: HARDWARE APPROACH

Design of SmaTra though hardware approach consists of use of an array of Infrared Radiation (IR) Module FC-51 to detect the vehicle density (Technologies, 2019). The operating

---

*Corresponding author: sachin.gajjar@nirmauni.ac.in

voltage of the module is 5 *Volts* DC and current rating is 20 mA. It consists of: (i) IR Emitter LED (700 nm – 1mm) which acts as IR transmitter, (ii) photodiode which acts as IR receiver, (iii) potentiometer to adjust the obstacle detection distance. The size of the module is 4.48 square centimeters. The IR rays from the module get reflected by vehicles travelling along the lane and are received by the receiver. For each successful reflection the module gives an output equivalent to logic 1. As shown in the Figure 1, each lane is covered using an array of IR sensor modules. The inputs from these modules indicate the traffic density. The timings of the traffic light at the junction are adjusted according to the input from the sensor modules.

2.1 *Counter circuit*

The timings of the traffic light are controlled with the help of a counter circuit consisting of a down counter. The output of the IR sensor module is in form of voltage which is digitally represented as logic 0 or logic 1. When an IR module is tripped, its output value changes.

It is considered that when an IR module is tripped the equivalent output value in range [0-1023], of its digital output voltage [0-5 V], is more than 511 i.e. more than 2.5 V in digital, and in other cases its value remains less than 500. The tripped IR sensor can be mapped to logic 1 and using this, a counter circuit is developed which controls the timings of traffic light based on the traffic density. Figure 2 shows the hardware setup of SmaTra. For experimental testing a prototype consisting of an array of four IR sensor modules that covers the entire lane is used. If the first IR sensor module is tripped and no other IR sensor module is tripped, it indicates low traffic density and wait timing of the counter circuit is kept as 15

Figure 1.   Pictorial representation of deployment of SmarTra system.

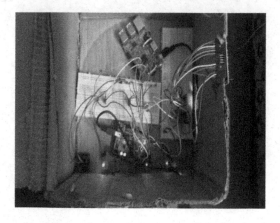

Figure 2.   Hardware setup of SmarTra.

secs. If only first and second IR sensor module trips it indicates moderate traffic and the wait timing of counter is set for 30 *secs.* When three IR sensor modules trip the waiting, time is kept as 45 *secs.* If all four IR sensor modules trip it indicates high traffic and the counter timing is set to 60 *secs.* The hardware of SmaTra works efficiently as long as there is no interruption in the pattern in which the IR sensor modules should be tripped. The latency of the system is 0.5 *secs* and majorly depends on response time of IR sensor module. Vehicle detection by IR sensor module in summer will be difficult as difference between surrounding temperature and vehicles would be less.

3 SMATRA DESIGN: SOFTWARE APPROACH

Figure 3 shows the software approach setup of SmaTra which uses Raspberry Pi (RPi) (Organization, 2019), Camera and an Intel Movidius Neural Compute Stick (NCS) (Intel, 2019. This setup will be placed at every side of traffic junction, to capture images of vehicles at the traffic junction. The software on RPi runs an image detection and image classification algorithm to detect the images of vehicles. The traffic density on the road is calculated based on the number of vehicles within the area covered in the image. Based on the density of vehicles, the waiting time of the vehicle at the traffic junction is varied. NCS consists of Intel Movidius Myriad 2 Vision Processing Unit (VPU) capable of running computationally inten-sive Convolutional Neural Network (CNN) algorithms. NCS supports Tensor-flow (Flow, 2019) and Caffe.

(Yangqing Jia, 2019) open source machine learning platforms for object detection, object classification, natural language processing and facial recognition. RPi-camera takes images of the vehicles at traffic signal and gives these images as input to NCS which detects number of vehicles. RPi in turn controls the vehicle waiting time based on the number of vehicles.

3.1 *Software design*

For SmaTra implementation through software approach, image classification is done using a pre-generated graph file which is generated by NCSDK. Next, it is required to load class labels and get the image input from RPi camera. The images are captured by RPi camera at intervals of every 5 *secs* and stored in RPi. The image capture interval should be greater than the total time required by NCS for image classification which in turn depends on the size of dataset. Class labels are used from text file synset_words (Azure, 2019). The dimensions for the images are 400 X 267 as per the requirements of CNN algorithm. SmaTra is tested using SqueezeNet, GoogLeNet and AlexNet (Simone Bianco, 2018) based network trained models that use mean subtraction technique. It is observed that best results are obtained on GoogLe-Net. As shown in the Figure 4, results show five predictions indicating probabilities of five

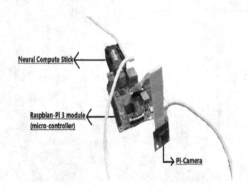

Figure 3.   SmaTra setup for software approach.

Figure 4. Image classification by NCS.

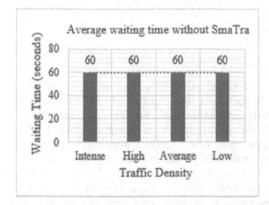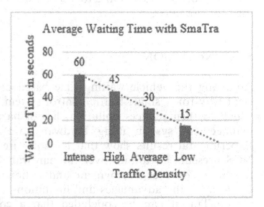

Figure 5. Average vehicle waiting time with and without SmaTra control system.

different objects. Highest probability is chosen as final result of classification. Figure 5 shows the average vehicle waiting time at the traffic junction with and without SmaTra control system. As seen in the figure the average vehicle waiting time at the traffic junction without SmaTra control system is fix irrespective of traffic density. With SmaTra control system, the average vehicle waiting time is less when traffic is low and high when traffic is high. This in turn saves fuel consumption at traffic junctions. Autoportal shows that the fuel consumption by a car standing at neutral gear is less than that by car that uses clutch pack and gears (Autoportal, 2019). It further estimates that fuel consumption for a car with 1000 cc of engine is 0.6 *liters* per hour and it is raised to a value of 1.2 *liters* per hour with AC turned ON. So, based on these values' consumption by a car per second with AC turned ON will be 0.00033 *liters*. As seen in the Figure 5, SmaTra control system saves a maximum of 45 *secs* of waiting time. This in turn leads to fuel saving equal of 45x0.00033 that is 0.015 *liters*.

4 COMPARISION OF HARDWARE AND SOFTWARE APPROACH

The SmaTra design through software approach is accurate, easy to deploy, requires less hardware and maintenance. The cost of implementation of SmaTra design through hardware approach is proportional to the length of road lane to be covered. In a typical Indian metro city, traffic signals are located at every *km*. This requires coverage of 150 *mts* of road side which can be done with five sensor modules to cover 30 *mts*. Thus, for a road stretch of 1 *km* the cost of implementation would be equal to cost of five IR sensor modules. The hardware approach can be scaled for any length of road but at cost of increase in number of IR module sensors. Further, placing IR sensors at close proximity will enhance the accuracy of traffic detection. The software approach would require RPi, camera and neural compute stick at each traffic junction. For road with a large length, cost of hardware approach would be directly proportional to the length of road whereas cost of software approach would be same irrespective of the length of the road. Further, the hardware approach requires to change the civil structure of the road to fix the IR sensor modules. The hardware approach works equally well during day as well as night as the IR sensor module shows less variance with surrounding light intensity. However, in software approach image classification accuracy is affected with variation of intensity of light during different times of day. For roads with sharp turn the camera will not be able to capture images accurately and in that case hardware approach proves out to be beneficial. The response time of software approach is higher as compared to hardware approach. The response time in hardware approach varies between 5 to 60 *secs* and increases as the number of sensors are increased. Whereas the response time in case of software approach varies from 2 to 5 *secs*. Further, the camera feed of the software design can be used for surveillance.

5 CONCLUSION

Adjusting the vehicle waiting time based on the vehicle density is one of the most efficient way for a smart traffic control system. It decreases the average vehicle waiting time which in turn reduces pollution, health hazards and saves fuel. SmaTra, a smart traffic management system using hardware and software approach is designed and tested. Experimental results show that the software approach using CNN algorithms running on NCS presents advantage over the hardware approach in terms of accuracy, cost, latency, flexibility, ease of deployment and maintenance requirement especially for long roads. Looking to the advantages and limitations of hardware and software approach for design of SmaTra, it can be concluded that a combination of both can further increase the benefits of SmaTra.

REFERENCES

Azure, M. (2019, July 14). *Microsoft Azure Notebook*. Retrieved from Neural Networks and Convolution Neural Networks Essential Training: https://notebooks.azure.com/jonfernandes/projects/NeuralNet works/html/data/synset_words.txt

Flow, T. (2019, July 14). *Tensor Flow Open Source Machine Learning Platform*. Retrieved from Tensor Flow: https://www.tensorflow.org/

Highways, M. O. (2019, July 14). *Ministry of Road Transport and Highways, Government of India*. Retrieved from Ministry of Road Transport and Highways, Government of India: http://morth.nic.in/index1.asp?lang=1&linkid=10&lid=144

Intel. (2019, July 14). *Intel Developer Zone*. Retrieved from Intel Software Developer Zone: https://soft ware.intel.com/en-us/openvino-toolkit

Intel. (2019, July 7). *Intel Software Developer Zone*. Retrieved from Intel Neural Computer Stick: https://software.intel.com/en-us/movidius-ncs

Kriesel, D. (2019, 7 11). *D. Kriesel, Data Science, Machine Learning, BBQ, Photos, and Ants in a Terrarium*. Retrieved from http://www. dkriesel. com/: http://www. dkriesel. com/en/science/neural_networks

Organization, R. P. (2019, July 14). *Raspberry Pi Foundation*. Retrieved from Raspberry Pi Organization: https://www.raspberrypi.org/

Prakash, A. (2019, July 14). *Deep Learning software installation guide on fresh Ubuntu*. Retrieved from Machine Learning on Ubuntu: https://iamaaditya. github.io/2016/01/Deep-Learning-software-installa tion-guide-on-fresh-Ubuntu/

Schutzberg, A. (2019, 7 11). *The Technology Behind Google Maps*. Retrieved from Directions Magazine: https://www.directionsmag.com/article/3323

Simone Bianco, R. C. (2018). Benchmark Analysis of Representative Deep Neural Network Architectures. *IEEE Access*, 1–8.

Technologies, R. (2019, July 13). *Rhydolabz*. Retrieved from FC-51: IR Infrared Obstacle Detection Sensor Module: https://www.rhydolabz.com/

Ubuntu. (2019, July 14). *Ubuntu Openstack*. Retrieved from Ubuntu Operating System: https://ubuntu.com/openstack

Yangqing Jia, E. S. (2019, July 14). *Caffe deep learning framework*. Retrieved from Caffe framework: https://caffe.berkeleyvision.org/

Zone, I. S. (2019, July 14). *Intel Movidius Neural Computer Stick SDK*. Retrieved from Intel Movidius Neural Computer Stick: https://movidius.github.io/ncsdk/

Autoportal. (2019, July 15). How much petrol consumed per hour when only AC is running but car is not moving?. Retrieved from Autoportal: https://autoportal.com/newcars/hyundai/eon/qna/354136.html

Technologies for Sustainable Development – Mahajan, Patel & Sharma (eds)
© 2020 Taylor & Francis Group, London, ISBN 978-0-367-33737-7

Implementation and defect analysis of QCA based reversible combinational circuit

Vaishali Dhare & Deeksha Agarwal
Institute of Technology, Nirma University, Ahmedabad, India

ABSTRACT: In the last four decades, Complementary Metal Oxide Semiconductor (CMOS) technology has achieved circuits and system with features like low power, high device density and high speed. But, heat dissipation issue affects the speed and performance in the computing. Reversible logic based circuits can overcome the problem of heat dissipation and improve the efficiency circuit and system. One of the viable implementations of reversible logic is Quantum-dot Cellular Automata (QCA) nanotechnology. Due to nano scale, occurrence of defect are present in QCA. Therefore, defect must be analyzed in QCA based circuits.

In this paper, combinational circuit, adder and substractor based on reversible logic is proposed. The proposed circuit is implemented using QCA devices Majority Voter (MV) and inverter. Further, the fault caused by multiple missing cells defect are analyzed.

1 INTRODUCTION

Since last five decades, the current silicon technology has made the Moore's law existent and offered high device density, low power and high speed implementations. The increase in device density by the scaling of transistor cause the heating problems [1]. Today's processors are designed using conventional (irreversible) logic circuits. In conventional logic circuits, the intermediate bits used to compute the final results are discarded unknowingly. This leads to energy dissipation which affects the speed and performance in the computing. The general processors dissipate more than 500 times the amount of heat, estimated by Landaeur [2] due to loss of 1 bit information. Thus, there is a great need of reversible logic circuits. In reversible computing bits are preserved and logical and physical reversibility guaranteed low energy dissipation. Hence, the processor with reversible logic will have less heating problem.

Further, reversible logic can be implemented using quantum mechanics based nanotechnology, Quantum-dot Cellular Automata (QCA) [3-6]. Being at nanoscale, occurrence of defects and faults cannot be ignored in this technology [7]. So, defect and fault analysis is important in QCA based reversible circuit.

The main contribution of paper are as follows:

1. Reversible adder substractor is implemented using QCA technology.
2. Further, multiple missing cell defect [8] analysis of implemented reversible circuit is carried out at device level using QCADesigner [9] tool.

The content of the paper is as follows: Section 2 presents preliminaries. Implementation of QCA based reversible adder/substractor is described in Section 3. Defect analysis is presented in Section 4. Paper concludes in Section 5.

2 PRELIMINARIES

In [2], Landauer have proven that for irreversible logic computations, each bit of information lost, generates heat energy given in equation 1.

$$E = KTln2 \text{ Joules} \tag{1}$$

Where k is Boltzmann's constant and T the absolute temperature at which computation is performed. Reversible circuits do not lose information and reversible computation in a system can be performed only when the system comprises of reversible gates. These circuits can generate unique output vector from each input vector, and vice versa, that is, there is a one-to-one mapping between input and output vectors. A k × k reversible gate is a k-input and k-output circuit as shown in Figure 1 that produces a unique output pattern for each possible input combination.

Reversible computing gates cannot be implemented with the traditional silicon technology like Complementary Metal Oxide Semiconductor (CMOS) but it can be implemented using QCA. QCA is one of the emerging nanotechnologies that exhibit a small feature size, high clock frequency, and ultra low-power consumption. QCA provides an alternative way of computation, in which the logic states ("0" and "1") are defined by the positions of the electrons. QCA is an array of cells in which each cell contains the binary information as shown in Figure 2.

QCA cells interact using Coulombic interaction with neighboring cells to generate polarization which in turn results in information transfer. A cell consists of four quantum wells on each corner of the substrate. Each cell has two electrons which are placed on either of the two diagonals of the cell substrate [3-6].

The main primitive of QCA are Majority Voter (MV) and inverter shown in Figure 3. (a) and (b) respectively [10]. A MV voter consists of five cells where the cells are positioned in a manner in which three cells act as inputs, one output cell and one central cell. The output of the MV is the majority of the three input cells. The Boolean expression that denotes the output of an MV is F = Maj(A,B,C) = A.B + B.C + C.A.

Figure 1.   K × k reversible gate.

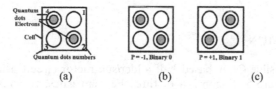

(a)   (b)   (c)

Figure 2.   QCA cell (a) schematic, (b) with binary 0 (c) with binary 1.

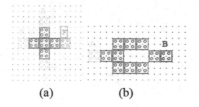

(a)   (b)

Figure 3.   QCA devices (a) majority voter (b) inverter.

245

The inverter gate is also used for designing various circuits in QCA. The quantum cells are aligned in such a manner so that the value driven from the input end inverts itself at the output end.

Defects are more likely occur in QCA devices. The detail classification is given in [11]. In this paper multiple missing cells defect and its corresponding faults [8] are considered.

Reversible Logic Gates

Reversible logic gates have same number of0inputs and outputs, if there are m number of inputs then the number of outputs should also be m. It can be called as one to one mapping system. In this system, outputs can be determined from inputs and vice versa. In these logic gates, fan out is not allowed0because such circuits do not0support one-to-many concept. There should be minimum number of reversible logic gates while designing the circuits. There are different parameters used for designing circuits, which are as follows:

• Constant input: For designing an 'AND' and an 'OR' logic from an MV, out of the three inputs available in an MV, one of the inputs should also be fixed to '0' and '1' respectively.
• Garbage outputs: Number of outputs that is not used in operation. i.e, input + constant input = output + garbage output.
• Quantum cost: This refers to the cost of the circuit in terms of cost primitive gates (1*1, 2*2). It is calculated knowing the number of primitive reversible logic gates required to realize the circuit. i.e, 1*1 cost = 0, 2*2 cost = 1.
• Gate level: Number of levels in a circuit to realize a function.

Density of circuit: Number of operations involved to realize function. i.e, AND, OR, NOT etc.

In this paper, reversible adder substractor circuit is implemented using Feyman reversible gate. This gate contains 2 inputs A, B and2 outputs P, Q from which one output is garbage output. Where P=A, Q=A⊕B. The quantum cost is 1. Block diagram and layout of Feyman gate is shown in Figure 4 (A) and (b) respectively.

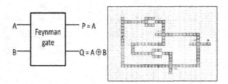

Figure 4.   Feyman reversible gate (a) block diagram (b) layout.

3   IMPLEMENTATION

The block level reversible QCA based half adder/subtractor circuit using reversible Feyman gate is shown in Figure 5. It consists of three Feyman gates, two majority voters and an

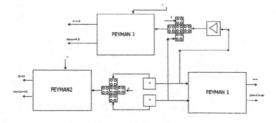

Figure 5.   Proposed reversible QCA based half adder/substractor.

inverter. Each Feyman gate consists of three majority voters and two inverters. So, in total the circuit consists of eleven majority voters, seven inverters and one crossover wire.

In half adder and subtractor sum and difference is implemented by exoring the two inputs. A Feyman gate with three MVs and two inverters has been used to implement the XOR logic. Here the two inputs are A and B, two of the three majority voters have their one of the input as logic 0 to get AND gate and the third majority voter has one of its input as logic 1 to get OR gate. Carry out is the ANDing of two inputs which is implemented with the help of a majority voter and a Feyman gate. The one input of the Fcyman gate is the AND of A and B with the help of a majority voter and the other input is logic 0. The configuration of other three majority voters is same as the first Feyman gate. Borrow is the AND operation of two inputs one of which is inverted. This is implemented with the help of a majority voter, an inverter and a Feyman gate. The one input of the Feyman gate is the AND of inverted A and B with the help of a majority voter and an inverter and the other input is logic 0. The configuration of other three majority voters is same as the first Feyman gate.

The block diagram is shown in Figure 5. implemented with the help QCA Designer tool to verify the functionality of it. The layout of the half adder/subtractor is shown in Figure 6(a). The eleven MVs are shown by red rectangles and the three Feyman gates are shown by black rectangles. The design shows two inputs to each of the three Feyman gates and two outputs from each of the three Feyman gates. The first Feyman has inputs A and B and outputs sum/difference and P (garbage output). The second Feyman has inputs A.B and logic 0 and outputs carry_out and Q (garbage output). The third Feyman has inputs A'.B and logic 0 and outputs borrow and R (garbage output).

The circuit is simulated using QCADesigner V 2.3.0 [9]. Bistable approximation simulation engine is used with cell size 18nmX18nm, dot diameter 5nm, cell to cell distance of 2.5nm and radius of effect 50nm. Other simulation parameters are kept as default. The simulation parameters are mentioned in Table 1.

The simulation waveform is shown in Figure 6(b). Output Sum/Diff is appearing after a delay of two clock cycles, output carry_out is appearing after a delay of two clock cycles and output borrow is appearing after a delay of four clock cycles as shown in the simulation in Figure 6(b). These outputs for the circuit have been highlighted rectangles.

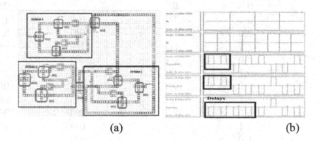

(a)   (b)

Figure 6.   QCA based reversible adder/substractor (a) device level layout (b) simulation result.

Table 1.   Simulation parameters.

Parameter	Simulation set up 1
Simulation Engine	Bistable approximation
Cell width	18nm
Cell height	18nm
Dot diameter	5nm
Cell to cell distance	2.5nm
Radius of effect	50nm

4 DEFECT ANALYSIS

Earlier single missing cell defect analysis for reversible SR latch is carried out in [12]. Here, multiple missing cell defect analysis is carried out using extensive simulation. QCADesigner tool is used with the same parameter as used earlier for implementation.

As given in [8], multiple missing cells defect causes single stuck at fault in QCA devices. The single stuck at fault set caused by multiple missing cells defect in MV is (S-a-A, S-a-A', S-a-B, S-a-C, S-a-C' and F(A' B C')). Where A, B and C are inputs of MV. Proposed reversible adder substractor is made of QCA MVs. So, effect of these faults in MV of reversible QCA based adder/substractor must be analyses. Also, need to see the effect on the output of adder/substractor. This section describe the detail analyses.

The effect on the output of the adder/substractor caused by faults in one of the MVs of first Feyman gate is listed in Table 2. Here the fault-free and faulty outputs have been listed for all the possible faults caused by the multiple missing cells defect.

First column of Table 1 is fault caused by multiple missing cell defect in MV1 of Feyman gate 1 shown in Figure 6. The faulty free and faulty values of output of adder/substractor namely Sum, Difference and Borrow are mentioned in the Table 1 for the corresponding values of primary inputs A and B. Same effects in terms of percentage in the graphical form is presented in Figure 8. The x-axis shows the type of defects due to an MV and the y-axis shows the percentage of error due to those defects in the outputs.

It is observed from Table 2 and Figure 7 (a) that only Sum output of Adder/substractor get affected by the all stuck at faults occurred in MV1 of Feyman 1 gate.

Similarly all the possibilities of faults in all the MVS of all Feyman gates are explored and analyzed. The percentage impact on the sum output due to the faults in MV2 of Feyman 1 is depicted in Figure 7(b). The graph shown in Figure 7 (a) & (b) depicts the same observation in a graphical manner. It also confirms that all the outputs except 'sum/diff' output have 0% defect and the defects for the output 'sum/diff' are non-zero.

Table 2.   Effect stuck at faults on output of adder/substractor.

Fault	Input	Sum/Diff	Carry	Borrow
S-a-B	0,0	0/0	0/0	0/0
	0,1	1/1	0/0	1/1
	1,0	1/1	0/0	0/0
	1,1	0/1	1/1	0/0
F(A' B C')	0,0	0/0	0/0	0/0
	0,1	1/1	0/0	1/1
	1,0	1/1	0/0	0/0
	1,1	0/1	1/1	0/0
S-a-A	0,0	0/1	0/0	0/0
	0,1	1/1	0/0	1/1
	1,0	1/1	0/0	0/0
	1,1	0/1	1/1	0/0
S-a-A'	0,0	0/0	0/0	0/0
	0,1	1/0	0/0	1/1
	1,0	1/1	0/0	0/0
	1,1	0/1	1/1	0/0
S-a-C	0,0	0/0	0/0	0/0
	0,1	1/0	0/0	1/1
	1,0	1/1	0/0	0/0
	1,1	0/0	1/1	0/0
S-a-C'	0,0	0/1	0/0	0/0
	0,1	1/1	0/0	1/1
	1,0	1/1	0/0	0/0
	1,1	0/1	1/1	0/0

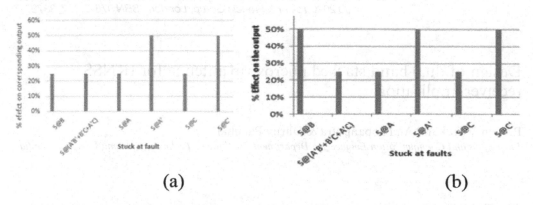

(a)   (b)

Figure 7. Fault effects caused by multiple missing cells defect on (a) output (Feyman 1, MV1) (b) output (Feyman 1, MV2).

5 CONCLUSION

In this paper, reversible logic based combinational circuit half adder and substractor is proposed. This circuit is implemented using QCA nanotechnology. The stuck at faults caused by multiple missing cells defect in QCA and hence reversible adder/substractor are analyzed.

The outcome of the proposed work may be treated as an importance for the fault detection of recent concept, reversible computing and nanotechnology. The same testing methods can be mapped to the other available nanotechnologies other than Quantum-dot Cellular Automata (QCA).

REFERENCES

[1] "International Technology Roadmap for Semiconductors (ITRS)", 2015 Edition, https://www.semicon ductors.org/clientuploads/Research_Technology/ITRS/2015/6_2015 ITRS 2.0 Beyond CMOS.pdf

[2] R. Landauer, "Irreversibility and Heat Generation in the Computational Process," IBM Journal of Research and Development, pp. 183–191, 1961.

[3] Lent, Craig S., P. Douglas Tougaw, Wolfgang Porod, and Gary H. Bernstein, "Quantum cellular automata," Nanotechnology, vol. 4, no. 1, 1993, pp. 49–57.

[4] C. S. Lent, P. D. Taugaw, and W. Porod, "Quantum cellular automata: The physics of computing with arrays of quantum dot molecules," in Proc. of the workshop on physics and computing, 1994, pp. 5–13.

[5] C. S. Lent and P. D. Taugaw, "Lines of interacting quantum-dot cells: A Binary wire," Journal of Applied Physics, 1993, 74, pp. 6227–6233.

[6] Mehta Usha, Vaishali Dhare, "Quantum-dot Cellular Automata (QCA): A Survey," arXiv preprint arXiv:1711.08153 (2017).

[7] Dysart, Timothy J., et al, "An analysis of missing cell defects in quantum-dot cellular automata," IEEE International Workshop on Design and Test of Defect-Tolerant Nanoscale Architectures (NANOARCH), 2005, 3, pp. 1–8.

[8] Vaishali Dhare, Usha Mehta, "Multiple Missing Cell Defect Modeling for QCA Devices" Journal of Electronic Testing, JETTA, vol. 34, no. 6, 2018, pp. 623–641.

[9] K. Walus, T.J. Dysart, G.A. Jullien, and R.A. Budiman, "Qcadesigner: a rapid design and simulation tool for quantum-dot cellular automata," in IEEE Trans. On Nanotechnology, vol. 3, no. 1, 2004, pp. 26–31.

[10] P. Tougaw and C. Lent, "Logical devices implemented using quantum cellular automata," Journal of Applied Physics, 1994, 75, (3), pp. 1818–1825.

[11] Vaishali Dhare, Usha Mehta, "Defect characterization and testing of QCA devices and circuits: A survey," 19th International symposium on VLSI Design and Test 2015, June 2015, pp. 1–2.

[12] Vaishali Dhare, Usha Mehta, "Single missing cell deposition defect analysis of sequential reversible circuit," NUiCONE 2017, Ahmedabad, India, November 2017, pp. 1–4.

Technologies for Sustainable Development – Mahajan, Patel & Sharma (eds)
© 2020 Taylor & Francis Group, London, ISBN 978-0-367-33737-7

Design of dual band stacked microstrip antenna for IRNSS receiver application

Bhavin V Kakani, Ami Jobanputra & Dhruv Panchal
Electronics and Communication Engineering Department, Institute of Technology, Nirma University, India

ABSTRACT: This article presents the design feature and simulation of a stacked microstrip patch antenna used for IRNSS receiver. Two antennas are sandwiched for the proposed idea; L5 band (1.17GHz) providing a gain of 4.5 dB and S band (2.48GHz) providing 7.66 dB gain for RHCP. The above simulated designs were fabricated, and results obtained were consistent with the simulations. For efficiency at these high frequencies RT/Duroid 5880 with a thickness of 3.175 mm is utilized.

1 INTRODUCTION

In the last two decades, there has been enormous growth in antenna designing for its promising application in the wide range of wireless communications incorporating Global Positioning System (GPS). GPS allows the user to locate its longitude, latitude and altitude position of any place across the globe by getting a signal from cluster of satellites with some finite accuracy. Global Navigation Satellite System (GNSS) is the standard generic term for radio navigation satellite system(Salvemini, 2015). GPS is an open source license to anyone carrying GPS receiver. The cons of GPS is that if the system of US fails then all over the world navigation accessibility will be affected severely. This leads to a requirement of independent navigation system as the fact that in critical situations access to foreign country controlled global navigation system is not guaranteed. (Tsui, 2000) (Wang, 2012)So, India developed in own domestic navigational system in 2010 named Indian regional Navigational Satellite System (IRNSS). The IRNSS receiver's antennae (Navigation using Indian Constellation NAVIC) to supplement the INS or Inertial Navigation System for en-route and terminal guidance. Satellite constellations are used to pinpoint the location of any receiver on the earth.(Thangadurai and Vasudha, 2017)

IRNSS communication subsystem consist of mainly 3 segments (Suryanarayan Rao K.N, 2010):

1. Ground segment
2. Space segment
3. User segment

The ground segment consist of central navigation center and control facility. This segment is responsible for maintenance, operation and controllability of IRNSS satellites. The space segment consist of constellation of 8 satellites. The user segment consist of IRNSS receiver capable of receiving standard positioning service (SPS) at L5 and S band. The IRNSS carrier frequencies for SPS is 2.48GHz.

2 DESIGN REQUIREMENTS

IRNSS navigation subsystem operates mainly in L5 and S frequency band. Dual band antenna is used to mitigate the effect of interference caused by earth's ionosphere on IRNSS

Table-1. IRNSS receiver antenna specification.

Design Parameter	S-band specifications	L5 band specifications
Centre frequency	2.493028 GHz	1.17645 GHz
Bandwidth requirement	24 MHz	8 MHz
Minimum & Maximum received power	-162.3 dBW	-162.3 dBW
	154.0 dBW	154.0 dBW
Gain	4 dBi	2 dBi
Polarization	RHCP	RHCP
Axial ratio	<3 dB	<3dB
Antenna Beam width	-60 to 60 degrees	-60 to 60degrees
Return Loss	<-10dB	<-10 dB

signal. The effect of ionosphere on this RF signal is frequency dependent phase shift commonly known as group delay caused by dispersive behavior of the environment at such height. Also, the effect of Faraday rotation may lead to use antenna that has circular property. This adversity can be effectively reduced by using two widely spaced frequency band(Anisha and Enoch, 2014).

Since it is dual band, where both are sufficiently separated to avoid any interference or intermodulation. As per the standard the minimum requirements for properly working of IRNSS antenna is listed in Table-1(Majithiya, Khatri and Hota, 2011).

The stacking can be then fed as per the three following cases; firstly, feed the upper patch only, so that the lower acts as parasitic and gets excited by its fringing field. Secondly, feed the lower patch only, so that upper patch gets excited by its radiation and lastly, feed both the patches with their centers displaced to coincide with their feed points. The Proposed idea used the last technique making the bottom patch designed for L5 band and top for S-band. RHCP with axial ratio less than 2dB is achieved by making square notched diagonally opposite to each other. Also, a return loss of high than -10dB is considered suitable for the antenna to have good radiation efficiency. The substrate used in designs is Rogers RT Duroid/5880 of thickness 3.175mm since it has a lower dielectric constant (of 2.2) as well as yields a very good radiation efficiency at high frequencies(Sidhu and Singh Sivia, 2015).

3 STACKED ANTENNA DESIGN METHODOLOGY

Helical antenna and microstrip patch antenna are the two most commonly used antenna for the purpose of GNSS receiver applications, but when it comes to design it for terminal end where there are several constraint in terms of shape, size, area, integration with other sub system, patch antenna win this bait(Chen *et al.*, 2012). The proposed design structure has multilayer square patches where their dimension are calculated using standard design equations(Balanis, 2007). The upper patch is designed to resonate at 2.4 GHz (S-band) while lower is designed for 1.17 GHz (L5 band) of operation. There are two 2.4 mm FR4 sheets stacked for L5 band to get a substrate of 4.8 mm thick and two 3.175 mm RT/ Duroid stacked to get 6.35 mm substrate. The ground plane is taken as 100x100 mm in dimension. Both the patches are connected with SMA-A connector and fed with a 50 ohm coaxial feed cable. Figure 1 shows the design made in simulation tool and actual design structure.

Some of the experimental methods to obtain circular polarization using single fed are being notched along the two diagonally edges of the patch, diagonally fed nearly square, square with stubs, corner shopped squares, diagonally triangular cut, diagonal slot etc. The proposed design uses square notch cuts at diagonally opposite corners of the patch on one end only.

Figure 1. Stacked IRNSS patch antenna design in simulation tool and actual fabrication.

Also, the feed point position is identified in such a way that it resonance two orthogonal modes with phase shift of ±45 degrees w.r.t the feed point, which leads to phase quadrature between two modes. This results in generation of circular polarization and limits the value of axial ratio below 2dB(Iwasaki, 1996).

Figure 2: shows the schematic layout of the proposed design and Table-2 gives details information regarding physical dimension of the proposed structure.

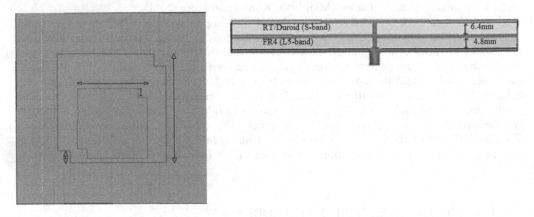

Figure 2. Schematic layout of stacked patch antenna.

Table 2. Proposed patch antenna specifications.

Patch dimensions	Upper Patch (S - band) - 34*32*.035 (mm³) Lower Patch (L5 - band) – 55*55*.035 (mm³)
Ground dimensions	100*100*.035 (mm³)
Substrate & Thickness	Upper Substrate (S - band) - RT/Duroid, 6.4 mm Lower Substrate (L5 - band) – FR4, 4.8 mm But fabricated thickness of Upper Substrate is 6.35 mm (two sheets of 3.175 mm are stacked vertically)
Feed point location	13.5 mm symmetrically from the centre, parallel to the radiating edge. (only to lower patch – L5 band)
Notch size	Upper Patch (S - band) – 6*6 (mm²) Lower Patch (L5 - band) - 6.75*6.75 (mm²)

4 SIMULATION RESULTS AND DISCUSSION

Simulation results are satisfying and matching with the design requirements as per Table-1. Figure 3 shows the measured return loss (S11) of the proposed structure. The above antenna radiates best at 1.1753 GHz where return loss is -19.5 dB. The measured return loss is below -10dB both at L5 and S band. Also, the axial ratio is found to be almost 0dB at 2.48GHz frequency, The L5 band didn't give complete circular bandwidth over the whole required range as axial ratio is slightly higher than 3dB limit but still in the tolerable limit to receive IRNSS signal as shown in Figure 4 which signifies circular polarization identity. VSWR is 1.02 at L5 band and 1.72 and S band shown in Figure 5. Figure 6 shows the 3D radiation pattern which is found to be hemispherical in shape and radiates maximum in the direction normal to the surface of the patch.

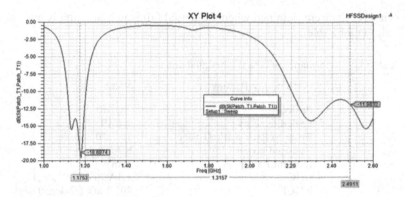

Figure 3.   Return loss of stacked IRNSS antenna.

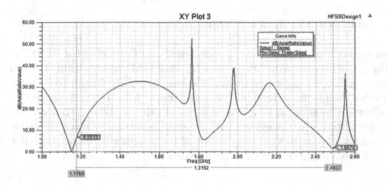

Figure 4.   Axial ratio of stacked IRNSS antenna.

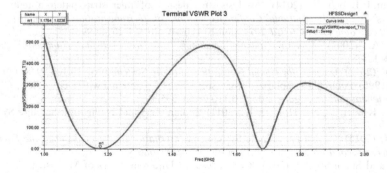

Figure 5.   VSWR of the stacked IRNSS antenna.

253

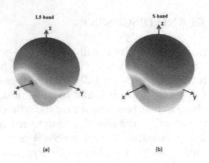

Figure 6.   3-D radiation pattern of the stacked antenna.

The following results were obtained when the actual fabricated antenna is put onto the test. The result is shown in Table 3.

Table 3.   Performance of actual fabricated antenna.

Sr. No.	Performance Parameter	Value
1.	Peak Gain	7.60 dB
2.	Peak directivity	4.4 dB
3.	Front to Back Ratio (FBR)	19.7 dB
4.	Radiated power	8.7 mW
5.	Incident Power	10mW
6.	LHCP	1.4 V (0 deg, 90 deg)
7.	RHCP	401.2 mV (-2 deg, 0 deg)

5   CONCLUSION AND FUTURE SCOPE

A rectangular microstrip stacked antenna is designed and fabricated for IRNSS receiver application and the results seems promising as per the standard requirements. The antenna system developed here needs to be improved in the following parameters for it to be accepted by commercial industries First is the reduction in size and mass of the antenna; second is the improvement in polarization of antenna at L5 band as axial ratio is above 3dB limit. Involvement of metamaterial may improvise the system performance as substrate dielectric value may be controlled and gain may be improved. Bandwidth of the antenna is restricted to few MHz, large application may have requirements of larger bandwidth.

REFERENCES

Anisha, A. and Enoch, S. (2014) 'Analysis and design of microstrip patch antenna for S-band applications', in *2014 International Conference on Control, Instrumentation, Communication and Computational Technologies, ICCICCT 2014*. doi: 10.1109/ICCICCT.2014.6993125.

Balanis, C. A. (2007) *Modern antenna handbook, Modern Antenna Handbook*. doi: 10.1002/9780470294154.

Chen, X. *et al.* (2012) *Antennas for Global Navigation Satellite Systems*. doi: 10.1002/9781119969518.

Iwasaki, H. (1996) 'A circularly polarized small-size microstrip antenna with a cross slot', *IEEE Transactions on Antennas and Propagation*. doi: 10.1109/8.537335.

Majithiya, P., Khatri, K. and Hota, J. K. (2011) 'Indian Regional Navigation Satellite System', *Inside GNSS*.

Salvemini, M. (2015) 'Global Positioning System', in *International Encyclopedia of the Social & Behavioral Sciences: Second Edition*. doi: 10.1016/B978-0-08-097086-8.72022-8.

Sidhu, S. K. and Singh Sivia, J. (2015) 'Comparison of Different Types of Microstrip Patch Antennas', *International Journal of Computer Applications*.

Thangadurai, N. and Vasudha, M. P. (2017) 'A review of antenna design and development for Indian regional navigational satellite system', in *Proceedings of 2016 International Conference on Advanced Communication Control and Computing Technologies, ICACCCT 2016.* doi: 10.1109/ICACCCT.2016.7831642.

Tsui, J. B.-Y. (2000) *Fundamentals of Global Positioning System Receivers, Fundamentals of Global Positioning System Receivers.* doi: 10.1002/0471200549.

Wang, J. J. H. (2012) 'Antennas for global navigation satellite system (GNSS)', in *Proceedings of the IEEE.* doi: 10.1109/JPROC.2011.2179630.

Suryanarayan Rao K.N., 2010, Project Director –IRNSS. ISAC Bangalore. Presentation on IRNSS.

Estimation of defocus blur radius using convolutional neural network

Ruchi Gajjar, Tanish Zaveri & Aastha Vaniawala
Electronics and Communication Engineering Department, Institute of Technology, Nirma University, Ahmedabad, Gujarat, India

ABSTRACT: Image blurring causes loss of information and restoring the degraded image requires knowledge about the original image and the blur kernel, which is usually unknown. However, if information about the blur kernel is available, the image can be deblurred. Degradation due to defocus blur is considered in this paper. To estimate the blur kernel and facilitate in the process of restoration, determination of amount of defocus radius is essential. Here, a convolutional neural network architecture is proposed to determine the defocus blur radius. The Fourier spectrum of defocus blurred images is fed as an input to the network for extraction of the features and classification of the defocus radius. The network is trained, validated and tested using images from standard datasets like USC-SIPI, Berkeley Segmentation dataset and Pascal VOC 2007. The experimental results show competitive results as compared to existing defocus radius estimation methods.

1 INTRODUCTION

An image is blurred when it gets convolved with a blur kernel called point spread function (PSF). Blurring acts as a low pass filter, discarding the high frequency information, thereby edges, in an image. Deblurring of an image can be categorized into blind and non-blind. If during restoration, both, the original image and the blur kernel are unknown, it is known as blind image restoration. In non-blind restoration, information about the blur kernel is priory available and the image is restored using it. However, in majority of the cases, the original image and the blur kernel, both are usually unknown and restoration becomes difficult. If the blur kernel can be estimated, image can be restored using any traditional deblurring technique as described by Kundur (1996).

An image is degraded by defocus blur when a cylindrical PSF of radius R gets convolved with the original image. Many methods reported in literature restore the image by estimating the defocus radius from the frequency spectrum of the blurred image. Mathematical modeling of the zero crossings obtained in the frequency response of the defocus blurred image was done by Moghaddam (2007). The relation between the radius of defocus and the spectral zeros was formulated by Gajjar (2017). Liang (2016) employed polar transformation on the logarithmic power spectrum of the blurred defocus image to estimate blur radius.

Along with determination of blur radius, certain approaches reported in literature, restore the image as well. Neural networks are used for the radius estimation. Su et.al (2008) estimated the defocus blur radius by using a Radial Basis Function network and restored the image using Weiner filtering. Chen ET. Al. (2012) used circular Hough transform along with neural network to determine the defocus radius. Yan (2013) employed deep belief network to first classify the type of blur and then to estimate the blur parameter. Yan (2016) used a General regression neural network (GRNN) for estimation of blur parameters. The network was trained by using preprocessed natural image patches obtained from the spectrum of the blurred image. The classification accuracy obtained was quite high, but the architecture was complex and computationally extensive. Apart from neural networks, decision trees were also used to determine the amount of defocus. Gajjar (2017) used Random Forest classifier to classify the defocus radius. The image was then restored by using Lucy Richardson (LR) algorithm.

In this paper, a convolutional neural network (CNN) is trained with the frequency spectrum of blurred images for classifying the defocus radius. The blur PSF is then constructed using this radius and the image is then restored using LR algorithm. The rest of the paper is organized as follows: Section 2 describes the image degradation model and the defocus PSF. The proposed CNN architecture is presented in Section 3. The dataset details are given in Section 4 and the results are discussed. Section 5 presents the conclusion.

2 BACKGROUND THEORY

2.1 *Image degradation model*

The image blurring process can be shown as follows:

$$g(x,y) = f(x,y) * h(x,y) + n(x,y) \tag{1}$$

Here, g(x,y) is the blurred image, f(x, y) is the latent clear image, h(x, y) is the blur kernel and n(x,y) denotes noise. Degradation due to blur only is considered in this paper and noise is assumed to be zero.

2.2 *Defocus blur*

Blurring due to defocus occurs if the scene to be captured is out of focus. This is usually depended on the distance between the lens of the camera and the object, the aperture of the lens and the focal length. Defocus blur can be defined as a cylindrical lens system having circular aperture. The defocus blur PSF can be mathematically expressed as:

$$h(x,y) = \begin{cases} \frac{1}{\pi R^2}, & x^2 + y^2 \leq R^2 \\ 0, & otherwise \end{cases} \tag{2}$$

The Fourier transform of PSF, h(x,y) yields

$$H(u, v) = \frac{J_1 \pi R r}{\pi R r} \tag{3}$$

where, J_1 is first order Bessel Function, R is the radius of the blur, r is the radius of periodic circles in the amplitude spectrum of H(u,v), where it takes value zero. These zero values would occur for $2\pi R r$ values at 3.83, 7.02, 10.2, 13.2, ...

In the frequency spectrum of H(u,v), if r_0 is the radius of the first zero crossing, i.e. the innermost circle, then the radius of defocus can be expressed as:

$$R = \frac{3.83 L_0}{2\pi r_0} \tag{4}$$

Here L_0 is the length of the image.

It is evident from Equation 4 that the radius of the defocus blur is inversely proportional to the amplitude spectrum of the defocused images. Hence, if the radius of the innermost circle in the frequency spectrum can be detected, the radius of defocus can be estimated. Figure 1 shows the frequency spectrum of images blurred with defocus of radius 5 and 20.

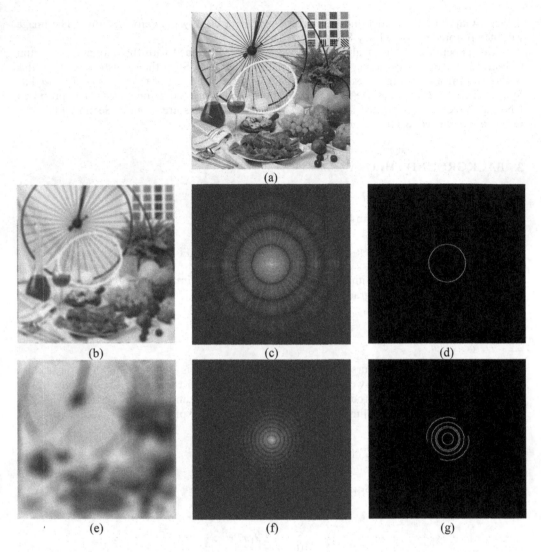

Figure 1. Spectrum of defocus blur images. (a) Original image, (b) Image blurred with defocus of radius 5, (c) Fourier spectrum of (b), (d) Edge detection on (c), (e) Image blurred with defocus of radius 20, (f) Fourier spectrum of (e), (g) Edge detection on (f).

3 PROPOSED METHOD

3.1 *Block diagram*

In order to determine the radius of defocus, the relation between the radius of the innermost circle in the spectrum of the blurred image and amount of defocus is explored. Figure 2 shows the block diagram of the proposed method. Fourier transform is applied on the input blurred images to extract the spectral features corresponding to the amount of defocus. The amplitude of this spectrum is then treated as an image and edge detection is applied to it using Canny edge detector, thereby achieving the zero crossing of the spectrum in a distinctly visible pattern. As the innermost zero crossing is inversely proportional to the radius of defocus, this is extracted using the First White method proposed by Gajjar (2016) and rest of the portion on the image is masked off. These images are now fed to the proposed convolutional neural

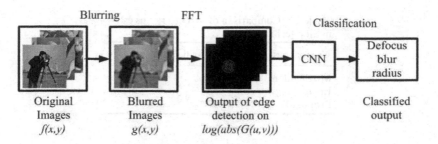

Figure 2. Block diagram of the proposed defocus blur radius estimation method.

network for classification. The output of the CNN is the class label corresponding to the defocus radius.

3.2 Proposed architecture

The proposed CNN architecture for classification of the defocus radius has following layers in its structure:

Layer 1: A convolutional layer with 16 kernels of size 9x9
Layer 2: A subsampling layer with a kernel size of 2x2
Layer 2: A convolutional layer with 32 kernels of size 3x3
Layer 4: A subsampling layer with kernel size of 2x2
Layer 5: A convolutional layer with 32 kernels of size 3x3
Layer 6: A subsampling layer with a rate of 2x2
Final Layer: Two fully-connected layers

4 EXPERIMENTAL RESULTS

4.1 Experimental set-up

The dataset comprises of 100 images of size 512x512 from USC-SIPI, 200 images from Berkeley and 200 images from Pascal VOC, all blurred with defocus of radius 3, 4, 5, 8, 10, 12, 15 and 20. Thus giving a total of 4,000 images. Out of these, 400 images are randomly selected for validation and 400 for testing, rest all are used for training the model. The total class labels are 8, corresponding to the blur radius.

As the variation in the frequency spectrum of the defocus blurred image is used to detect the radius of defocusing, these images are pre-processed before feeding them to the CNN. To detect the zero crossings in the frequency spectrum of the input blurred image, edge detection is applied. The resultant image contains concentric circles, where the radius of the innermost circle is related to the radius of defocus based on Equation 4. This innermost circle is extracted using the First White method. This resultant image is then resized to 28x28 and then fed to the CNN for training.

4.2 Results

The estimated defocus blur radius using the proposed CNN is compared with the radius estimation by other state of art techniques available. Table 1 shows the comparison of radius classification accuracy. It is evident from this table that the classification accuracy achieved by the proposed CNN architecture is 97%, which is high as compared to that achieved by Moghaddam (2007) and Liang (2016).

Once the defocus radius is estimated by the proposed CNN, the blur kernel can be created. This kernel can be used as an input to non-blind image restoration algorithms for image

Table 1. Comparison of defocus radius estimation techniques.

Approaches	Classification accuracy
Moghaddam	87 %
Liang	90 %
Proposed	97 %

 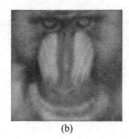 

(a)   (b)   (c)

Figure 3. Restoration results. (a) Original image, (b) image degraded with defocus blur of radius 10, (c) Image restored with estimated radius estimated by the proposed method.

restoration. Figure 3 shows the restored output, when the radius of image blurred with defocus of 10 was correctly classified by the proposed CNN. This radius was used to restore the image using Lucy Richardson algorithm.

5 CONCLUSION

In this paper, an approach to determine the defocus blur parameter using convolutional neural network is proposed. The radius of defocus blur was estimated by the convolutional neural network which was trained with edge detected images obtained from the Fourier spectrum of the blurred image. The proposed network was trained and tested on images from standard datasets. The obtained defocus radius classification accuracy is 97% which is high as compared to other existing methods.

REFERENCES

Arbelaez, P., Fowlkes, C., Martin, D., 2007. The Berkeley Segmentation Dataset and Benchmark UC Berkeley Comput. Vis. Group. Available:http://www.eecs.berkeley.edu/Research/Projects/CS/vision/bsd.

Chen, H.C., Yen, J.C. and Chen, H.C., 2012, August. Restoration of out of focus images using neural network. In IEEE 2012 International Conference on Information Security and Intelligent Control: 226–229.

Everingham, M., Van Gool, L., Williams, C.K., Winn, J. and Zisserman, A., 2007. The PASCAL visual object classes challenge 2007 (VOC2007) results.

Gajjar, R., Pathak, A. and Zaveri, T., 2016. Defocus Blur Parameter Estimation Technique. *International Journal of Electronics and Communication Engineering and Technology*, 7(4): 85–90.

Gajjar, R. and Zaveri, T., 2017, February. Defocus blur parameter estimation using polynomial expression and signature based methods. In *2017 IEEE 4th International Conference on Signal Processing and Integrated Networks (SPIN)*: 71–75.

Gajjar, R. and Zaveri, T., 2017. Defocus blur radius classification using random forest classifier. In *2017 IEEE International Conference on Innovations in Electronics*, Signal *Processing and Communication (IESC)*: 219-223.

Jiang, Y., Wu, Q. and Guo, P., 2005, October. Defocused image restoration using RBF network and kalman filter. In *2005 IEEE International Conference on Systems, Man and Cybernetics*, 3: 2507–2511.

Kundur, D. and Hatzinakos, D., 1996. Blind image deconvolution. *IEEE signal processing magazine*, 13 (3): 43–64.

Liang, M., 2016. Parameter Estimation for Defocus Blurred Image Based on Polar Transformation. *Rev. Tec. Ing. Univ. Zulia.*, 39(1): 333–338.

Moghaddam, M.E., 2007. A mathematical model to estimate out of focus blur. In *2007 5th IEEE International Symposium on Image and Signal Processing and Analysis*: 278–281.

Su, L.Y., Li, F.L., Xu, F. and Liu, Y.R., 2008, May. Defocused image restoration using RBF network and iterative Wiener filter in wavelet domain. In *2008 IEEE Congress on Image and Signal Processing*, 3: 311–315.

The USC-SIPI Image Database, Available: http://decsai.ugr.es/cvg/dbimagenes/

Yan, R. and Shao, L., 2013. Image Blur Classification and Parameter Identification Using Two-stage Deep Belief Networks. *In BMVC*: 70–1.

Yan, R. and Shao, L., 2016. Blind image blur estimation via deep learning. *IEEE Transactions on Image Processing*, 25(4): 1910–1921.

Instrumentation and control engineering track

Technologies for Sustainable Development – Mahajan, Patel & Sharma (eds)
© 2020 Taylor & Francis Group, London, ISBN 978-0-367-33737-7

Powertrain modelling and range analysis for all terrain electric vehicles

Keval Parmar
Electrical Engineering, Nirma University, Ahmedabad, India

Jay Desai, Jatin Patel & Nital Patel
Instrumentation and Control Engineering, Nirma University, Ahmedabad, India

ABSTRACT: Electric vehicles (EV) are the future of transportation, they have lots of advantages compared to their internal combustion engine (ICE) counterparts. Their popularity is also widely spreading across the motorsport community with different competitions already being held and more coming in the future. This poses a challenge for engineers to design efficient EVs with the help of simulation software. A new reverse engineered powertrain method is discussed in this paper to design a model which calculates the range of an EV. This model is focused towards the EVs that are meant for motorsport racing. The velocity and change of slope over the course of a lap is used to simulate the range. This ensures accurate range for the vehicle on a particular track that is used for reference.

1 INTRODUCTION

Electric vehicles are steadily gaining popularity and are bound to overtake internal combustion engine vehicles in the coming future. EVs have an obvious advantage over the ICE vehicles as they produce no emissions; they produce little to no noise; they have better performance; they have very high efficiencies and so on. With these advantages there also comes some obvious disadvantages; one of them is that the charging time of an EV is considerably higher than just filling a tank with petrol[3]. This defect can easily be tackled by knowing the accurate range of the vehicle.

The popularity of electric vehicles in day to day life has also seen an increase in popularity of EVs in motorsport racing. All electric motorsport series like Formula E and World Rallycross championship are spear heading the EV revolution. The student competitions like SAE eBAJA India and Formula Electric also promote all electric racing.

The range analysis of the vehicle is very crucial as it helps the engineers to design and manufacture their vehicles in a more efficient way. Components like the gearbox, motor, wheels, tyres, track conditions all play a key role in the range of an EV. The estimation of range by simulation will directly help the engineers to decide all of the above components in a way that their vehicle would last for the span of the race.

There are different approaches to range estimation. The algorithms existing today rely on the state of charge (SoC) data of the battery pack obtained from field testing[5]. This field testing includes running of the vehicle for multiple charge-discharge cycles and averaging the range obtained[2]. Whereas unconventional methods include simulations using softwares like MATLAB Simulink. One such method is the "contour positioning system (CPS)"[6]. CPS takes into account travel on plane as well as travel on slope. But it takes a constant driving speed which is not a realistic simulation. Another method of advanced CPS takes a drivers approach of accelerating and braking to estimate the range. This method is also known as "dynamic range estimator(DRE)"[6].

The model discussed in this paper, takes a novel reverse engineered approach that is not explored before. It's called "Reverse Engineered Powertrain (REP)". REP uses the varying velocity as well as varying slope of the vehicle on a single lap of a racetrack and calculates the forces applied on the vehicle at every instant. These forces are used to calculate electric charge that is being drawn by the motor controller from the battery pack at that instance. This will continue till the battery discharges. This helps us to find the electrical energy drawn from the battery for an entire race, a single lap and also the energy drawn at various parts of the race track. This helps engineers design the vehicle more efficiently and also according to the race track that they're about to race on.

2 REP MODEL

The model is divided into 4 parts which are interconnected in reverse fashion than how the power is delivered in an actual powertrain[3]. The four parts are as follows:

1. Force model
2. Gearbox model
3. Motor controller model
4. Battery model

In REP the velocity and angle of slope at every instant of the lap is taken as an input of the force model. These forces are then used to calculate the electrical power that will be drawn from the battery pack with the help of a gearbox model and a motor controller model.

The functioning of every section here is in reverse order of an actual powertrain i.e. the power flow in a functioning powertrain is such that the motor controller draws power from the battery pack which then drives the motor, the shaft of the motor then delivers power to the gearbox which in turn delivers it to the wheels after a reduction in rotations and increment in torque. The models here have an exact opposite power flow. The model runs until the SoC reaches a defined low point i.e. when battery discharges, the resulting time for which the model runs is effectively the range of the vehicle.

For better understanding of the model of a single seater ATV is taken as an example.

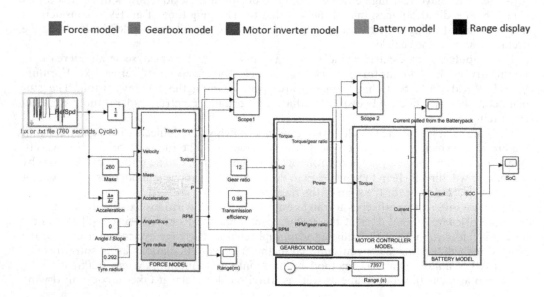

Figure 1. REP model.

2.1 Force model

The force model calculates various forces that are applied on the vehicle resulting in tractive force required to propel the vehicle in forward or reverse directon [1].

It uses 5 inputs to do so.

1) Velocity (V) - It uses predefined values of velocities of the vehicle over the course of one lap of a race track which are measured using a rotary sensor. It uses a loop of such values to simulate a race like environment. Each value of velocity is measured at an interval of 0.5s which can be lowered according to the sensor used. The given model has a lap of around 12.6 mins around a dirt track.
2) Instantaneous acceleration (a) – The above stated values of velocity are differentiated w. r.t time in order to find out the instantaneous acceleration.
3) Mass of the vehicle (m)
4) Angle of slope (θ) - The slope used here for reference is 0 for simplicity but can also be fed in the same manner as velocity.
5) Tyre radius(r)

The force model feeds the torque (τ), tractive force (F_T), angular velocity (Ω) and mechanical power (P) induced on the wheels to the gearbox model[1].

The equations used to calculate the said forces are as follows:

Drag Force: $F_D = \frac{C_D A \rho V^2}{2}$

Rolling Resistance: $F_{RR} = C_{RR} m g$

Accelerating Force: $F_a = a(m + 0.05m)$

Angular velocity: $\omega = \frac{60}{2\pi} \frac{v}{r}$

Tractive Force: $F_T = F_a - F_{RR} - F_D$

Torque: $\tau = F_T r$

Power: $P = F_T v$

2.2 Gearbox model

The torque (τ), mechanical power (P) and angular velocity (ω) of the vehicle received from the force model are used as inputs here. The torque (τ_g) is decreased according to the defined gear ratio (G_r) and the angular velocity (ω) is increased accordingly as well. The efficiency of the gearbox is also considered here Larminie & Lowry, 2003).

Torque = $\tau_g = \frac{\tau}{G_r \mu_T}$

Output Power of Gearbox: $P_g = \frac{\tau_g \omega G_r \mu_T}{9.5488}$

2.3 Motor controller model

The motor that has been used for the tests is a 5kW Permenant magnet brushless motor (PMBLDC) with peak power of 8kW, maximum angular velocity of 3500rpm and a maximum torque of 26Nm. The value of current (I) drawn from the battery pack by the motor controller is obtained against respective values of power (P) and torque (τ_g) obtained from the gearbox model[3]. The manufacturers data sheets that provide torque vs current curves are used as a reference. These values of current vary with various motors and motor controllers[4]. These values of current that are calculated by the motor controller model are further fed into the battery model.

2.4 *Battery model*

The battery model uses the current values obtained from the motor controller model and mathematically integrates it w.r.t time to obtain electric charge. The charge obtained is subtracted from the total charge stored in the battery pack. This process continues until the SoC reaches a defined low point i.e. battery discharges, the model stops, henceforth giving the time for which the vehicle ran on its defined course[7]. The battery pack considered for example here has a capacity of 110 Ah at 48V nominal. Characteristics other than discharging characteristics of the battery have not been considered in this model[1].

3 SIMULATION RESULTS AND DISCUSSION

The model has been made using Simulink and the results are discussed with the help of graphs below.

3.1 *Force outcomes*

As discussed in topic 2.1 the values of forces induced on the vehicle are calculated by the force model on the basis of velocity and slope of the vehicle.

The model generates values of torque and rotational velocity of the wheels which are fed into the gearbox to be reduced and increased respectively. This can be seen clearly in the Figure 2 and 3 respectively. In Figure 2 the graph at the top shows that the torque on the wheel reaches a peak value of 312 Nm and at the same instant it can be seen in the graph below, that after the reduction of gearbox by factor of 12 the torque obtained is 26Nm.

Similarly the angular velocity of the wheels increases from 266rpm to 3200rpm which is shown in the Figure 3.

Figure 4 shows the mechanical power that is being generated on the wheels of the vehicle. It can be seen clearly that as the vehicle accelerates and its velocity increases the value of mechanical power increases.

3.2 *Current drawn from the pack*

Figure 5 below shows the electric current that the motor controller draws from the battery pack at every instant. Figure 5 shows that the value of current increases when the vehicle accelerates. Figure 5 shows the peak and nominal values of current drawn from the battery pack.

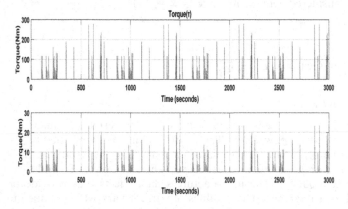

Figure 2.   Torque graph.

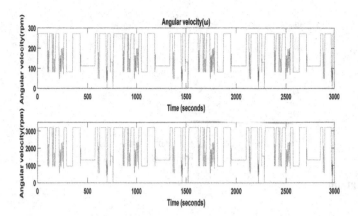

Figure 3. Angular velocity graphs.

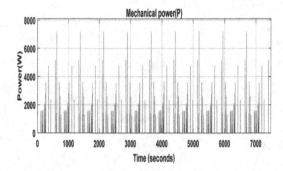

Figure 4. Mechanical power graph.

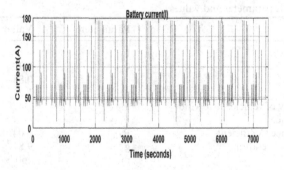

Figure 5. Current drawn from the battery pack (A).

3.3 *The range*

There are two ways to analyze the range of a vehicle, in terms of time and in terms of distance. Both of these entities depend highly on the way a race track is designed. The model calculates both of them.

The distance is calculated by continuously integrating the velocity w.r.t to time and adding it.

As shown in Figure 6 the model runs until the SoC of the battery pack reaches a predefined low value which is 10% in this case. This will generate the time for which the vehicle ran. The Figure 7 shows the value of the distance that the vehicle has covered. In this case it is 88kms.

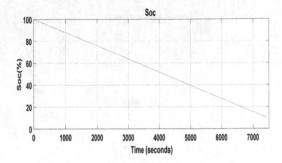

Figure 6.   SoC graph.

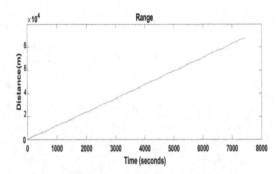

Figure 7.   Distance covered.

Table 1.   Vehicle parameter and values.

Name	Parameter	Value	Unit
Drag Force	F_D	-	N
Drag Coefficient	C_D	0.3	-
Frontal Area	A	0.85	m^2
Fluid Density	ρ	1.2	Kg/m^3
Relative Velocity of vehicle w.r.t wind	V	-	m/s
Rolling Resistance	F_{CC}	-	N
Rolling Resistance Coefficient	R_{CC}	0.04	-
Gravitational Acceleration	g	9.81	m/s^2
Mass of Vehicle	m	260	kg
Acceleration of vehicle	a	-	m/s^2
Velocity of vehicle	v	-	m/s
Tractive Force	F_T	-	N
Torque	τ	-	Nm
Power	P	-	Nm/s
Angular Velocity	ω	-	*rpm
Output Torque from Gearbox	τ_g	-	Nm
Output Power from Gearbox	P_g	-	Nm/s

* rpm is rotations per minute

Table 2. Other parameter and values.

Name	Parameter	Value	Unit
Tyre Radius	R	0.292	m
Transmission Efficiency	μ	0.98	-
Gear Ratio	G_r	12	-

Table 3. Change in outputs by Change in Parameters.

	Max. Torque τ_g (Nm)	Max. Angular Velocity (rpm)	Range (Second)
Gear Ration			
12	26	3200	7397
11.56	27	3082	7119
Mass(kg)			
260	26	3200	7397
240	24.3	3086	7629

4 CONCLUSION

The model developed is mainly focused in helping engineers design the vehicle in a more dynamic manner. As discussed, the REP method is focused on EVs that are used in motorsport racing categories like rallycross racing or BAJA racing; for this reason electric components like HVAC (heating, ventilation and air conditioning) system that are found in commercial EVs, which draw a lot of electrical power from the battery pack, haven't been considered here [5].

This model though is very modular as compared to other methods as the data given in Table 3 proves it; Varying certain parameters have a definite change in the range of the vehicle. You can see that changing the gear ratio brings obvious change into the angular velocity and torque and in turn also brings an effective decrease in the range of the vehicle. Whereas, a decrease in the weight of the vehicle also increases its range hence, validating the model [6].

This helps the engineers in deciding what components to use in the vehicle and to help find performance enhancements. REPs approach to range calculation is such that the range can be calculated instantly over the course of a race if a problem occurs with the vehicle, the driver has to drive the vehicle in a different manner. This is very helpful in long endurance races where drivers fatigue towards the end of the race causes them to slow down and has a considerable effect on range of the EV. Hence such an algorithm helps the engineers to design a reliable model to build their EV around.

REFERENCES

1. Larminie, J., Lowry, J. (2003). Electric vehicle technology. England: John Wiley & Sons LTD.
2. Laurikko, J., Granstrom, R., Haakana, A. (2012). Assessing range and performance of electric vehicles in. California: World Electric Vehicle Journal Vol. 5.
3. Sarrafan, K., Sutanto, D., Muttaqi, K. M., Town, G. (2017). Accurate range estimation for an electric vvehicle. Australia: IET Electrical Systems in Transportation.
4. Schreiber, V., Wodtke, A., Augsburg, k. (2014). Range prediction of electric vehicles. Ilmenau: ILME NAU SCIENTIFIC COLLOQUIUM.
5. Varga, B.O., Sagoian, A., Mariasiu, F. (2019). Prediction of Electirc Vehicle Range: A Comprehensive. Cluj Napoca: energies.
6. Wai, C.K., Rong, Y. Y., Morris, S. (2015). Simulation of a distance estimator for battery. Malaysia: Alexandra Engineering Journal.
7. Yin, J. Shen, Y., Liu3, X. T., Zeng, G.J. Liu, D.C. (2017). Research on State-of-Charge (SOC) estima tion using current. China: IOP Publishing.

Technologies for Sustainable Development – Mahajan, Patel & Sharma (eds)
© 2020 Taylor & Francis Group, London, ISBN 978-0-367-33737-7

Input fault detection using ensemble kalman filter for nonlinear descriptor systems

Tigmanshu J. Patel, M.S. Rao, Jalesh L. Purohit & Vipul A. Shah
Faculty of Technology, Dharmsinh Desai University, Nadiad, Gujarat, India

ABSTRACT: Model based fault diagnosis relies on process models and differs in terms of adopted model as well as algorithms used. The system model expressed in descriptor form is a natural starting point for modeling of complex industrial processes (Nikoukhah 1992). Recent studies (Darouach 1992, Chisci 1992, Mandela 2010, Puranik 2012) demonstrate state estimation of descriptor systems by Kalman filters and its variants. However, the application of such filters to fault detection of descriptor systems has not been assessed. This paper aims to analyse fault detection (FD) of nonlinear descriptor system (NLDS) by implementation of such filters. An Ensemble Kalman filter (EnKF) is evaluated for state estimation and filter performance is examined. Further, the residuals are generated and analysed by various residual evaluation functions. The approach for state estimation, input fault detection and performance analysis of residual evaluation functions is demonstrated on a slow electrochemical descriptor system.

1 INTRODUCTION

1.1 *Motivation*

Model based fault diagnosis has been a keen area of investigation since many years whose central idea is to substitute the hardware redundancy by a process model via software implementation. The techniques used for fault detection and diagnosis need a specific representation of the system. In contrast, a large class of engineering systems can be modeled in descriptor/DAE (Differential Algebraic Equations) form as a natural starting point. The differential equations describe process dynamics while algebraic constraints arise from equilibrium conditions. DAE's were regarded as implicitly written ODE's until 1980's, when several numerical integration methods failed after which it received considerable attention. It is hence important to work directly with systems in descriptor form for FDD (Fault detection and Diagnosis) and FTC (Fault tolerant control).

Several researchers (Chisci 1992, Nikoukhah 1992, Darouach 1995, Mandela 2010, Puranik 2012) have evaluated state estimation of descriptor systems using Kalman filters and its variants of which only few (Mandela 2010, Puranik 2012) have applied Kalman filter based approaches to NLDS. In (Mandela 2010), the state estimation of NLDS is studied using a DAE EKF (Extended Kalman filter) which overcomes the shortcomings of work done by (Beccera 2001). In (Puranik 2012), a particle filter is applied to NLDS. Recently (Purohit 2015), a multi-rate DAE EKF is applied to an irregularly sampled large scale DAE system. However, prior works did not address fault detection of NLDS based on such filters. Moreover, a qualitative approach for comparative studies of computationally inexpensive residual evaluation methods is by and large missing in literature. In (Puranik 2012), a particle filter with 40 particles was implemented for state estimation. However, the rationale for choosing a certain number of particles with regards to considered case study are missing. Also, these algorithms shall be applied to real time systems where 'time' becomes a major constraint. There has been almost no quantitative analysis for efficacy of such algorithms before real time implementation for descriptor systems.

1.2 Scope of work

This work critically examines an approach for fault detection of NLDS using a Kalman filter variant. The residual generation is accomplished by using observed states from Ensemble Kalman Filter (EnKF). The rationale behind the choice for a certain number of ensemble members is investigated and clarified. The residuals are evaluated using computationally inexpensive residual evaluation functions. This paper also attempts to delve into the problem of algorithm efficacy by analysis of computational cost and RMSE (Root Mean Square Error).

2 STATE ESTIMATION

2.1 System representation

A myriad of engineering systems such as chemical process, electrical, thermodynamic systems etc. lead to descriptor form of system representation. Descriptor systems are also referred to as implicit systems, singular systems, generalized state-space systems, semi-state systems and differential algebraic equation (DAE) systems. The process model to describe a nonlinear DAE/NLDS as a function of differential and algebraic variables can be given as

$$\dot{x} = F(x, z, u); \; g(x, z, u) = 0; \; y(t) = h(x, z) \tag{1}$$

A discrete representation of NLDS can be given by

$$x_{k+1} = x_k + \int_{kT}^{(k+1)T} f(x(\tau), z(\tau), u_k) + w_k; \; g(x(\tau), z(\tau), u) = 0; \; y_k = h(x_k, z_k) + v_k \tag{2}$$

where $x \in R^d, z \in R^a, y \in R^q$ and $u \in R^p$ are differential states, algebraic states, system outputs and inputs respectively and under the assumptions that: 1) w_k and v_k are mutually uncorrelated zero mean white Gaussian noise processes with known distribution 2) measurements are available at regular intervals with sampling time T 3) input/s as well as disturbances are piecewise constant over the sampling interval which ensures that the system is regular and thus solvable 4) Initial values be consistent. The consistency of initial conditions may depend on the derivatives of input function u (Scholz 2015).

2.2 Ensemble Kalman filter

An Ensemble Kalman filter (EnKF) which belongs to a class of particle filters is implemented for state estimation of NLDS (Puranik 2012). For the system given by eq. 2, the process and measurement noise have zero mean with known covariance matrices Q and R respectively. The EnKF relies on a certain number of particles to facilitate state estimation. Initially, N_e particles (differential states) are drawn from a known distribution with mean $\hat{x}_0$ and corresponding algebraic states are obtained by solving constraint equation $g(x, z, u) = 0$. Each ensemble member at a particular time instant is a set of consistent differential and algebraic states and can be visualized as,

$$\{e_1, e_2, e_3, .., e_{N_e}\} = \left\{ \begin{bmatrix} x_1 \\ z_1 \end{bmatrix}, \begin{bmatrix} x_2 \\ z_2 \end{bmatrix}, \begin{bmatrix} x_3 \\ z_3 \end{bmatrix}, \ldots, \begin{bmatrix} x_{N_e} \\ z_{N_e} \end{bmatrix} \right\} \tag{3}$$

where x and z are differential and algebraic vectors respectively. Each ensemble member with consistent initial values for DAE is propagated through system model and is the first step during each iteration of estimation algorithm. Since the algebraic equation/s arise from equilibrium conditions, it is exact and hence the process noise is only added to differential states.

$$\widehat{x}_{k+1|k} = \widehat{x}_{k|k} + w_k \tag{4}$$

For noisy differential states, the algebraic states are computed to ensure consistency. This is carried out because process noise already added to differential states will contribute in a manner that algebraic constraints may not be satisfied. Further, the output is predicted as,

$$\widehat{y}_{k+1|k} = h\left(\widehat{x}_{k+1|k}, \ z_{k+1|k}\right) + v_k \tag{5}$$

In addition, the mean of differential states, algebraic states and predicted output are deduced by,

$$\overline{x}_{k+1|k} = \sum_{i=1}^{N_e} \left(\widehat{x}_{k+1|k}^i\right); \ \overline{z}_{k+1|k} = \sum_{i=1}^{N_e} \left(\widehat{z}_{k+1|k}^i\right); \ \overline{y}_{k+1|k} = \sum_{i=1}^{N_e} \left(\widehat{y}_{k+1|k}^i\right) \tag{6}$$

where i relates to i^{th} ensemble member. The error covariance matrix P can then be obtained as

$$P_{k+1|k} = \frac{1}{N_e - 1} \left\{ \begin{bmatrix} \widehat{x}_{k+1|k}^i \\ \widehat{z}_{k+1|k}^i \end{bmatrix} - \begin{bmatrix} \overline{x}_{k+1|k} \\ \overline{z}_{k+1|k} \end{bmatrix} \right\} \left\{ \begin{bmatrix} \widehat{x}_{k+1|k}^i \\ \widehat{z}_{k+1|k}^i \end{bmatrix} - \begin{bmatrix} \overline{x}_{k+1|k} \\ \overline{z}_{k+1|k} \end{bmatrix} \right\}^T \tag{7}$$

The ensemble members are further updated by solving an optimization problem iteratively for each ensemble member. The optimization function is posed as

$$\left(\widehat{x}_{k+1|k}, \widehat{z}_{k+1|k}\right) = \min_{x_{k+1}, z_{k+1}} \ \varepsilon_{k+1}^T P_{k+1|k}^{-1} \varepsilon_{k+1} + e^T_{k+1} R^{-1} e_{k+1} \tag{8}$$

subject to the constraints $g\left(x_{k+1|k}, \ z_{k+1|k}, \ u_k\right) = 0$; $x_L \leq x_{k+1} \leq x_H$; $z_L \leq z_{k+1} \leq z_H$, where

$$\varepsilon_{k+1|k} = \begin{bmatrix} x_{k+1|k} \\ z_{k+1|k} \end{bmatrix} - \begin{bmatrix} \widehat{x}_{k+1|k} \\ \widehat{z}_{k+1|k} \end{bmatrix}; \ e_{k+1|k} = y_{k+1} - \widehat{y}_{k+1|k} \tag{9}$$

The updated ensemble members are used to determine the state estimates. First, the mean of the differential states is calculated as

$$\overline{x}_{k+1|k+1} = \sum_{i=1}^{N_e} \left(\widehat{x}_{k+1|k+1}^i\right) \tag{10}$$

The scalar value obtained from eq. 10 is estimated differential state which is further utilized to obtain the algebraic state by solving algebraic constraint equation. With this, the iterative algorithm propagates the estimated states through system model again and starts its next iteration.

3 RESIDUAL GENERATION AND EVALUATION

3.1 Residual generation

The estimated states of nonlinear descriptor system obtained by employing an EnKF facilitates to determine predicted output. In order to detect a fault, the first step is residual generation. The residuals are generated by

$$r_i = y_n - \widehat{y} \tag{11}$$

where $i \in [1, q]$, $y_n \in R^q$ is measured output/s and $\widehat{y} \in R^q$ is predicted output/s. The residual generation is followed by extracting features or statistics of the residual signal. In this step, the residual signal is evaluated beforehand during fault free conditions by residual evaluation function to determine thresholds. These thresholds allow to distinguish system operation between fault free and faulty conditions. During system operation, the state estimator is implemented online and residuals are relayed to fault detection module which maps the residuals via residual evaluation function continuously and checks for any threshold violations.

3.2 *Residual evaluation methods*

Some computationally expensive methods for residual evaluation are also employed across literature which determines statistical parameters of residual signal or adaptive thresholds. In the light of implementation issues for any algorithm to a real time system, time is a major constraint. The state estimation described in Section 3 or any other observer based methods shall consume some time to arrive at a solution. Considering this fact, computationally inexpensive residual evaluation methods are chosen as follows: 1) Trend Analysis (rate of change) 2) RMS (Root mean square) function 3) Average function 4) Peak value function.

4 RESULTS AND DISCUSSION

For the purpose of simulation, a case study of galvanostatic process of nickel hydroxide electrode shown in (Mandela 2010) is considered where mole fraction of nickel hydroxide and potential difference at solid-liquid interface are differential and algebraic states respectively. The system model captures charging, open circuit and discharging processes. It is assumed that differential state is corrupted by process noise. The system output is potential difference which is affected by measurement noise. Covariance matrices are Q = [0.00001] and R = [0.0001]. The sampling time considered is $\Delta t = 15s$ while initial state for the system as well as estimator is chosen to be [0.35024 0.4071]^T. The results of state estimation shown in Figure 1 are generated with 5 ensemble members. RMSE analysis for different no. of ensemble members is tabulated in Table 1. As evident, RMSE increases progressively with increase in number of ensemble members except from $N_e = 3$ to $N_e = 5$. It also advocates a proportional relationship between ensemble size and time consumed. An input (abrupt) fault is introduced at around time t = 3000s which changes the state of electrochemical cell from charging (u = 0.00001A) to open circuit condition (u = 0.00000A). The residuals are evaluated and resultant evaluation function value is compared with respective pre-

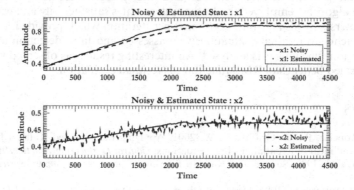

Figure 1. Estimation results for $N_e = 5$: Differential state x_1 (mole fraction) & algebraic state x_2 (potential difference).

Table 1. Filter performance analysis.

Ensemble size	3	5	10	15	20	30	40
RMSE(differential state)	0.0362	0.0076	0.0579	0.0806	0.0714	0.1040	0.1206
RMSE(algebraic state)	0.0061	0.0061	0.0117	0.0149	0.0134	0.0182	0.0197
Time (per iteration)	01.03s	02.80s	04.46s	06.30s	08.84s	12.94s	16.96s

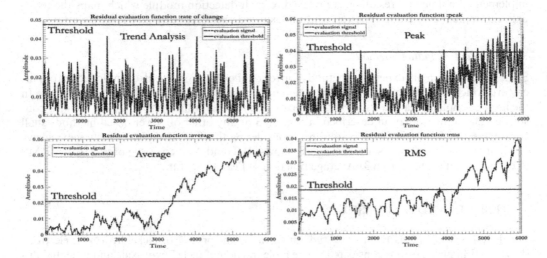

Figure 2. Fault detection: Residual evaluation functions vs. thresholds.

determined thresholds depicted in Figure 2. Since the electrochemical system is a slow system an abrupt fault would change system states slowly. Hence, the Trend Analysis function is unable to detect a fault. An evaluation window of 10 samples is con-sidered for RMS and Average functions. The residual signal gradually accumulates after t = 3000 s and convincingly violates thresholds in case of RMS and Average methods unlike Trend Analysis as well as Peak function. The peak function does not provide confident fault occurrence information. The Average function comparatively performs better than RMS function.

5 CONCLUSION

In this study, the problem of fault detection and filter performance analysis for NLDS was addressed. The algorithm employed tracks the system states satisfactorily and was investigated by two parameters, RMSE and time consumed per iteration which provided a better insight. Various approaches to residual evaluation were assessed and fault detection ability of these residual evaluation functions was analysed. An interesting result was observed where Average function performed better than RMS function for residual evaluation.

REFERENCES

Becerra,V.M., Roberts, P.D & Griffiths, G.W. 2001. Applying the extended Kalman filter to systems described by nonlinear differential-algebraic equations, *Control Engineering Practice*, 9(3):267.
Chisci L & Zappa G. 1992. Square-root Kalman filtering of descriptor systems, *Systems and Control Letters*.
Darouach M. & Boutayeb M. 1995. Design of Observers for Descriptor Systems, *IEEE Transactions on Automatic Control*, 40(7):1323–1327.

Gao Z. & Ding S X. 2007. Actuator fault robust estimation and fault-tolerant control for a class of non-linear descriptor systems. *Automatica*, 43(5):912–920.

Hamdi H, Rodrigues M, Mechmeche, Theilliol D & Benhadj N. 2012. Fault detection and isolation in linear parameter-varying descriptor systems via proportional integral observer, *IJACS*, 224–240.

Mandela RK, Rengaswamy R, Narasimhan S. & Sridhar L.N. 2010. Recursive state estimation techniques for nonlinear differential algebraic systems. *Chemical Engineering Science*, 65(16):4548–4556.

Nikoukhah R, Willsky A.S. & Levy B.C. 1992. Kalman Filtering and Riccati Equations for Descriptor ystems, *IEEE Transactions on Automatic Control*, 37(9):1325–1342.

Puranik Y, Bavdekar V, Patwardhan S C & Shah S. 2012. An ensemble kalman filter for systems governed by differential algebraic equations (DAEs), volume 8, *IFAC Proceeding Volumes*, 531–536.

Purohit J L, Patwardhan S C & Mahajani S M. 2015. State Estimation of a Reactive Distillation System using Multi-rate DAE EKF, *Indian Control Conference*.

Scholz L. 2015. *Control Theory of Descriptor Systems Lecture Notes*.

Technologies for Sustainable Development – Mahajan, Patel & Sharma (eds)
© 2020 Taylor & Francis Group, London, ISBN 978-0-367-33737-7

Smart irrigation system through mobile app

Sandip A Mehta
Assistant Professor, Institute of Technology, Nirma University, India

Akshat Bhutiani
Student, Institute of Technology, Nirma University, India

ABSTRACT: Traditionally, mobile apps have been used to book tickets, view videos and listen to music podcasts and other things. However, it has not been used to control any connected devices in a feasible manner. As the concept of the Internet of Things grows, it is envisioned that mobile devices and ubiquitous computing will be common place. This technology will allow us to remotely monitor & control various hardware devices. Using the Internet of Things, our proposed method aims to develop a Smart Irrigation System controlled by a mobile device.

This system makes use of a WiFi module integrated with a microcontroller unit. Data will be sent wirelessly to the Android Device. The Android Device has a custom app that is built specifically for the purpose of monitoring and control of various appliances. This proposed method makes use of various sensors such as humidity, moisture and temperature sensors mounted at various places to measure the various process parameters. For the purpose of control, a pump and an LED is also installed.

1 INTRODUCTION

India is a land of agriculture. Indian agriculture largely has been historically rain dependent. In recent times, with modern technology agriculture is not reducing dependence on rain by resorting to elaborate canal systems, bore well systems and drip irrigation. In agriculture, irrigation is an important process that influences the production of the crop, the quality of the harvest and various other parameters. Due to inefficient use of water resources, India loses around 50% of its received rainfall due to inefficiencies in storage. Further as India generates power using thermal and gas based plants, around 60-70% of the received rainfall is used up in these activities thereby leaving it contaminated and unfit for drinking. Therefore to reduce the problem of water wastage it is proposed to use drip irrigation.

Drip irrigation is aimed at significant increase in efficiency of available water for farming/agriculture. Drip irrigation is a micro-irrigation system that has the potential to save water and nutrients by allowing water to drip slowly to the roots of plants, either from soil surface or below. The goal is to place water directly into the root zone and minimize evaporation. Drip irrigation systems distribute water through a network of valves, pipes, tubing, and emitters.

Drip irrigation in current form has the limitation of providing water to plant roots as per a fixed pre-defined volume and timing. It does not check the true requirement of water and therefore does not optimize the amount of drip water or the frequency of drip water. This not only leads to an inefficient use of water but also has impact on the crop yield as optimized water delivery is not being managed.

In Drip irrigation, the quantum of discharge through the drip system is fixed on the basis of the size of the drip holes and the fixed frequency of water release.

This leads to -

- Inefficient use of water resource (wastage)
- Inefficiency in crop yield due to either excess moisture/deficient moisture

1.1 *Motivation*

Our main motivation is to create an instrumentation based solution to monitor the soil moisture at frequent intervals and switch on drip irrigation system only when soil moisture drops below pre-defined level. The other motivation is to make the solution very user friendly and accessible on the ubiquitous mobile phone. The mobile app will have dials, gauges and buttons to make the UI simple and user friendly. It is further proposed that remote data logging and other features will also be provided.

Various sensors such as soil moisture sensor, humidity sensor and air temperature sensors are interfaced with the controller unit. The readings will be taken at some particular frequency of certain interval and according to pre-programmed logic, appropriate action will be taken.

1.2 *Literature survey for the mobile app*

Some efforts have been made for the mobile app development for irrigation system, like in [1] the soil water measurement has been done. In [2] smart irrigation system for the strawberries has been proposed. Similar attempts are made using mobile app for irrigation [3–6] and home automation [7–13] system. A good literature review is available in [14,15] for home automation. In most of these methods mobile phone app which works on the suitable hardware are device dependent. i.e. they need dedicated hardware platform for the proper function. Further most of the app are developed using readily available app software and they used normally LabView platform for interfacing the data.

1.3 *Proposed method*

It is proposed that the system will be able to remotely monitor and control. Various sensors interfaced will send data through internet to the mobile device. The mobile device will be able to control the actuators and store data for further action. It is also proposed that the user will be able to remotely monitor the system. The proposed method is shown Figure 1.

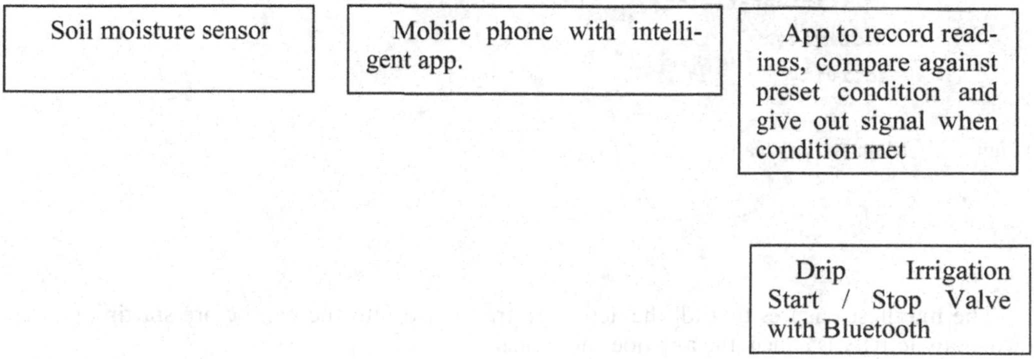

Soil moisture sensor	Mobile phone with intelligent app.	App to record readings, compare against preset condition and give out signal when condition met
		Drip Irrigation Start / Stop Valve with Bluetooth

Figure 1. Proposed method.

For the purpose of demonstration the following things will be used –

1) Soil Moisture Sensor
2) Humidity sensor
3) Air Temperature Sensor
4) LED

The app will have a screen for the purpose of monitoring the devices.

It will also have a secure Login & Logout facility to prevent misuse by unauthorized personnel.

2   VARIOUS ACTIVITIES IN THE APP

The app consists of various activities. The first and most important activity is the Main Activity. It is loaded as soon as the app is started. To ensure that the main activity is loaded, it has to be declared in the Manifest of the App. This is done in the following way as shown in Figure 2

```xml
<?xml version="1.0" encoding="utf-8"?>
<manifest xmlns:android="http://schemas.android.com/apk/res/android"
    package="com.androidtutorialshub.loginregisterkotlin">
<uses-permission android:name="android.permission.INTERNET"/>
    <application
        android:allowBackup="true"
        android:icon="@mipmap/ic_launcher"
        android:label="Login Register Kotlin"
        android:supportsRtl="true"
        android:theme="@style/AppTheme">
        <activity
            android:name=".activities.LoginActivity"
            android:screenOrientation="portrait">
            <intent-filter>
                <action android:name="android.intent.action.MAIN" />

                <category android:name="android.intent.category.LAUNCHER" />
            </intent-filter>
        </activity>
        <activity
```

Figure 2.   Main activity map.

The manifest ensures that all the activities are loaded into the app before starting so that when an activity is called, the app does not crash.

2.1   *Login activity*

The Login Activity is called as soon as the user touches the login Button. The Login Activity passes the user entered data to a helpers package known as input validation. Within the input validation class, there are several methods to check whether the input entered is correct or not. If the entered input is correct only then the app will allow the user to access the control panels and monitoring views.

Figure 3. Login activity map.

2.2 *The user activity class*

The user activity class is the class where the user performs all control and monitoring function.

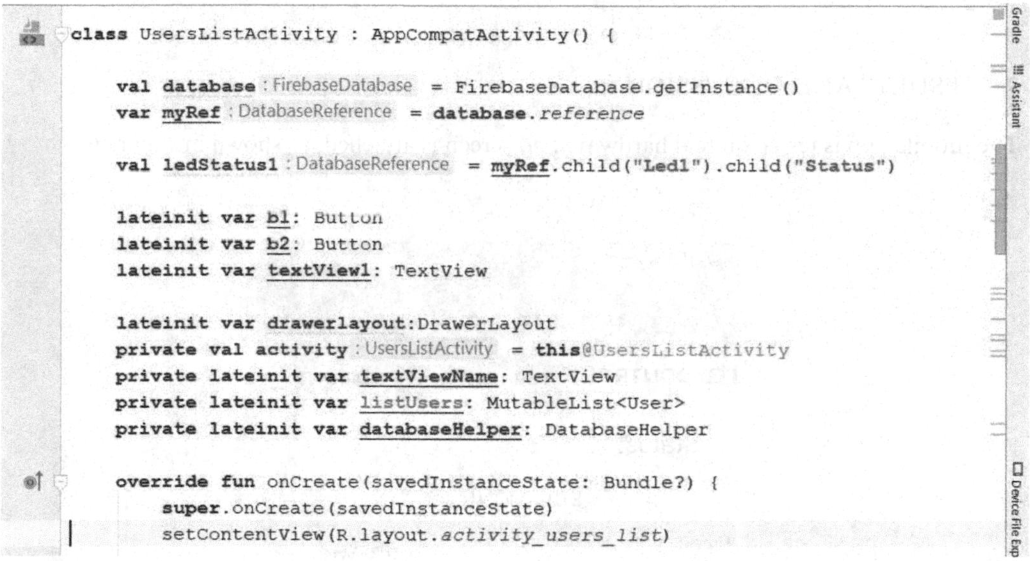

Figure 4. User activity map.

The code in the Node MCU unit that is used to send and receive data through the mobile app –

The hardware interfacing set up is as shown in Figure 5

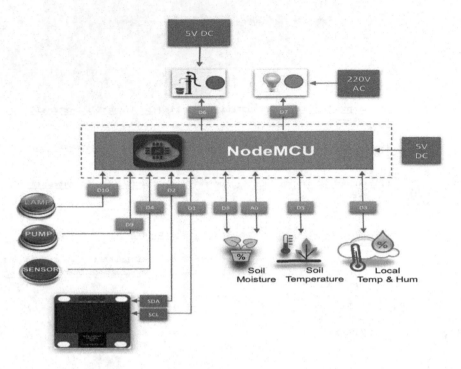

Figure 5.   Hardware interface for the proposed method.

3   RESULTS AND DISCUSSION

The mobile app is tested on real hardware and screen is attached, as shown in Figure 6.

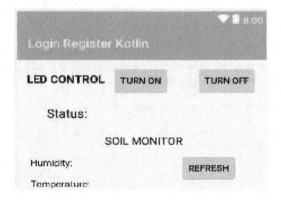

Figure 6.   Screenshot of mobile app.

4 CONCLUSION

This paper demonstrates that it is possible to make an Android App for control and monitoring of process parameters. It is particularly beneficial to the plant manager as he/she can remotely view and control various sensors and actuators in the plant.

Since overall implementation of the app is not only affordable but also it can be widely adapted by many industries. This app is also useful in places where there is a scarcity of water and water is to be managed efficiently. Further this app doesn't require any dedicated hardware support and hence it can be configured for remote monitoring using any hardware device available for the use.

REFERENCES

[1] D.M. Freebairn, A. Ghahramani, J.B. Robinson, D.J. McClymont, A tool for monitoring soil water using modelling, on-farm data, and mobile technology, Environ. Model. Softw. (2018). doi:10.1016/j.envsoft.2018.03.010.

[2] R. González Perea, I. Fernández García, M. Martin Arroyo, J.A. Rodríguez Díaz, E. Camacho Poyato, P. Montesinos, Multiplatform application for precision irrigation scheduling in strawberries, Agric. Water Manag. (2017). doi:10.1016/j.agwat.2016.07.017.

[3] D. Kissoon, H. Deerpaul, A. Mungur, A Smart Irrigation and Monitoring System, Int. J. Comput. Appl. (2017). doi:10.5120/ijca2017913688.

[4] N. Mhaisen, O. Abazeed, Y. Al Hariri, A. Alsalemi, O. Halabi, Self-Powered IoT-Enabled Water Monitoring System, in: 2018 Int. Conf. Comput. Appl. ICCA 2018, 2018. doi:10.1109/COMAPP.2018.8460401.

[5] A. Kumar, A. Surendra, H. Mohan, K. Muthu Valliappan, N. Kirthika, Internet of things based smart irrigation using regression algorithm, in: 2017 Int. Conf. Intell. Comput. Instrum. Control Technol. ICICICT 2017, 2018. doi:10.1109/ICICICT1.2017.8342819.

[6] A. Laamrani, R.P. Lara, A.A. Berg, D. Branson, P. Joosse, Using a mobile device "app" and proximal remote sensing technologies to assess soil cover fractions on agricultural fields, Sensors (Switzerland). (2018). doi:10.3390/s18030708.

[7] K. Mandula, R. Parupalli, C.H.A.S. Murty, E. Magesh, R. Lunagariya, Mobile based home automation using Internet of Things(IoT), in: 2015 Int. Conf. Control Instrum. Commun. Comput. Technol. ICCICCT 2015, 2016. doi:10.1109/ICCICCT.2015.7475301.

[8] V. Lohan, R.P. Singh, Home Automation Using Internet of Things, in: Lect. Notes Networks Syst., 2019. doi:10.1007/978-981-13-0277-0_24.

[9] N. David, A. Chima, A. Ugochukwu, E. Obinna, Design of a Home Automation System Using Arduino, Int. J. Sci. Eng. Res. (2015).

[10] N. Sriskanthan, F. Tan, A. Karande, Bluetooth based home automation system, Microprocess. Microsyst. (2002). doi:10.1016/S0141-9331(02)00039-X.

[11] R. Teymourzadeh, S.A. Ahmed, K.W. Chan, M.V. Hoong, Smart GSM based home automation system, in: Proc. - 2013 IEEE Conf. Syst. Process Control. ICSPC 2013, 2013. doi:10.1109/SPC.2013.6735152.

[12] B. Yuksekkaya, A.A. Kayalar, M.B. Tosun, M.K. Ozcan, A.Z. Alkar, A GSM, internet and speech controlled wireless interactive home automation system, IEEE Trans. Consum. Electron. (2006). doi:10.1109/TCE.2006.1706478.

[13] R.K. Kodali, V. Jain, S. Bose, L. Boppana, IoT based smart security and home automation system, in: Proceeding - IEEE Int. Conf. Comput. Commun. Autom. ICCCA 2016, 2017. doi:10.1109/CCAA.2016.7813916.

[14] A. Cyril Jose, R. Malekian, Smart Home Automation Security: A Literature Review, Smart Comput. Rev. (2015). doi:10.6029/smartcr.2015.04.004.

[15] V.S. Gunge, P.S. Yalagi, Smart Home Automation: A Literature Review, Int. J. Comput. Appl. (2016). doi:10.6029/smartcr.2015.04.004.

Mechanical engineering track

Study on microstructural and mechanical properties of dissimilar metals joining using FSW

Devang K. Sisodiya

M.Tech Student, Mechanical Engineering Department, School of Engineering, Institute of Technology, Nirma University, Ahmedabad, Gujarat, India

Nilesh D. Ghetiya

Associate Professor, Mechanical Engineering Department, School of Engineering, Institute of Technology, Nirma University, Ahmedabad, Gujarat, India

Kaushik M. Patel

Professor, Mechanical Engineering Department, School of Engineering, Institute of Technology, Nirma University, Ahmedabad, Gujarat, India

ABSTRACT: Friction Stir Welding (FSW) is used in the welding of dissimilar materials. Joining of dissimilar material is required in chemical, nuclear, power generation, transportation and electronic industry. In the present study, an attempt was made to analyse the microstructure and mechanical properties of welded aluminium alloy AA6061 and copper alloy using FSW. The FSW was carried out using varying welding speed and tool offset. A mixture of recrystallized aluminium matrix and deformed copper particles were visible in nugget zone. Microstructure variation significantly affects the tensile strength and micro hardness distribution in the cross-section of the joint. Maximum tensile strength of 186 MPa at 40 mm/min welding speed and 1mm tool offset was obtained towards Al side.

Keywords: FSW, dissimilar joint, mechanical Properties, macrostructure, microstructure

1   INTRODUCTION

Joining of dissimilar materials like aluminum and copper is of great interest in power generation, transportation, chemical, nuclear and electronic industries [1,2]. FSW is most suitable processes to join Al-Cu joint as compared to the other fusion welding processes [3]. Important parameters of a FSW process are rotational speed, welding speed, tool offset and tilt angle. In FSW a non-consumable tool is used which consists of a pin and shoulder geometry [4]. Ouyang et al [5] concluded that the FSW of AA6061 to Cu is difficult to produce higher strength joint but recently it was reported that the offsetting of tool pin towards Al side provides the better joint strength. Xueet al. [6] found that the defect free joints are produced when copper is fixed at advancing side and Al is placed towards retreating side. Tool pin offset towards aluminium side produces more stirring action at aluminium side and few particles detached at Cu side which leads to make good bonding between Cu particles and aluminium matrix, which ultimately helps to make a defect free joint [7]. The objective of the present investigation was to study the effect of welding speed and tool offset on the microstructure and mechanical properties of FSW between AA6061 and Cu.

2   EXPERIMENTAL PROCEDURE

FSW were carried out on a vertical milling machine with indigenous designed fixture used to hold the work piece having dimensions of $300 \times 100 \times 5$ mm^3. AA6061 was placed towards advancing side (AS) and pure Cu on retreating side during welding. The tool was made of high speed steel having 20 mm shoulder diameter, 4.5 mm mean pin diameter and 4.7 mm pin height. The tool was tilted at 2° angle. Welding was done using welding speed of 40 mm/min and 63 mm/min and at constant rotational speed of 1400 rpm/min. The tool offset 0 mm and 1mm towards Al was used during welding. The cross-sections of metallographic samples for optical microscopy were polished to a mirror finish by using a diamond paste. The samples were etched with Keller's reagent and observed using optical microscopy. Vickers micro hardness was measured under a load of 9.81 N for a 10s dwell time. All tensile tests were performed as per the ASTM E8M-04 standards. Tensile strength of each joint was evaluated through an average value of three tensile specimens being cut from the same joint.

3 RESULTS AND DISCUSSION

3.1 *Macrostructure visualization*

The macrostructural cross section of the weld specimen as well as top surface of the welded specimen is shown in Figure 1. It was observed that the most of the surface defects were present in NZ. A void was observed at NZ of RS with tool rotational speed (TRS) of 1400 rpm and welding speed (WS) of 40 mm/min with both having 0 mm and 1 mm tool offset condition towards Al side. At TRS of 1400 rpm, WS of 63mm/min and 0mm tool offset condition, the

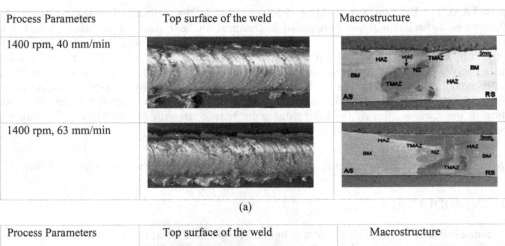

(a)

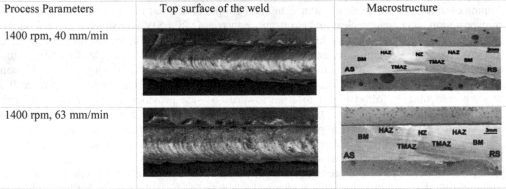

(b)

Figure 1. Macrostructure and top surface of weld surface obtained at various welding speed and tool offset (a) 0 mm tool offset (b) 1 mm tool offset at Al side.

top surface of the welded quality was very smooth and shining however a slight crack was present in between TMAZ and NZ at RS.

Due to low hardness value of Copper, it's deposition at AS is higher. It can be seen that the burst or material deformation around the weld line is more at WS of 63 mm/min compared to the WS of 40 mm/min at constant TRS of 1400 rpm.

3.2 *Microstructure evolution*

Figure 2 & 3 (a) shows the NZ of Al-Cu FSW joint which was obtained at 1400 rpm and 40 mm/min with 0 mm and 1 mm tool offset condition respectively. It was observed that due to the thermo-mechanical action of tool, NZ consists of a mixture of recrystallized Al matrix and the more deformed twinned Cu particles due to its lower strength. Cu particles were distributed in an irregular shapes and size due to which NZ appeared to be inhomogeneous. In this portion average grain size were found to be 19.05 μm and 16.50 μm respectively.

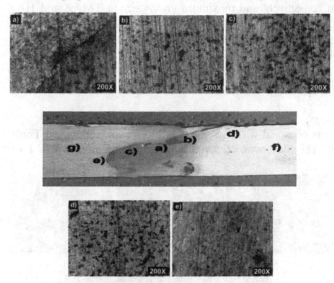

Figure 2.   Microstructure for N = 1400 rpm, WS = 40 mm/min and 0 mm tool offset with condition of (a) NZ (b) TMAZ on Cu side, (c) TMAZ on Al side, (d) HAZ on Cu side, and (e) HAZ on Al side.

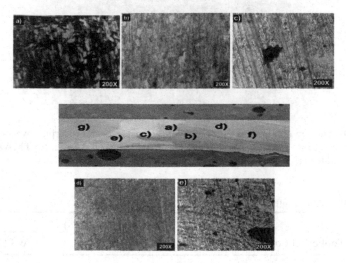

Figure 3.   Microstructure for N = 1400 rpm, WS = 40 mm/min and 1 mm tool offset at Al side with condition of (a) NZ, (b) TMAZ on Cu side, (c) TMAZ on Al side, (d) HAZ on Cu side,and (e) HAZ on Al side.

Figure 2 & 3 (b) and 2 & 3 (c) shows TMAZ on Cu and AA6061 side respectively. Transition between TMAZ and HAZ can be characterized by observing the slight increase in grain size. Maximum grain size in this region was found to be 22.55 µm on Al side and 37.50 µm on Cu side. The formation of fine void defect also called as groove defect were observed at high rotation speed which was occurred due to the excess heat input during the welding in the NZ. In this zone Cu and Al experiences severe plastic flow due to the stirring action of the tool. A slight crack was present in this region due to the insufficient mixing of AA6061 and Cu.

Figure 4(a) shows the NZ of Al-Cu FSW joint which was obtained at 1400 rpm and 63 mm/min with 1mm tool offset at Al side. When the tool was rotated, the material flows towards the rotating direction of the tool and same is visible in the macrostructure. It is observed that a Cu lining was created which deposited upto the TMAZ on the Al side. It was also seen that the thickness of Cu lining was more towards the Al side as compared to the Cu side. Maximum grains size in NZ were found to be 13.45 µm while maximum grains size in TMAZ and HAZ on Al side were 33.8 µm and 38.05 µm respectively and maximum grains size in TMAZ and HAZ on Cu side were 28.90 µm and 34.5 µm respectively. A number of voids were formed in the HAZ on both sides of AA6061 and Cu. In both the cases, 1 mm tool offset at Al side provided the better microstructure results as compared to the 0 mm tool offset.

3.3   Mechanical properties

3.3.1 Tensile strength
Tensile strength was obtained with varying welding speed as shown in Table 1. The maximum tensile strength of 186 MPa was achieved at WS of 40 mm/min with 1 mm tool offset and

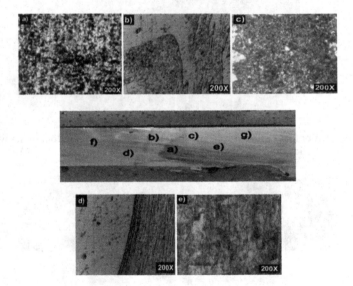

Figure 4.   Microstructure obtained at 200X magnification for N = 1400 rpm, WS = 63 mm/min and 1 mm tool offset at Al side with condition of (a) NZ, (b) TMAZ on Al side, (c) TMAZ on Cu side, (d) HAZ on Al side and (e) HAZ on Cu side.

Table 1.   Tensile strength at various welding speed and tool offset.

Sr. No.	Tool offset Condition (mm)	Tensile Strength (MPa)	
		WS 40 (mm/min)	WS 63 (mm/min)
1	0	123	89
2	1 at Al side	186	104

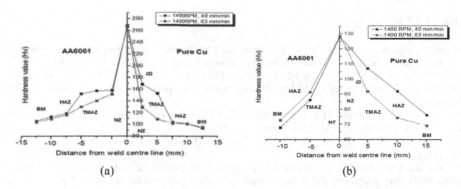

Figure 5. Vickers microhardness distribution at variation of welding speed keeping (a) 0 mm tool offset (b) 1 mm tool offset at Al side.

minimum tensile strength of 89MPa was achieved at WS of 63 mm/min with 0 mm tool offset. The reason for the higher strength is due to the proper mixing between two materials. It is observed that the WS as well as tool offset affects the joint strength; higher strength is observed when the tool offset was toward the harder material. This is due to the proper material mixing and sufficient heat generation.

3.3.2 *Micro-hardness*
Micro hardness values mainly depend on the formation of precipitates. Measurement of hardness of the FSW specimens welded were carried out using Vickers hardness tester. In FSW, hardness values of BM is different from that of HAZ and TMAZ because these zones experiences temperature cycles during the process. NZ shows variation in hardness values because of the mixing of two materials. Micro hardness distribution was observed by using vicker-hardness machine having diamond indenter of square base. Vickers micro hardness measurements were done along the cross section of the specimen and is shown in Figure 5. It was observed that the hardness values were increased from Al base to NZ and then decreased from NZ to the base of Cu for 0 mm tool offset condition and various other process parameters. Since, dissimilar materials have different mechanical properties therefore an asymmetric micro hardness curve was obtained while calculating hardness values for all of the welding parameters. Maximum of 268 Hv hardness value was measured at NZ with 63 mm/min welding speed and 0 mm tool offset whereas minimum hardness value of 40 mm/min was measured at NZ at 1mm tool offset toward Al side. that is 49 Hv.

4 CONCLUSIONS

From this investigation, the following conclusions were derived:

➢ Higher welding speed resulted in poor plastic flow, insufficient heat input and caused some voids in the joints which reduced the tensile strength of the joints.
➢ Maximum grain size was obtained in 1mm tool offset at Al side as compared to the 0 mm tool offset condition.
➢ The joints were fabricated using 20 mm tool shoulder diameter with a rotational speed of 1400 rpm and welding speed of 40 mm/min, which exhibited superior tensile strength as compared to the other joints.

REFERENCES

1. Ahmed O; Saja M; Andre D (2014), Experimental and theoretical analysis of friction stir welding of Al–Cu joints, *International journal of advanced manufacturing technology*, 71:1631–1642.
2. Kallee W.S.; Dave E.N.; Wayne M. T. (2001), Friction Stir Welding- Invention, Innovation and Applications. Proceedings of INALCO, *8th International Conference on Joints in Aluminum*, Munich, Germany.
3. Rai, R.; De, A.; Bhadeshia, H.K.D.H.; Debroy T. (2011), Review: friction stir welding tools. *Journal: Science and technology of welding and joining*, 16, 325–342.
4. Mishra R and Mahoney M. (2007), *Friction Stir Welding and Processing*, ASM International Publishing.
5. Ouyang, J.; Yarrapareddy E.; Kovacevic R. (2006), Microstructural evolution in the friction stir welded 6061 aluminum alloy (T6-temper condition) to copper. *Journal of materials processing technology*, 172: 110–122.
6. Xue, P.; Ni, D.R.; Wang, D.; Xiao, B.L.; Ma, Z.Y. (2011), Effect of friction stir welding parameters on the microstructure and mechanical properties of the dissimilar Al–Cu joints. *Materials science and engineering*: A, 528 4683–4689.
7. Galvao, I.; Loureiro, A.; Verdera, D.; Gesto, V.; Rodrigues, D.M. (2012), Influence of tool offsetting on the structure and morphology of dissimilar aluminum to copper friction stir welds. *Metallurgical and materials transactions*: A, 43 5096–5105.

Technologies for Sustainable Development – Mahajan, Patel & Sharma (eds)
© 2020 Taylor & Francis Group, London, ISBN 978-0-367-33737-7

Design of automatic vertical storage system

Nishant Vyas, B.K. Mawandiya & Mayur A. Makhesana
Mechanical Engineering Department, Institute of Technology, Nirma University, Ahmedabad, India

ABSTRACT: Increase in range of products has created the need and importance of the scientific vertical storage systems. Factors like reduction in inventory, mass production, easy storage/retrieval, and workers' safety show the necessity of storage systems. Because of its high storage/retrieval capacity, popularity and importance of the vertical storage system are increasing in the industries. Vertical storage is essential for companies as constant performance and accuracy in terms of picking of orders is required. In the printing industry, wide ranges of rolls are being stored on vertical racks or floor when they are not in use which occupies the floor space and causes stock inaccuracies. Thus, to solve the storage problem as well as to utilize the vertical space, the automatic vertical storage system is preferable. This works aims to design the automatic vertical storage system for printing rolls which utilize the vertical space in the industry, obtain better inventory control, ensure the safety of the workers and in turn reduces the loading and unloading time. In the present study, a conceptual design of the carousel has been done using Solid Edge ST9 software. Analytical calculations of the critical parts have been carried out.

1 INTRODUCTION

The automatic vertical storage system is one type of AS/RS. AS/RS is composed of a variety of systems controlled by computers for automatic load placements and retrieval of loads from preset storage locations. AS/RS is generally used where a very high volume of load or product is being moved in and out of storage. The main elements of an AS/RS are cranes, input/output points, racks, aisles and pick up positions. Cranes are automated machines that can move autonomously while picking up or dropping off loads. Location where loads are picked up or dropped off can be called as Input/output points. Racks are metal structures, providing a location for storage of loads. Empty spaces between racks, for movement of cranes can be called as Aisles. Pick positions are places where an individual item can be removed from already retrieved load before the load is sent back to the system (Roodbergen et al., 2009).

Important variables for such systems are columns, number of bins, service time, rows and operator utilization. The popularity of horizontal as well as vertical storage systems is increasing due to their easy material handling, storage capacity, and easy computer control adaption by industry and complete automation. The bins move in the horizontal plane for the horizontal storage systems, whereas bins move in the vertical plane for vertical storage systems. Utilization of vertical space and minimization of floor space occupied is the main concept of vertical storage system/vertical carousel. It also reduces the effort for the workers as the load is delivered at a desired location or height (Gabriel-Santos et al., 2016).

The work aims to design the automatic vertical storage system for printing rolls which utilize the vertical space in the industry, obtain better inventory control, ensure the safety of the workers and in turn reduces the loading and unloading time.

- It will increase stock-keeping capacity within limited space
- It protects from damage for storage goods
- Automation of storage and retrieval process reducing labor
- Increases accuracy as human intervention is minimum

2 CONCEPTUAL DESIGN OF AUTOMATIC VERTICAL STORAGE SYSTEM

The vertical storage system consists of bins or gripper types of arrangement to hold the material, which is hinged at both ends to a chain or a belt, which in turn moves over the fixed path along the drive for conveying purpose. A multiple numbers of such bins or grippers are used for storage purpose. The chain is driven by a brake gear-motor. As the motor rotates the chain arrangement along with bins or grippers also move such that the material stored is not dropped. Loading/Unloading point is provided at a convenient height for easy material handling by workers (Fenner, 2009).

Main components of the vertical storage system are:

1. Frame (Structure)
2. Driver sprocket assembly
3. Driven sprocket assembly
4. Hook rod assembly (Gripper assembly)
5. Chain
6. Motor

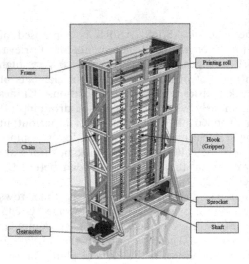

Figure 1. Conceptual 3D model of vertical storage system.

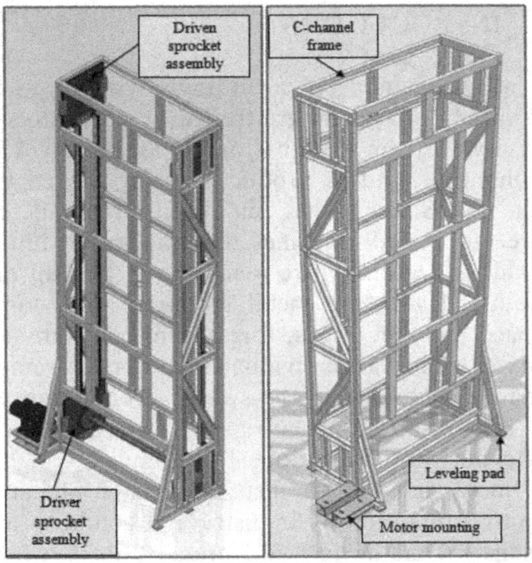

Figure 2. Structure with various assemblies mounted.

The Storage system is conceptualized for the maximum capacity of 62 printing rolls. Design is done such that roll of maximum diameter 250 mm, a roll of maximum length of 2200 mm can be stored. The maximum mass of single printing roll is considered 50kg.

2.1 Frame (structure)

Frame or structure of the storage system supports the sub-assemblies and components of the storage system. It also deals with static and dynamic loads without any deflection or distortion. Sub-assemblies, driver sprocket assembly and driven sprocket assembly are mounted at the bottom and top of the frame respectively. On one side motor mounting is done to support the motor.

2.2 Driver sprocket assembly

Driver sprocket assembly or Driver assembly is connected with the motor and is responsible for the motion of the system. Assembly consists of driver sprockets, driver shaft, flanges and support plate for the sub-assembly. Driver assembly is mounted on a frame with the help of support plates on both sides.

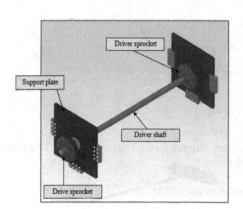

Figure 3.   Driver sprocket assembly.   Figure 4.   Driven sprocket assembly.

Sprockets are mounted on driver shaft with the help of keys and are held at a position with the help of grub screws. Motion is provided by motor to drive sprocket, located outside, which in turn rotates driver shaft. Along with driver shaft driver sprockets also rotates. Flanges are mounted on support plates for support to the driver shaft. Arrangement of double bearing support is provided through flanges at both ends.

2.3 Driven sprocket assembly

Driven sprocket assembly or driven assembly is connected with driver assembly through chains on both sides. Assembly consists of driven sprockets, driven shaft and support plate for the sub-assembly. Driven assembly is mounted on the frame with the help of support plates on both sides.

2.4 Hook rod assembly (Gripper assembly)

Hook or Gripper assembly is the one that is in direct contact with the printing rollers. It consists of attachment links, bush, rod, stoppers, and hook. Attachment links are used to attach assembly with the chain on both sides, so it moves along with the chain. Bush is used to constraining the movement of attachment links. The Rod is a cylindrical component which is a joining component between chain on both sides. Stoppers are mounted over the rod over which the hook is mounted. This whole assembly of stoppers and hook can move in linear as well as rotary motion, thus providing hinge effect for the stored printing roll and it does not allow the roll to fall while moving along the path of chain. The position of hooks can be adjusted as per the requirement of rolls by moving it along the rod.

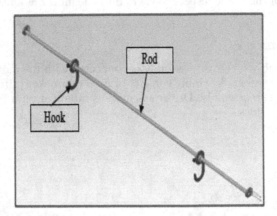

Figure 5.   Hook rod/Gripper assembly.

3   ANALYTICAL CALCULATIONS

The storage system is conceptualized for the maximum capacity of 62 printing rolls. The maximum mass of single printing roll is considered 50kg.
Chain selection
Chain pull or Force required to pull rolls F (Jagtap et al., 2014)

$$F = [(n_r \times m_r \times 9.81) + (n_a \times m_a \times 9.81) + (m_c \times 9.81)] \qquad (1)$$

Where,
n_r = No. of rolls in system
m_r = Mass of single roll
n_a = No. of hook assembly
m_a = Mass of single rod
m_c = Approx mass of chain

The no. of chains used is 2, so the load is equally distributed on each side, select a chain with minimum breaking load more than the calculated load. The selection of chain should be done while keeping Factor of Safety (FOS) in mind.

Motor selection
The motor used in vertical storage must be able to transmit power even at maximum loading condition i.e. completed filled storage system. The motor must be able to transmit power at constant low speed and high torque as required by the system. And motor must also stop at any instance and should sustain the position at any loading condition. Thus in order to fulfill the above criteria, a Brake geared motor is to be used (Anath et al., 2013). The torque required for carrying the rolls at maximum loading conditions can be given by,

$$T = F \times r \tag{2}$$

Where,

T = Torque required (Nm)

F = maximum load

r = radius of sprocket= $\frac{PCD \ of \ sprocket}{2}$

Maximum velocity at which the system should rotate,

$$rpm(n) = \frac{v \times 60}{2\pi r} \tag{3}$$

Power required can be given by,

$$P = \frac{2\pi n T}{60} \tag{4}$$

Where,

P = power required (watts)

n = output rpm of motor

T = torque required

The motor should be selected such that its output rpm and power are in close agreement of the calculated values.

4 APPLICATION OF THE CONCEPT

The accurate application of AS-RS affords an extensive list of user benefits. They can be employed in all those places where there is a requirement of handling materials by storing them at defined locations. It can be seen that automated vertical storage systems are capable of effective material handling. They handle raw materials, products where work is in process, control inventories, finished goods of all kinds and make it possible to retrieve them where placed. This project can be applied for storage and retrieval of spare parts in various industries. Its application includes controlling of small inventories by automated storing and retrieving them when required. Only the design of the gripper is to be changed as per the requirement of storage material. Different types of grippers are required for handling materials with different shapes and sizes.

5 FUTURE SCOPE

In this study, a conceptual model for storage of printing rolls is presented, one can design gripper required for storage of products of different shapes and sizes. Further, one can fabricate a prototype according to the model presented and the same can be tested for industrial use. Also, the program (code) for automation can be developed for the vertical storage system.

6 CONCLUSION

An Automatic Vertical Storage System is conceptualized and designed for the storage of various sizes of the printing rolls. Analytical calculations for the critical parts and load calculation for motor and chain selection is carried out. Implementation of this system storage can yield better space utilization and save time from sorting out the product without any damage to products.

This system eradicates manual work and can easily be controlled by any common man. It is certainly a good way to make storage secure, is reliable and allows quick retrieval of goods.

REFERENCES

Anath, K.N., Rakesh, V. & Visweswarao P.K. 2013. Design and Selecting Proper Conveyor Belt. *International Journal of Advanced Technology*. 4 (2): 43–49.

Fenner Dunlop Conveyor Handbook, Conveyor Belting Australia, (2009), pp. 1–70.

Gabriel-Santos, A., Rolla, J., Martinho, A., Fradinho, J., Gonçalves-Coelho, A. and Mourão, A., 2016. The rational footsteps for the design of the mechanism of a vertical carousel-type storage device. *Procedia CIRP*, 53: 193–197.

Jagtap, A.A., Vaidya, S.D., Samrutwar, A.R., Kamadi, R.G. and Bhende, N.V., 2015. Design of material handling equipment: belt conveyor system for crushed Biomass wood using v merge conveying system. *International Journal of Mechanical Engineering and Robotics Research*, 4(2): 38–48.

Roodbergen, K.J., & Vis, I.F.A.2009. A survey of literature on automated storage and retrieval systems. *European Journal of Operational Research*. 194 (2): 343–362.

Technologies for Sustainable Development – Mahajan, Patel & Sharma (eds)
© 2020 Taylor & Francis Group, London, ISBN 978-0-367-33737-7

Single layer graphene as nanofiller to enhance mechanical properties of matrix material

Shyam Tharu & Mitesh Panchal
Mechanical Engineering Department, Institute of Technology, Nirma University Ahmedabad, India

ABSTRACT: Graphene is one of the most suitable nanomaterials to enhance the mechanical properties of the base material. In the present work graphene has been considered as a filler material to enhance the mechanical properties of different base materials namely; gold (Au), aluminum (Al), copper (Cu), iron (Fe) and magnesium (Mg). The analytical as well as 3-dimensional modeling based finite element analysis based simulation approaches are considered to estimate the elastic modulus of graphene based nanocomposites. The square representative volume element (RVE) with variation in width as well as length of graphene has been considered for the analysis. The effect of variation in width of graphene on elastic modulus without in terphase is not much significant irrespective of matrix materials, when interphase is considered between nanofiller and matrix material the elastic modulus increases with increase in width. Similarly, as length of graphene increases elastic modulus increases as. Overall consideration of interphase between nanofiller and matrix increases the elastic modulus of nanocomposites.

Keywords: Graphene, nanocomposite, elastic modulus

1 INTRODUCTION

Nanomaterials have enormous mechanical properties to make nanocomposites with vast useful applications in different sectors like, aviation, defense, health, etc. Nanocomposites have benefits of superior mechanical properties like higher elastic modulus and strength, high noise damping capacity, lower density than other ceramic materials and metallic materials [Ijima (1991), Zhao-yong (2011)]. The graphene sheet is a new evaluation of material science field. The material research field is now a days focusing on ultra-thin materials like carbon films, and one of the best example of carbon films is graphene sheets. The graphene is a two dimensional nanostructure with thickness ranging 1 to 100 nm. The graphene sheet consist of a layer of carbon atoms with hexagonal structure has been described by Ramanathan (2008). The carbon nanotubes and graphene sheets have higher surface areas as they are nanometer scaled structures. The higher surface areas of CNTs and graphene sheets can be utilized to have nanocomposites of excellent mechanical properties.

The interphase between base material (matrix material) and filler material, and its properties play an important role to derive the properties of the nanocomposites. Thus, it is important to estimate the properties of the nanocomposites based on the properties of the matrix material, nano-filler material as well as the interphase between matrix and filler materials [Singh (2016), Hortarangel (2013)]. Shaikh and Gohil (2010) have developed the equation of interphase by considering three phase representative volume element (RVE), and they analyzed the effect of interfacial bonding on the effective mechanical behavior of nanocomposites materials. Spanos et al. (2015) had described about the graphene based nanocomposites with the use of micromechanical theory, in which the nanofiller was evaluated using atomistic approach and the matrix material modeled using continuum approach to have more efficient computational work to analyze nanocomposites. Mohammadpour et al. (2014) had studied the large strain behavior of nanocomposites using FEM and predicted the Young's modulus and ultimate strength of nanocomposites under axial loading. Srivastava and Kumar (2017) had studied the effect of the interface using FE model and concluded that the thickness as well as elastic modulus of the interphase zone behaves different from the general consideration and it is highly affected by the matrix material of nanocomposite. Sears and Batra (2004) used the molecular mechanics simulation to predict the Yong's modulus, Poisson's ratio and shear modulus, of nanocomposites and results are compare with the results obtained based on continuum modeling, in which CNTs have as continuum tube. Huang et al. (2008) established the cohesive law for CNT based nanocomposites, the cohesive law obtain using van der Waals force theory.

Montazeri and Rafii-Tabar (2011) employ a molecular structural mechanics (MSM)/finite element (FE)-based multiscale modeling approach to compute the reinforcement effects of carbon nanostructures embedded in polymers to conclude, at a low nanofiller content the graphene perform significantly better than carbon nanotubes in terms of enhancing the Young modulus of the nanocomposite.

2 ANALYTICAL FORMULATION FOR MECHANICAL PROPERTIES

The analytical formulation to estimate the elastic modulus of nanocomposite has been formulate by different approaches, Tsai-Pagano based approach (not considering interphase properties between nanofiller and matrix material) and continuum model based Rule of Mixture (ROM) approach (considering interphase properties between nanofiller and matrix material).

The formulation developed by Tsai-Pagano approach, estimate the elastic modulus of nanocomposites without presence of interphase between nanofiller (single layer graphene sheet) and considered matrix material. Tsai-Pagano approach is given by Kunanopparat (2008) as,

$$E_c = \frac{3}{8}E_l + \frac{5}{8}E_t \tag{1}$$

where, E_l = longitudinal elastic modulus; E_t = transverse elastic modulus; and E_c = elastic modulus of composite.

$$E_l = \frac{1 + 2(l/w)BV_n}{1 - BV_n}E_m \ and \ E_t = \frac{1 + 2CV_n}{1 - CV_n}E_m \tag{2}$$

where, l = length of single layer graphene; w = width of single layer graphene; E_m = elastic modulus of matrix material; and V_n = volume fraction of nanofiller.

$$B = \frac{E_a - 1}{E_a + 2(l/w)} \ and \ C = \frac{E_a - 1}{E_a + 2} \tag{3}$$

where, E_a = ratio of elastic modulus of nanofiller to matrix material.

$$E_a = \frac{E_n}{E_m} \tag{4}$$

where, E_n = elastic modulus of nanofiller; and E_m = elastic modulus of matrix material.

The continuum model based Rule of Mixture approach is considered to estimate the elastic modulus of nanocomposites with presence of interphase properties between the nanofiller and matrix material. For the evaluation of interphase properties, cohesive energy law ($\emptyset$) between nanofiller and matrix material for unit length is estimated from expression given by Jiang et al. (2008),

$$\emptyset = \frac{2\pi}{3}\rho_m\rho_g Q_I P_I^3 \left(\frac{2P_I^9}{15r^9} - \frac{P_I^3}{r^3}\right)$$

(5)

where, Q_I and P_I = intermolecular potential parameter ; ρ_m= density of matrix material; ρ_g= density of graphene sheet; and r = distance between a point in the matrix material from surface of single layer graphene sheet.

LB mixing rule by Srivastava and Kumar (2017),

$$P_I = \frac{P_c + P_m}{2} \ and \ Q_I = \sqrt{Q_c . Q_m}$$

(6)

The equation of cohesive energy law equation (7) is used for calculating stiffness of interphase material, ratio with total strain e.g., nothing but interphase potential as equation (8) and (9) by Sears and Batra (2004),

$$E_I = \frac{\partial^2 \emptyset}{\partial^2 S}$$

(7)

The strain S, at any distance in the matrix material, $S = \frac{r - t_I}{t_I}$, also t_I is the equilibrium distance of nanofillers (single layer graphene sheet) and matrix material.

$$t_I = 0.8584 P_I$$

(8)

From the equation of cohesive energy law, the interphase elastic modulus has been formulated as given below,

$$E_I = 8\pi t_I \rho_m \rho_g Q_I \left[\left(\frac{P_I}{r}\right)^{11} - \left(\frac{P_I}{r}\right)^5\right]$$

(9)

For interphase zone modeling between the nanofiller and matrix material for given volume fraction using square RVE is utilized to evaluated the effect of vdW interaction on elastic properties of nanocomposites material.

The volume fraction of nanofiller (V_n) in matrix material into square RVE is given as,

$$V_n = \frac{w.t}{a^2}$$

(10)

The volume fraction of interphase (V_I) in matrix material into square RVE is given as,

$$V_I = \frac{(w + 2t_I).(t + 2t_I)}{a^2}$$

(11)

where, t_I = interphase thickness ; w = width of single layer graphene sheet; t = thickness of single layer graphene sheet; and a = size of side square RVE.

The elastic modulus of nanocomposites with presence of interphase, elastic modulus of single layer graphene based nanocomposites is given by Srivastava and Kumar (2017) as,

$$E_c = V_n E_n + V_I E_I + (1 - V_n - V_I)E_m \qquad (12)$$

where, V_n and V_I = volume fraction of nanofillers and interphase material respectively; E_n, E_I and E_m = elastic modulus of nanofillers, interphase and matrix material respectively.

3 FEM BASED SIMULATION FOR ESTIMATE THE ELASTIC MODULUS OF NANOCOMPOSITE

For FEM based simulation, 3-dimensional model of nanocomposite is developed as per considered dimensions of single layer graphene sheet, interphase thickness tI and square RVEs of nanocomposites. The thicknesses of interphase for different matrix materials, tI

Table 1. Elastic modulus and density of different considered matrix material.

	Elastic Modulus, GPa	Density, kg.m^{-3}
Gold, Au	78	19320
Aluminum, Al	70	2702
Copper, Cu	130	8960
Iron, Fe	211	7860
Magnesium, Mg	45	1738
Graphene	1000	1320

are calculated as per equation (8), which are used for 3-dimensional model of nanocomposites with interphase. To estimate the elastic modulus, boundary conditions applied as one of the ends of the 3-dimensional model is fixed and on the other end of the component is subjected to axial tensile load of 10 N. The elastic modulus of nanocomposite is estimated based on Hook's law.

$$\sigma = E\varepsilon \qquad (13)$$

where, σ = stress ; ε = strain; E = elastic modulus.

4 RESULTS AND DISCUSSION

Initially, the effect of presence of interphase on elastic modulus of nanocomposite has been analyzed for different considered widths of the graphene sheet for nanocomposites of all the considered matrix materials. The obtained (Figure 1) results depict that as the width of graphene sheet increases the elastic modulus increases. Also, the presence of interphase between nanofiller and matrix materials increases the elastic modulus of nanocomposite. Also, the analysis has been performed to analyze the effect of size variation of the nanofiller on elastic modulus of nanocomposites. The obtained results (Figure 2) depict that as the width of the graphene sheet increases the elastic modulus of nanocomposite increases irrespective of the matrix materials, and as the length of graphene sheet increases the elastic modulus increases but the variation is not much significant.

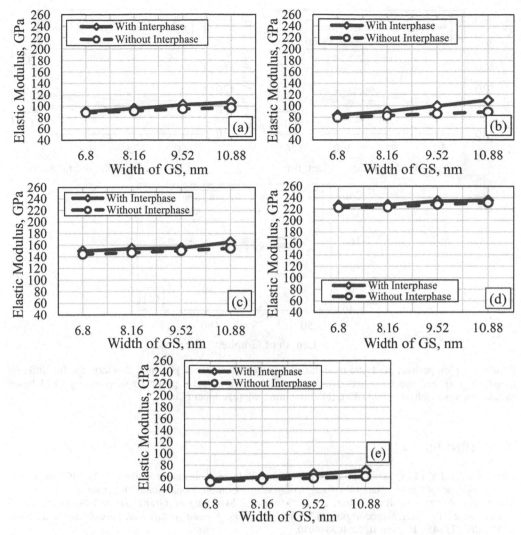

Figure 1. Comparison of elastic modulus of nanocomposites with presence of interphase and without presence of interphase for length of graphene sheet 50 nm against variation in width of single layer graphene. (a) gold (Au); (b) aluminum (Al); (c) copper (Cu); (d) iron (Fe); and (e) magnesium (Mg).

5 CONCLUSION

The presence of interphase between single layer graphene sheet and base material (matrix material) increases the overall elastic modulus of nanocomposites. The variation in size of graphene sheet is also analyzed and obtained results depict that as the width and length of graphene sheet increases elastic modulus of nanocomposites increases, due to increase in size results in to more and more contribution of interphase and which is ultimately increases the resultant elastic modulus of nanocomposites. The effect of variation in width of graphene sheet on elastic modulus of nanocomposite is more significant compare to the variation in length of graphene sheet. The presented analysis is found to be more useful to estimate the elastic modulus of nanocomposites.

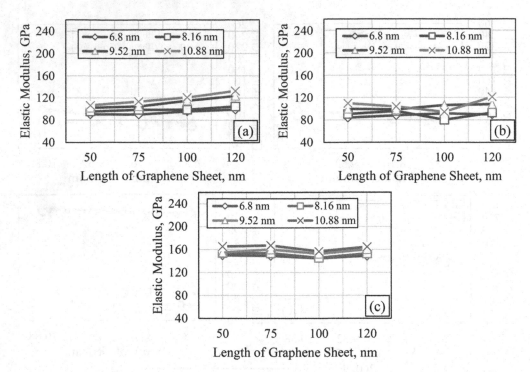

Figure 2. Comparison of elastic modulus of nanocomposites with presence of interphase for different length of graphene sheet against variation in width of single layer graphene sheet (using FEM based simulation approach). (a) gold (Au); (b) aluminum (Al); (c) copper (Cu).

REFERENCES

Chen, X L, and Y J Liu. 2004. "Square Representative Volume Elements for Evaluating the Effective Material Properties of Carbon Nanotube-Based Composites." Computational Materials Science 29 (1): 1–11.

Montazeri, A., and H. Rafii-Tabar. 2011. "Multiscale Modeling of Graphene- and Nanotube-Based Reinforced Polymer Nanocomposites." *Physics Letters, Section A: General, Atomic and Solid State Physics* 375 (45). Elsevier B.V.: 4034–4040.

Ijima. "1991 Nature Publishing Group."

Hortarangel, Jaime, Witold Brostow, and Victor M Castano. 2013. "Mechanical Modeling of a Single-Walled Carbon Nanotube Using the Finite Element Approach." *Polimery/Polymers* 58 (4): 276–281.

Singh, Nakshatra Bahadur, and Sonal Agarwal. 2016. "Nanocomposites: An Overview." *Emerging Materials Research* 5 (1): 5–43.

Kunanopparat, Thiranan, Paul Menut, Marie Hélène Morel, and Stéphane Guilbert. 2008. "Plasticized Wheat Gluten Reinforcement with Natural Fibers: Effect of Thermal Treatment on the Fiber/Matrix Adhesion." *Composites Part A: Applied Science and Manufacturing* 39 (12). Elsevier Ltd: 1787–1792.

Gohil, Piyush P, and A A Shaikh. 2010. "Analytical Investigation and Comparative Assessment of Interphase Influence on Elastic Behavior of Fiber Reinforced Composites." *Journal of Reinforced Plastics and Composites* 29 (5): 685–699.

Srivastava, Ashish, and Dinesh Kumar. 2017. "A Continuum Model to Study Interphase Effects on Elastic Properties of CNT/GS-Nanocomposite." *Materials Research Express* 4 (2). IOP Publishing: aa5dd2.

Lu, W. B., J. Wu, J. Song, K. C. Hwang, L. Y. Jiang, and Y. Huang. 2008. "A Cohesive Law for Interfaces between Multi-Wall Carbon Nanotubes and Polymers Due to the van Der Waals Interactions." *Computer Methods in Applied Mechanics and Engineering* 197 (41–42): 3261–3267.

Sears, A., and R. C. Batra. 2004. "Macroscopic Properties of Carbon Nanotubes from Molecular-Mechanics Simulations." *Physical Review B - Condensed Matter and Materials Physics* 69 (23): 1–10.

Mohammadpour, Ehsan, Mokhtar Awang, Saeid Kakooei, and Hazizan Md Akil. 2014. "Modeling the Tensile Stress-Strain Response of Carbon Nanotube/Polypropylene Nanocomposites Using Nonlinear Representative Volume Element." *Materials and Design* 58. Elsevier Ltd: 36–42.

Spanos, K. N., S. K. Georgantzinos, and N. K. Anifantis. 2015. "Mechanical Properties of Graphene Nanocomposites: A Multiscale Finite Element Prediction." *Composite Structures* 132. Elsevier Ltd: 536–544.

Ramanathan, T., A. A. Abdala, S. Stankovich, D. A. Dikin, M. Herrera-Alonso, R. D. Piner, D. H. Adamson, et al. 2008. "Functionalized Graphene Sheets for Polymer Nanocomposites." *Nature Nanotechnology* 3 (6): 327–331.

Zhao-yong, Ding, S U N Bao-min, J I A Bin, and Ding Xiao-lian. 2011. "Flame Synthesis of Carbon Nanotubes and Nanocapsules" 688: 116–121.

Synthesis, characterization and applications of magnetorheological fluid

Hiren Prajapati* & Absar Lakdawala
Assistant Professor, Mechanical Engineering Department, Institute of Technology, Nirma University, India

ABSTRACT: Magnetorheological Fluid (MRF) consists of micron sized iron particles dispersed in nonmagnetic viscous carrier fluid. MRF changes its apparent viscosity in the presence of external magnetic field. This rheological behavior of MRF categorizes it as a smart fluid. This rheological characteristic can be used to broaden working range of any fluid based mechanical systems like isolators, tuned vibration absorber, shock absorbers, clutches, engine mounts, alternators, power steering pumps, control valves, brakes, dampers etc. But it also has some issues which creates hurdles in its successful use in these applications. This article discusses the most challenging issues and methods used to overcome the same. It also includes characterization at different stages of MRF synthesis. At the end some MRF based applications are discussed.

Keywords: Smart fluids, Magnetorheological Fluid (MRF), ferro fluids, sedimentation, in use thickening, hard caking, characterization

1 INTRODUCTION TO MAGNETORHEOLOGICAL FLUIDS (MRF)

The material which changes its behavior under influence of any external agent are termed as smart materials. Some smart materials are shape memory alloys, piezo-electric materials, Electrorheological Fluids (ERFs) and Magnetorheological fluids (MRFs). (Ashtiani 2015) The suspended magnetic particles lined up in the magnetic flux lines increase viscosity and again particles disperses in absence of magnetic field. The characteristics of MRF like high apparent yield stress, quick transformation from liquid to semisolid and semisolid to liquid, simple interface between electrical power input and mechanical power output etc., have drawn attention of many researchers and made MRF suitable to develop various mechanical systems like isolators, shock absorbers, clutches, engine mounts, alternators, power steering pumps, control valves, brakes, dampers etc. The MRF consists of micron sized magnetic particles, nonmagnetic carrier fluid and other additives. (Premalatha 2012), (Shetty 2011), (Sidpara 2009) The magnetic particles are generally carbonyl iron (CI), Electrolytic Iron (EI), highly pure iron or iron/cobalt alloy, nickel/zinc ferrite. (Ashtiani 2015),(Olabi 2007) To achieve MRF with high yields stress, the particles should have high magnetic saturation and less coercivity. (Ashtiani 2015), (Lim 2004), (Sidpara 2009). Generally, the MRF should have less off state viscosity and large variation in viscosity under application of magnetic field. (Yang 2009) MRF is synthesized using various carrier fluids like polyvinylnbuty and naphthol-thickened kerosene, sunflower oil, silicon oil, white and light grade mineral oils, a combination of synthetic oil, water and organic liquids, polyester, polyether, hydrocarbonic oils etc. (Ashtiani 2015) Commonly used carrier fluids are silicon oil, synthetic oil and mineral oil. (Kumbhar 2015) Selection of carrier fluid based on its viscosity is really tricky, as low viscosity can lead to instability and

---

*Corresponding author: E-mail: hiren.prajapati@nirmauni.ac.in

sedimentation problems, whereas high viscosity increases undesirable off state viscosity. The third constituent of MRF is different additives. These additives include stabilizer and surfactant. To reduce particle settling or sedimentation different stabilizers (e.g. carbon fibers, silica nanoparticles (Lim 2004), grease (Shetty 2011)) are added.

2 ISSUES WITH MAGNETORHEOLOGICAL FLUID (MRF) AND THEIR REMEDIES

2.1 Sedimentation

The settling of the particles in MRF results in poor Magnetorheological response. Although, small particle size and high viscosity carrier fluids are recommended to reduce sedimentation, low off state viscosity is preferred for better response. Different approaches are attempted to control sedimentation: (a) adding thixotropic agents (carbon fibers, silica nanoparticles (Lopez 2005), grease (Shetty 2011)) (b) adding surfactants (oleic (Lim 2004), stearic acid) (c) adding magnetic nanoparticles (Yang 2009) (d) the use of viscoplastic media as continuous phase (Carlson 2002) (e) water-in-oil emulsions as continuous phase. (Lopez 2005), (Olabi 2007) The most popular stabilizers are grease (Shetty 2011) and fumed silica (Ashtiani 2015), (Lopez 2005). To study the effect of additives on sedimentation rate, nano particles of fumed silica and oleic acid are added in silicon oil and mineral oil respectively. (Lopez 2005). No change in off state viscosity of different sample were observed with silica in Silicon oil. The stability of the MRF is achieved at lower percentage of silica (say 1.17 g/L). In a similar study (Lim 2004) mineral oil based MRF was synthesized with varying percentage of fused silica and reduction in sedimentation was observed. Contradictory to above, (Carlson 2002; Premalatha 2012) have observed reduction in MR effect with addition of surfactant. As discussed earlier, Ferro-Fluid (FF) can solve the issue of sedimentation albeit with lower magnitude of yield stress as compared to MRF (Olabi 2007). In an interesting study (Lopez 2006) an optimum solution was proposed by mixing micron sized carbonyl powder in magnetite ferro-fluids. It is observed that addition of magnetite do not cause sudden yielding of MRF although the yielding behavior was non-Newtonian. Yang et al. (Yang 2009) observed very slow settling rate of ferro-fluid based MRF (FMRF). The deformation of FMRF is also smooth and continuous in the transition from pre to post yield region while brittle collapse was observed in few case. With addition of ferro-fluid in MRF, a clear reduction in yield stress was observed by Yang 2009. Carlson (Carlson 2002) observed increase in off state viscosity by factor 3 with addition of 1% nano sized ferrite particle in his study.

2.2 Effect of off state viscosity

Off state viscosity does not scale with particle concentration at different shear rate. Most of the MRF fluids exhibit shear thinning but at higher shear rate, rate of shear thinning also decreases. If particle concentration is increased to achieve higher yield stress then off state viscosity of MRF is also increased. (Olabi 2007) However Jolly et al. (Jolly 1999) observed that viscosity in absence of field is not scaled with particle concentration but off state viscosity is function of composition and chemistry of carrier fluid. If more viscous carrier fluid is used to control particle settling then off state viscosity increases and this makes MR system operation difficult in normal conditions (i.e in absence of field). In devices like dampers and brakes settled particles can be redisposed easily by normal motion of moving components as long as MRF does not settle in to hard sediment. MRF which forms soft sediment is sufficient for such applications. Attempt to make MRF absolutely stable may reduce performance of device. These devices are operating with very high frequency which causes shear rate of approximately 1 to 4 104 sec -1. MRFs which are producing soft sediments are easily redispersible at such a high speed (Carlson 2002), (Jolly 1999).

2.3. In Use Thickening (IUT) of MRF

Carlson has discussed more critical issue than yield strength and settling in MRF. It is easy to redisperse soft sediments in high shear rate applications like dampers and brakes but later in practical application they found more serious problems. One of such problem is "In Use Thickening" (IUT). When MRF, with low off state viscosity, subjected to such high shear rate over a long period of time, the fluid will thicken. This change in viscosity is believed to be due to fracture of micron sized carbonyl particles in to nano sized particle due to inter particle stresses. This increase in viscosity causes increase in shear force by factor of 2.5 in off state. MRF progressively thicker and thicker until it becomes an unmanageable paste. (Carlson 2002).

2.4 Hard cake formation

Once particles are settled due to density difference between particles and carrier fluid, as top layered particles produces more gravitational forces on bottom layered particles. This cause tight packing of particles. Particles also have some residual magnetism in absence of external magnetic field. This will cause agglomeration. This will formulate hard cake after settling which hinders in redispersion of MRF. Proper surfactant is to be used to reduce residual magnetism in particles.(Wahid 2016) The effective use of surfactant may reduce hard cake formation and make easy redispersion in low shear rate based application. Iglesias et al. (Iglesias 2012) have dispersed micro sized iron particle in ferro-fluid to study the effect on redispersibility. Only 3% of magnetite addition gives good reduction in settling rate and easy redispersion but contradictory to this benefit they observed reduction in the rheological response.

3 SYNTHESIS AND CHARACTERIZATION OF MAGNETORHEOLOGICAL FLUID (MRF)

Magnetorheology is the branch of rheology that deals with the flow and deformation of materials under an applied magnetic field.(Sarkar 2015) The rheological properties of MRF depends on concentration and density of particles, particle size and shape distribution, properties of carrier fluid, additional additives, applied field, temperature and other factors.(Carlson 2002), (Jolly 1999), (Lim 2004) (Lopez 2005), (Saraswathamma 2015), (Sidpara 2009) The yield stress is increased with increase in size of particle or increased particle concentration. (Carlson 2002) (Jolly 1999), (Olabi 2007), (Saraswathamma 2015), (Sidpara 2009) At low to moderate magnetic flux densities, the fluid stress and magnetic flux have quadratic relation (Jolly 1999), (Yang 2009) and at low shear rate of 26 s-1, follow power law index 1.75. Beyond flux density 0.2 to 0.3, there is no change in fluid stress which reveals saturation of MRF. (Jolly 1999) One of the most effective method to produce stable MRF is to use additives (surfactant and stabilizers) in carrier fluid first and then use suspending agent. (Ashtiani 2015)

a) Additives are to be dispersed in carrier fluid and get homogeneous mixture using mechanical stirrer (Premalatha 2012) ultrasonic bath (Lopez 2005)
b) Particles are to be dispersed with the appropriate additive and carrier oil mixture in a polyethylene container
c) This mixture should be stirred by mechanical stirrer at 400 rpm (Premalatha 2012) or first by hand and then put in an ultrasonic bath (Lopez 2005)

3.1 Magnetorheometer

Magnetorheometers are available in various configurations like parallel plat, cone and plat and concentric cylinder rheometer. Cone and plat rheometer maintains constant strain rate throughout the gape but results are highly sensitive to correct positioning (gap) of cone and plate. Concentric cylinder rheometer allows testing at very high shear rate ($> 10^3$ s^{-1}) without losing sample but large sample volume is required (> 20ml). The gap distance of parallel plat rheometer can be varied. It can be used for highly viscous samples (> 10 mPa:s). The sample is placed in gap between two parallel plates. The bottom plate is fixed and top plate is motor driven. The plates are sand blasted which avoid slippage of sample during testing.(Saraswathamma 2015), (Sidpara 2009) The solenoid is kept below bottom plate which produces magnetic field perpendicular to shear flow direction of MRF. Shear rate of sample is adjusted using gap distance and upper plate speed. It is also proved that more promising results are offered by rotational disk rheometers. (Ashtiani M, 2015) When sample is loaded between two parallel plates under magnetic field then it exhibits yield stress. Due to this yield stress motor driven shaft of upper plate experiences twisting. This torque is measured using torque sensors and later converted into shear stress of MRF. (Saraswathamma 2015)

3.2 Study of sedimentation rate

The MRF looks dark gray colored but when particles settling starts then a clear boundary is formed between clear and turbid (carrier with suspended particles) part. The simplest way of sedimentation study is using kynch theory of sedimentation. According to kynch, it is assumed that initially all the particles are homogeneously suspended in carrier fluid. A clear line of separation is formed once settling starts. The sedimentation is measured by visual observation of the position changes of boundary between clear and turbid part of MRF. Generally MRF is filled in graduated cylinder and with time height of boundary separation is measured. Sedimentation ratio is simply defined as a proportion between length of clear and turbid part of MRF. (Turczyn 2008) If particle concentration is high then more is the upward fluid flow, which will keep particles suspended. (Richardson 2003) This physics works well for the non-stabilizers containing MRFs but observed to have contrast response in the presence of stabilizers. (Turczyn 2008) The adsorption isotherm of surfactant on dispersed particle can be used for sedimentation study. First using Beer law, optical absorbance of surfactant is calibrated in terms of particle concentration in suspended form. Then with time optical adsorption of the MRF is to be performed. Every time adsorption density of solution at some fixed concentration is to be calculated and obtained results are to be converted in form of particle suspension using Beer calibration curve. (Lopez 2005) This methodology could give more accuracy and precision compared to simple visual observation based kynch theory.

3.3 Study of redispersion

It is impossible to get completely stable MRF. This issue can be acceptable if sediment is soft and gets redispersed quickly with less shear rate. Lord Motion Master RD-1005 damper requires two to three strokes to redisperse MRF once damper has not been operated for several months. (Carlson 2002) For redispersibility measurement the MRF is to be filled in glass cylinder and once it gets settled, a standard needle is penetrated to measure penetration force with depth. Penetration force is measured at different time interval. Higher the penetration force, more stiff the settled layer and difficult to redisperse again. (Iglesias 2012).

4 APPLICATIONS OF MRF

The rheological nature of MRF converts the passive system into semi active or active system. The working range of the system becomes dynamic with the use of MRF. Most common applications Magnetorheological Dampers (MR Damper) are in automobile suspension, washing machine, prosthetic knee, gun recoiling protection pad, seismic dampers, cable stayed bridge (Dongting Lake) etc. (Carlson 2002), (Gadekar 2017) Magnetorheological fluids also find their application in brakes. Magnetorheological brakes consist stator (Hollow circular casing), rotor (Non-magnetic Rotating disc) and current coils on periphery of rotating disc. When current is flowing through current carrying coils then it will generate magnetic field lines and suspended magnetic particles in MRF will align themselves. This alignment will raise the viscosity and provide braking torque. (Kumbhar 2015) Processing of materials like optical glass and sapphire is difficult. However causing flow of MRF along with abrasive particles through rotating electromagnetic wheel and brittle material gives good surface finish. This has made MRF applicable in superfinishing processes. (Sidpara 2009).

5 CONCLUSION

The performance of MRF based system highly depends upon quality of MRF which have been used in system. There is nothing like ideal MRF for every system. If we compare performance requirement of automobile damper and earthquake dampers, then they are totally different. Yield stress is not only dependent on magnetic particles but it is also influenced by type of carrier fluid. It is evident from the literature that, the clear picture of effect of various parameters on the synthesis and characterization of MRF is not available. In reference (Lopez 2005) no significant reduction in magnetic property of MRF with presence of fused silica is observed. However, contradictor of Lopez et. al. (Lopez 2005), Calson et. al. (Carlson 2002) and Premalatha et al. (Premalatha 2012) have observed reduction in MR effect. Another issue, the use of only 1% nano magnetite particles as continuous phase reduces sedimentation, (Lopez 2006) but it also increases off state viscosity by factor 3 (Carlson 2002) and such MRF behaves as Non-Newtonian in absence of magnetic field as well. (Ashtiani M, 2015) Similarly, the sedimentation reducing methods also reduces rheological effects of MRF. Hence field for selecting proper additives is still open for stable and redispersible MRF.

REFERENCES

Ashtiani M, Hashemabadi SH, and Ghaffari A. A review on the magnetorheo-logical fluid preparation and stabilization. Journal of magnetism and Magnetic Materials, 374:716–730, 2015.

Carlson J D. What makes a good MR fluid? Journal of intelligent material systems and structures, 13 (7-8):431–435, 2002.

Gadekar P, Kanthaleand VS and Khaire ND. Magnetorheological fluid and its applications. International Journal of Current Engineering and Technologyl, (2017).

Iglesias GR, Lopez-Lopez MT, Duran JDG, Gonzalez-Caballero F, and Delgado AV. Dynamic characterization of extremely bidisperse magnetorheological Fluids. Journal of colloid and interface science, 377(1):153–159, 2012.

Jolly M R, Bender J W, and Carlson J D. Properties and applications of commercial magnetorheological fluids. Journal of intelligent material systems and structures, 10(1):5–13, 1999.

Kumbhar B K, Patil S R, and Sawant S M. Synthesis and characterization of magnetorheological (MR) fluids for MR brake application. Engineering Science and Technology, an International Journal, 18 (3):432–438, 2015.

Lim S T, Cho M S, Jang I B, and Choi H J. Magnetorheological characterization of carbonyl iron based suspension stabilized by fumed silica. Journal of magnetism and magnetic materials, 282:170-173, 2004.

Lopez-Lopez MT, Vicente J D, Caballero F G, and Duran JDG. Stability of magnetizable colloidal suspensions by addition of oleic acid and silica nanoparticles. Colloids and Surfaces A: Physicochemical and Engineering Aspects, 264(1-3):75–81, 2005.

Lopez MT, Kuzhir P, Lacis S, Bossis G, Caballero FG, and Duran JDG. Magnetorheology for suspensions of solid particles dispersed in ferrofluids. Journal of Physics: Condensed Matter, 18(38):S2803, 2006.

Olabi AG and Grunwald A. Design and application of magnetorheological fluid. Materials & design, 28 (10):2658–2664, 2007.

Premalatha SE, Chokkalingam R, and Mahendran M. Magneto mechanical properties of iron based MR fluids. American journal of polymer science, 2(4):50–55, 2012.

Richardson CJF. Chemical Engineering Particle technology and separation processes, volume 2. Elsevier, 5 edition, 2003.

Saraswathamma K, Jha Sunil, and Rao P V. Rheological characterization of MR polishing fluid used for silicon polishing in bemrf process. Materials and Manufacturing Processes, 30(5):661–668, 2015.

Sarkar C and Hirani H. Effect of particle size on shear stress of magnetorheological fluids. Smart Science, 3(2):65–73, 2015.

Shetty BG and Prasad P.S.S. Rheological properties of a honge oil based magnetorheological fluid used as carrier liquid. Defence Science Journal, 61(6):583–589, 2011.

Sidpara A, Das M, and Jain VK. Rheological characterization of magnetorheological finishing fluid. Materials and Manufacturing Processes, 24(12):1467–1478, 2009.

Turczyn R and Kciuk M. Preparation and study of model magnetorheological fluids. Journal of Achievements in Materials and Manufacturing Engineering, 27(2):131–134, 2008.

Wahid SA, Ismail I, Aid S, and Rahim MSA. Magneto-rheological defects and failures: A review. In IOP Conference Series: Materials Science and Engineer-ing, volume 114, page 012101. IOP Publishing, 2016.

Yang Y, Li L, and Chen G. Static yield stress of ferro-fluid based magnetorheological fluids. Rheologica acta, 48(4):457–466, 2009.

Technologies for Sustainable Development – Mahajan, Patel & Sharma (eds)
© 2020 Taylor & Francis Group, London, ISBN 978-0-367-33737-7

FE simulation of metal spinning process using LS-DYNA software

K.V. Sheth, B.A. Modi & A.M. Gohil
Mechanical Engineering Department, School of Engineering under Institute of Technology, Nirma University, Ahmedabad, India

ABSTRACT: Metal spinning is one of the ancient metal forming processes. Today, it has gained popularity due to its immense advantages over other metal forming processes. There are several complexities associated with the spin forming process. Different metal requires different forces and different procedures to generate the desired shape of the product. The uniformity in the thickness is also an important aspect to consider in manufacturing with the spinning process. The present work aims to perform the FE simulation of the metal spinning process using explicit dynamics solver LS-DYNA. To study the metal spin forming process resultant force and stress obtained through FE simulation are used as response variables.

1 INTRODUCTION

Metal spinning is treated as the new edge process with several benefits over other metal forming processes like deep drawing, blanking, collar forming, etc. It is basically classified as the tensile and compressive forming type manufacturing process. Various types of shapes are produced with metal spinning process. Generally, metal spinning is used to produce symmetrical shapes like a cone, bowl, dish, etc. Shell and tube spinning process is used with spinning (Tschaetsch, 2007), when the wall thickness is intended. The metal spinning process connects the craftsman and modern machine tool of present and futures. Automation in spinning is preferred for the mass production of the identical symmetric part. Conventional spin forming (manual spin forming) is applied for the intricate architectural and decorative products. Manual spin forming creates the smooth finishing product which has good mechanical strength and fine surface finish. Spinnability of the metal is an important parameter and is defined as the ability of sheet metal to undergo deformation by spinning without the wrinkles in the flange and no fractures on the wall.

Several researchers investigated the various parameters and complexities associated with metal spin forming process. During spin forming process radial tensile stresses and tangential compressive stresses act upon the blank due to dynamic force conditions. So the resultant force acting upon the blank is important parameter to consider for successful spin forming of parts. Semiatin et al. (2006) describes the basic terminology and essential values of spin forming methods. Moreover, types of spin forming processes and their application is also described. Liu (2007) performed the FE modelling of conventional spin forming process. Music et al. (2010) worked with different parameters to identify the important parameters affecting spin forming process. Marghmaleki et al. (2011) done the FE simulation using ABAQUAS software by taking different shapes of mandrel and records the thermomechanical effect due to shape variation using FE simulation approach. Watson et al.(2015) investigates the metal spin forming process by taking six key parameters and found that the roller feed per pass and feed rate affects the wrinkle formation. Sangkharat and Dechjarern (2017) perform the FE simulation by takin SPCC aluminum blank and validate it by experiments performed by Taguchi method. Härtel et al. (2017) shows the different approach for non-circular spinning process and observed the thinning with the use of numerical simulation and found that with tight roller feed pass and tight tool angle gives increment in thinning of the blank. Xia et al.(2014) reviews the spin forming process and associated data from which they found the resultant force acting upon the conical mandrel and location of its intensity. They also found out the effect of roller nose radius change and stress distribution. Ahmed et al.(2015) performed the experimental investigation of the deep spinning process for single pass spinning operation and found that with the increase of the clearance between mandrel and chuck, the Limiting Spinning Ratio(LSR) would increase drastically.

In this paper, there is FE simulation of metal spin forming process using LS-DYNA software. Relationship is established between resultant force and thickness of the sheet metal blank using linear regression analysis which can be used to identify the force required to spin the sheet metal.

2 FE SIMULATION

The simulation requires the step by step procedure which includes functions, behaviors and the main characteristics of spin forming process to be idealized without losing the important information and approximating the actual condition prevailing in the process. The simulation process represents the operating condition of the products with time domain relation. For the said purpose, pre-processing and post-processing is carried out in LS-Pre Post 4.1 and the solver used for the process is LS-DYNA. Mandrel is idealized as conical frustum as per the cone geometry dimension to be produced. Sheet metal blank is represented as circular plate, tool as sphere and tail stock end as cylinder (Figure 1a). All the parts are discretized using BT-shell type element which is widely accepted for FE simulation of metal forming processes. Tool, tail stock and mandrel have been defined using RIGID material. Sheet metal blank is defined as deformable body and properties are defined using MAT_POWER_LAW_PLASTI-CITY keyword. Motion is given to the tool using BOUNDARY_PRESCRIBED_MO-TION_RIGID keyword. Using BOUNDARY_SPC keyword sheet metal blank is constrained. Contact between mandrel & sheet metal blank and tool & sheet metal blank have been established using CONTACT_FORMING_ONE_WAY_SURFACE_TO_SURFACE keyword. Simulation is carried out in single pass forming giving motion to the tool using DEFINE_CURVE keyword. Initial velocity to the tool is imparted using INITIAL_VELOCI-TY_GENERATION keyword.

The other parameters and material properties are taken as: mandrel rotational speed = 280 rpm, federate 0.2 mm/rev, and no lubrication condition. Tool tip is considered as the sphere of 15 mm and tool angle is set to be 45 degrees from rotational axis. The blank

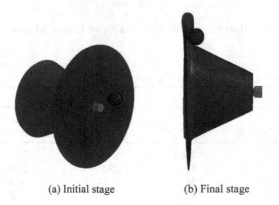

(a) Initial stage   (b) Final stage

Figure 1.   Various stages seen during spin forming simulation.

thickness variation is taken as 0.9 mm thick to 1.3 mm. The mandrel, backplate and tool have been given the material properties of stainless steel and the blank materials and their properties are as per the Table 1.

For analyzing the results, force is considered as a parameter of interest. Figure 2 gives the time vs. resultant force curve obtained from the FE analysis of Al 1060 Aluminum Alloy of 1 mm thickness. Force curve is showing initial upward trend and then reaching to the maximum as the contact is developed gradually. Once the contact is established force curve stabilizes. Maximum force obtained in the analysis is used for comparison and is summarized in Table 2. In most of the cases, the maximum resultant force is obtained when there is an initial contact between blank and the tool with

Table 1.   Material properties of various materials of aluminum alloys.

Material	Modulus Of Elasticity (GPa)	Poisson's Ratio (μ)	Strength Coefficient(K) (MPa)	Strain Hardening Exponent(n)	Initial Yield Stress (MPa)	Ultimate Tensile Strength (MPa)
Al 1060	70	0.33	613.8016	0.5637	123.20	162.8
Al 1100 - O	68.9	0.333	159.5374	0.0676	92	110
Annealed Aluminum	70	0.3	113.10	0.230	27.1	46.2
SPCC Aluminum	200	0.29	454.4201	0.1128	210	400
Hard Aluminum	70	0.33	132.50	0.0260	107.20	112.7

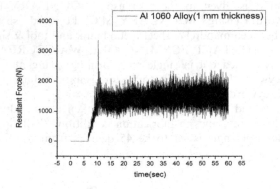

Figure 2.   Resultant force vs time graph with Al 1060 alloy of 1 mm thickness.

Table 2.   Resultant force values in different aluminum alloy materials.

Thickness(mm)	Resultant Force(N)				
	Al 1060	Al 1100	Annealed Aluminum	Hard Aluminum	SPCC Aluminum
0.9	2805	2860	2377	2590	4863
1	3361	3368	2824	3140	5926
1.1	3725	3826	3264	3605	6775
1.2	4298	4327	3786	4229	7813
1.3	4886	4905	4308	4819	9137

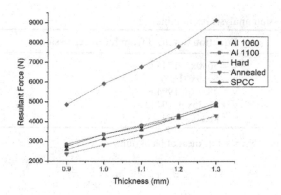

Figure 3.   Resultant force graph.

the back support of backplate. For the case of the AA 1060 material with blank thickness 1 mm, the maximum resultant force occurred at 10.4 seconds. Figure 3 gives the variation in the maximum force with change in material and thickness. As the thickness of the material changes, maximum force increases as more material is required to be deformed by the tool. Also moving from Annealed aluminium to SPCC aluminium force increases which can be attributed to higher ultimate tensile strength.

3   REGRESSION ANALYSIS

To find out the relation between thickness and the resultant force, regression analysis is carried out. For the said purpose to obtain the best fit polynomial and linear regression analysis is performed on the force results as shown in Figure 4.

From the case of polynomial regression analysis as shown in Figure 4 a, the Pearson's R value obtained as 0.9928 while in linear regression the Pearson's R value is 0.9967, hence, it is better to get results with linear regression. For different material of aluminum Alloy the relationship between thickness and resultant force are shown in the Table 3 Established relationship can be used to predict the forces involved in the metal spin forming process and to design the tool and other components of fixture and machine.

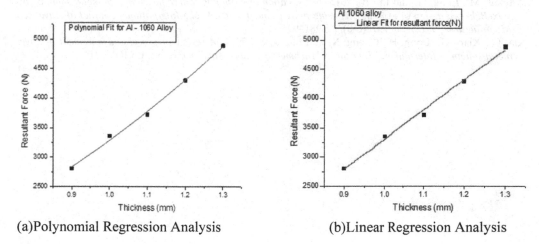

(a)Polynomial Regression Analysis   (b)Linear Regression Analysis

Figure 4.   Polynomial and linear regression analysis.

315

Table 3. Linear regression analysis result table.

Material	Equation obtained from linear regression	Pearson's R- Value
Al 1060	y = 5030x – 1745	0.9967
Al 1100 - O	y = 5049x -1696	0.999
Annealed Aluminum	y = 4824x – 1994.6	0.99908
SPCC Aluminum	y = -4575x + 10430	0.99719
Hard Aluminum	y = 4824x – 1994.6	0.9990

Where, y = Resultant Force(N), x = thickness of blank(mm).

4 CONCLUSION

In the present work, the FE simulation setup for single pass spin forming operation is settled out in the LS-DYNA software to study the resultant forces acting upon the blank during the metal spin forming operation. From the regression analysis it is established that the resultant forces follows the linear relationship with the blank thickness. Thickness and ultimate tensile strength have marked effect on the resultant force.

REFERENCES

Ahmed, K.I., Gadala, M.S. and El-Sebaie, M.G., 2015. *Deep spinning of sheet metals. International Journal of Machine Tools and Manufacture*, 97, pp.72–85.
Härtel, S. and Laue, R., 2016. *An optimization approach in non-circular spinning. Journal of Materials Processing Technology*, 229, pp.417–430.
Liu, C.H., 2007. *Dynamic finite element modeling for the conventional spinning process. Journal of the Chinese institute of engineers*, 30(5),pp.911–916.
Marghmaleki, I.S., Beni, Y.T., Noghrehabadi, A.R., Kazemi, A.S. and Abadyan, M., 2011. *Finite element simulation of thermomechanical spinning process. Procedia Engineering*, 10, pp.3769–3774.
Music, O., Allwood, J.M. and Kawai, K., 2010. *A review of the mechanics of metal spinning. Journal of materials processing technology*, 210(1),pp.3–23.
Sangkharat, T. and Dechjarern, S., 2017. *Spinning Process Design Using Finite Element Analysis and Taguchi Method. Procedia Engineering*, 207, pp.1713–1718.
Semiatin, S.L., Marquard, E., Lampman, H., Karcher, C. and Musgrove, B., 2006. *ASM handbook, vol 14B: Metalworking: sheet forming. ASM International*, Materials Park, Ohio, USA, pp.656–669.
Tschaetsch, H., 2007. *Metal forming practise: Processes-machines-tools. Springer Science & Business Media.*
Watson, M., Long, H. and Lu, B., 2015. *Investigation of wrinkling failure mechanics in metal spinning by Box-Behnken design of experiments using finite element method. The International Journal of Advanced Manufacturing Technology*, 78(5-8), pp.981–995.
Xia, Q., Xiao, G., Long, H., Cheng, X. and Sheng, X., 2014. *A review of process advancement of novel metal spinning. International Journal of Machine Tools and Manufacture*, 85, pp.100–121.

Technologies for Sustainable Development – Mahajan, Patel & Sharma (eds)
© 2020 Taylor & Francis Group, London, ISBN 978-0-367-33737-7

Strain based and stress based fracture forming line for single point incremental forming

Y.R. Prajapati, A.M. Gohil, B.A. Modi & K.V. Sheth
Mechanical Engineering Department, School of Engineering under Institute of Technology, Nirma University, Ahmedabad, India

ABSTRACT: Strain based Forming Limit Diagrams (FLD) are used to predict a failure in sheet metal forming. The main problem for strain based forming limit diagrams is its inability to predict failures accurately in various sheet forming operations where there is a change in strain path and mode of deformation. Whereas the stress based forming limit diagram can reveal formability limit of material more precisely. Strain based criterion is valid only when the strain path is linear throughout the whole deformation process. Stress based forming limit criterion is not dependant on strain path. So stress based criterion can be used for predicting forming limits in forming operations where strain ratio does not remain constant. There are several situations in forming operations where stress based FLD can be applied to predict the failure correctly like incremental forming, hydroforming, multistage forming, deep drawing of pockets etc. Different yield criterion is used to transform strain based FLD into stress based FLD. For incremental sheet forming process conventional Forming Limit Curve(FLC) is not used as it is prepared considering necking limit whereas ISF process is limited by fracture. Hence, in literature it is termed as Fracture Forming Line(FFL). In this paper, first strain based Fracture Forming Line(FFL) is established by generating different shapes by incremental sheet forming process. From strain based FFL, stress based FFL is prepared using Hills quadratic criterion.

1 INTRODUCTION

A Forming Limit Diagram (FLD) is used in metal forming processes to predict forming behaviour of a sheet metal. The Forming Limit Diagram attempts to expose a graphical description of material failure tests. To prepare the FLD, different geometrical shapes which are capable to generate different straining conditions, are prepared using single point incremental forming process. In order to determine that a given region has failed, circle grid is marked on a sheet metal blank before deformation. After deformation through the process, deformed circle which takes the shape of ellipse are measured. Measured points are marked on FLD. From a number of such points forming limit curve(FLC) is generated, which divides the FLD into safe and failure region. Generated FLC can be used to predict the failure of the newly designed component for a given material and thickness of the blank. Stress based forming limit diagram is obtained from the results of strain based Forming Limit Curve. When strain-based FLD is converted into the forming limit stress diagram, the resulting stress based diagram is negligibly affected by changes to the various strain paths. This phenomena makes stress based Forming Limit Stress Diagram (FLSDs) an alternative to strain based FLDs for the prediction of failure.

Incremental sheet metal forming is a forming process where a sheet is formed into the final workpiece of desired shape by a sequence of small incremental deformations (Jeswiet et al., 2005b). Process has gained the attention of researchers because it does not require dedicated tooling and hence suitable even for single piece production. However, the process takes longer processing time. Looking at the research potential, the process has been studied by various researchers from the point of thinning band (Hussain and Gao, 2008), force (Filice et al., 2006), power consumption (Ingarao et al.,2012), accuracy (Duflou et al.,2007), and surface finish (Attansio et al.,2006).

Jeswiet and Young (2005a) observed very high strain of the order of 300% which is the main advantage of the incremental sheet forming process. The reason behind this high formability is that in incremental sheet forming process necking is suppressed and failure is limited by fracture (Malhotra et al., 2012). Conventional processes are limited by necking and to predict the failure Forming Limit Diagram and Forming Limit Curve is used. However, the same cannot be used for incremental sheet forming process. To predict failure, Fracture Forming Line is used for incremental sheet forming process (Martins et al., 2008).

Prajeesh et al.(2014) have developed new correlation among the parameters of formability which include normal anisotropy, strain hardening exponent & thickness and plane strain intercept in the FLD. Strain based FLD is having the limitation that it predict failure well when strain path is linear. Hence, Forming Limit Stress Diagram(FLSD) is also proposed which overcomes this limitation(Stoughton, 2000) (Stoughton, 2001) (Stoughton and Zhu, 2004) and predicted failure even when the strain path is non-linear. Safdarian(2016) & Safari et al.(2011) have established that Ductile Fracture Criterion is more useful in generating FLD than the conventional Forming Limit Diagram approach. Panich et al.(2011) obtained Forming Limit Stress Diagram (FLSD) of steel sheet of SPCE270 grade using FE method.

In this paper strain based FLC is obtained by preparing three shapes namely straight groove, circular groove and cone. These shapes produce different straining conditions which are studied through circle grid analysis. First, strain based FLD is prepared which is used to prepare Stress based FLC using Hills quadratic criterion.

2 STRAIN BASED FLD

Experimental setup used for single point incremental forming is shown in Figure 1. Experiments are carried out on VMC Jyoti CNC PX-10 having a Siemens controller. Aluminium AA-1060 sheet is used in experiment and the properties of which is shown in Table 1. For the SPIF process, sheet blank is required to be prepared as per ASTM E2218 02 standard. Blank

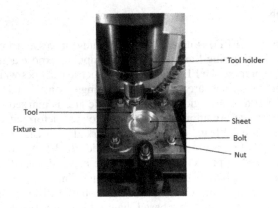

Figure 1. Experimental set-up.

Table 1. Material properties of various materials of aluminum alloys.

Material	Modulus Of Elasticity (GPa)	Poisson's Ratio (μ)	Strength Coefficient(K) (MPa)	Strain Hardening Exponent (n)	Initial Yield Stress (MPa)	Ultimate Tensile Strength (MPa)
Al 1060	70	0.33	613.8016	0.5637	123.20	162.8

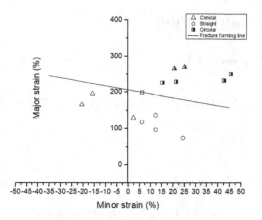

Figure 2.   Strain distribution of straight groove, circular groove and cone.

of 100 X 100 mm^2 size are cut from the sheet. After that, a grid pattern having 1 mm diameter circles are marked by laser engraving technique. Three different shapes namely straight groove, circular groove & cone have been prepared by incremental sheet forming to obtain strain based FLD.

Circle grid measurement is carried out using portable digital microscope. Figure 2 shows the strain distribution obtained for all the three shapes.

A fracture forming line is drawn from a safe point obtained from circular groove test which is dividing the safe points from the fractured points. A line is traced from this point slope of which is obtained using the equation established by Martins et al.[9].

Major strain= slope * minor strain + constants

$$\text{Where, Slope} = -[5(r_{tool}/t) - 2]/[3(r_{tool}/t) + 6]$$
$$= -[5(5/1) - 2]/[3(5/1) + 6]$$
$$= -1.095$$

(r_{tool} = Radius of tool and t = thickness of sheet metal blank)

• Constant = Major strain − (slope * minor strain)

$$= 199.9 - (-1.095 * 6.1)$$
$$= 206.57$$

This strain based Fracture Forming Line (FFL) is used to generate stress based Fracture Forming Line using equations mentioned in Section 3.

3   STRESS BASED FLD

Stress based FLD obtained by using Hills quadratic criterion. By solving hills criterion following equations are established [17].

• Stress ratio, $\alpha = \{(1 + R)\beta + R\}/\{1 + R + R\beta\}$ (1)

• Anisotropy ratio, $R = (R_0 + 2 \times R_{45} + R_{90})/4$ (2)

• Major strain, $\varepsilon_1 = \log(1 + e_1/100)$ (3)

- Minor strain, $\varepsilon_2 = \log(1 + e_2/100)$ (4)

Where, Strain ratio, $\beta = \varepsilon_2/\varepsilon_1$ (5)

Effective strain, $\bar{\varepsilon} = ((1 + R))/\sqrt{(1 + 2R)} * \sqrt{(e_1{}^{\wedge}2 + e_2{}^{\wedge}2 + (2R/(1 + R))\varepsilon_1\varepsilon_2)}$ (6)

- Effective stress, $(\bar{\sigma}) = K \times \bar{\varepsilon}^n$ (7)

- Ratio of effective stress $(\bar{\sigma})$ and major principal stress (σ_1) is given by,

$$\xi = \sqrt{(1 + \alpha^{\wedge}2 - [(2R/(1 + R)) \times \alpha])}$$ (8)

So Major stress is, $\sigma_1 = (\bar{\sigma})/(\xi)$ (9)

Minor stress $\sigma_2 = \alpha \times \text{major stress} = \alpha \times (\bar{\sigma})/(\xi)$ (10)

The plot of major stress (σ_1) & minor stress σ_2 for each point of strain based FLD gives forming limit stress diagram (FLSD). The points on fracture forming line is converted into stress using hills quadratic criterion as shown in Figure 3. This stress based fracture forming line can be used to predict the failure of a component along with FE simulation.

4 FE SIMULATION OF STRAIGHT GROOVE SHAPE

FE simulation of straight groove shape have been performed using LS-DYNA solver. From the simulation fracture point is identified (Figure 4a). Strain points have been obtained at three different Z-depth: 1. Preceding Z-depth 2. Z-depth at the point of failure and 3. Succeeding Z-depth. Strain data is converted in to the stress points and is represented in Figure 4b for all the cases. It is observed that stress points in Case 1 is completely below the stress based Fracture Forming Line and hence no failure at this Z-depth. In Case 2 stress points have moved close to stress based Fracture Forming Line, and one of the points is crossing the

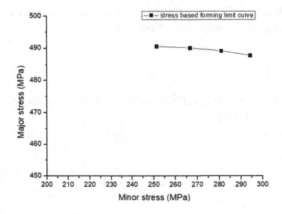

Figure 3.   Stress based forming limit curve.

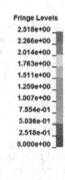

Fringe Levels
2.518e+00
2.266e+00
2.014e+00
1.763e+00
1.511e+00
1.259e+00
1.007e+00
7.554e-01
5.036e-01
2.518e-01
0.000e+00

Figure 4a.   FE simulation of straight groove.

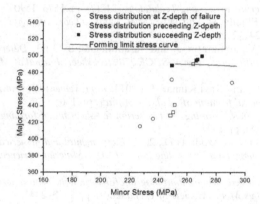

Figure 4b.   Stress distribution of straight groove.

fracture forming line indicating failure of the component. In Case 3 more points are crossing the fracture forming line confirming the failure of the component.

5   CONCLUSION

- By using Hills quadratic criteria strain based FFL is converted to stress based FFL for AA1060 material of 1 mm thickness.
- Equation of strain based fracture forming line is, Major strain = [(-1.095) * Minor strain] + 206.57.
- Straight groove have been simulated using LD-DYNA software and results of stress are mapped on stress based fracture forming diagram. Developed stress based fracture forming line predicts the failure of straight groove and can be used to predict the failure of other shapes also.

REFERENCES

Attanasio, A., Ceretti, E. and Giardini, C., 2006. *Optimization of tool path in two points incremental forming. Journal of Materials Processing Technology*, 177(1-3), pp.409–412.

Duflou, J.R., Callebaut, B., Verbert, J. and De Baerdemaeker, H., 2007. *Laser assisted incremental forming: formability and accuracy improvement. CIRP annals*, 56(1), pp.273–276.

Filice, L., Ambrogio, G. and Micari, F., 2006. *On-line control of single point incremental forming operations through punch force monitoring. CIRP annals*, 55(1), pp.245–248.

Gajjar, Nilesh, 2012. *Investigation of formability of alluminum alloy in hydroforming square cup*, IITD, Dissertation.

Hussain, G., Hayat, N. and Gao, L., 2008. *An experimental study on the effect of thinning band on the sheet formability in negative incremental forming. International Journal of Machine Tools and Manufacture*, 48(10), pp.1170–1178.

Ingarao, G., Ambrogio, G., Gagliardi, F. and Di Lorenzo, R., 2012. *A sustainability point of view on sheet metal forming operations: material wasting and energy consumption in incremental forming and stamping processes. Journal of Cleaner Production*, 29, pp.255–268.

Jeswiet, J. and Young, D., 2005a. *Forming limit diagrams for single-point incremental forming of aluminium sheet. Proceedings of the Institution of Mechanical Engineers, Part B: Journal of Engineering Manufacture*, 219(4), pp.359–364.

Jeswiet, J., Micari, F., Hirt, G., Bramley, A., Duflou, J. and Allwood, J., 2005b. *Asymmetric single point incremental forming of sheet metal. CIRP annals*, 54(2), pp.88–114.

Malhotra, R., Xue, L., Belytschko, T. and Cao, J., 2012. *Mechanics of fracture in single point incremental forming. Journal of Materials Processing Technology*, 212(7), pp.1573–1590.

Martins, P.A.F., Bay, N., Skjødt, M. and Silva, M.B., 2008. *Theory of single point incremental forming. CIRP annals*, 57(1), pp.247–252.

Panich, S., Uthaisangsuk, V., Juntaratin, J. and Suranuntchai, S., 2011. *Determination of forming limit stress diagram for formability prediction of SPCE 270 steel sheet. Journal of Metals, Materials and Minerals*, 21(1), pp.19–27.

Prajeesh, B., Raja Satish, D. and Ravi Kumar, D., 2014. *Improvement in accuracy of failure prediction in FE simulations of sheet metal forming of Al alloys. Aimtdr*, pp.1–6.

Stoughton, T.B., 2000. *A general forming limit criterion for sheet metal forming. International Journal of Mechanical Sciences*, 42(1), pp.1–27.

Safari, M., Hosseinipour, S.J. and Azodi, H.D., 2011. *Experimental and numerical analysis of forming limit diagram (FLD) and forming limit stress diagram (FLSD). Materials Sciences and Applications*, 2(05), p.496.

Safdarian, R., 2016. *Stress based forming limit diagram for formability characterization of 6061 aluminum. Transactions of Nonferrous Metals Society of China*, 26(9), pp.2433–2441.

Stoughton, T.B., 2001. *Stress-based forming limits in sheet-metal forming. Journal of Engineering Materials and Technology*, 123(4), pp.417–422.

Stoughton, T.B. and Zhu, X., 2004. *Review of theoretical models of the strain-based FLD and their relevance to the stress-based FLD. International Journal of plasticity*, 20(8-9), pp.1463–1486.

Technologies for Sustainable Development – Mahajan, Patel & Sharma (eds)
© 2020 Taylor & Francis Group, London, ISBN 978-0-367-33737-7

Experimental investigation of nucleate boiling over a smooth and micro-grooved cylindrical surface

Balkrushna A. Shah
Assistant Professor, Mechanical Engineering Department, Institute of Technology, Nirma University, Gujarat, India

Mohommed Naseem Quanungo
PG Student, Mechanical Engineering Department, Institute of Technology, Nirma University, Gujarat, India

Mohommed Naseem Quanungo, Manish Sumera & Shahil Chaudhary
UH Student, Mechanical Engineering Department, Institute of Technology, Nirma University, Gujarat, India

Vikas Lakhera
Professor, Mechanical Engineering Department, Institute of Technology, Nirma University, Gujarat, India

ABSTRACT: This paper presents the experimental results of nucleate pool boiling heat transfer for R141b over a smooth and micro-grooved cylindrical copper surface. The experiments have been conducted at a saturation temperature of 32°C and for heat fluxes ranging from 20 to 150 kW/m^2. A copper tube of outside diameter 20 mm and length of 50 mm was used as the test surface. The experimental heat transfer coefficients for pool boiling over smooth cylindrical copper tube have been compared with the empirical correlations and comply within ± 15%. The boiling heat transfer coefficient on the micro-grooved surface was found to be 63.4% to 82.3% higher in comparison to that of plain surface for the wall superheat range from 6 to 8.5 K.

1 INTRODUCTION

The development of smaller and energy-efficient thermal devices remains the objective of the industry worldwide. Boiling heat transfer has been studied since long by many researchers as it is an effective mechanism that can remove a huge quantity of thermal energy from a surface due to a very high heat transfer coefficient which results in maintaining lower surface temperature. The boiling heat transfer has been used in several engineering and industrial fields that need high heat flux removals, such as electronic chip cooling, flooded evaporators, kettle reboilers, and marine ship power generation. Presently boiling heat transfer performance over various types of microstructures and nanostructures has gained interest and further work is carried out by many researchers.

The effects of the surface roughness on pool boiling of R-134a and R-123 over various material have been studied by Jabardo et al (Ribatski & Jabardo, 2003). The study found that very rough cylindrical surface results in reduced thermal performance at higher heat flux and gives better performance at lower heat flux in comparison to smooth surface.

Rocha et al (Rocha, Kannengieser, Cardoso, & Passos, 2013) measured the temperature distribution along the circumference of micro-finned tube surface. The study compared the experimental results for the R134a pool boiling over the micro-finned tube with the available Gorenflo-Kenning equation. Chen (Chen, 2013) investigated the pool boiling of R123 on the enhanced cylindrical tube and used hot water flow inside the tube instead of heater for heating the refrigerant which gets boiled outside the tube. The study found that HTC on the enhanced tube was about 6 to 10 times higher than that on the plain tube within the heat flux range from 12 to 62 kW/m^2.

Mehta and Kandlikar (Mehta & Kandlikar, 2013a) carried out the investigations over cylindrical microchannel copper tubes with water. The study involved the effect of geometric parameters like fin depth, find width and pitch on the pool boiling of open cylindrical microchannel copper tubes. The work also studied the effect of orientation on heat transfer augmentation with cylindrical microchannel test surfaces. According to the study, the bubble dynamics along with the rewetting phenomenon results in overall heat transfer augmentation along with area enhancement in microchannel test surfaces. The work found that microchannel surfaces give higher heat transfer enhancement in a horizontal orientation as compared to vertical orientation. The study also found that the rectangular cross-section groove channel remains flooded with water and hence it gives better performance than the V-groove microchannel surface (Mehta & Kandlikar, 2013b).

2 EXPERIMENTAL APPARATUS AND TEST PROCEDURE

2.1 *Apparatus details*

An experimental setup was designed and developed to investigate the nucleate pool boiling heat transfer over cylindrical surfaces in a horizontal orientation at atmospheric pressure. The wall of the boiling vessel is made of glass, allowing visualization of the boiling process. To ensure minimum heat loss from the glass during the boiling process, thermal insulation is applied over the glass surface before initiating the experimentation process.

Figure 1 shows the schematic of the test setup used for pool boiling investigations. The boiling vessel is made from borosilicate glass and its dimensions were 400 mm (height) x 90 mm (diameter). The setup was made leak-proof by using O-ring along with vacuum grease in between steel plates, used at the top and bottom, to make the boiling vessel air-tight. The test surface assembly houses the cartridge heater of 50 mm (length) x 9.5 mm (diameter) with a rated heat output of 500 W. Heating is provided to the copper tube through this heater. The inner and outer diameter of the copper tube test section was 10 mm and 20 mm respectively. To ensure negligible thermal contact resistance, A thermal conducting paste (Anabond 652c)

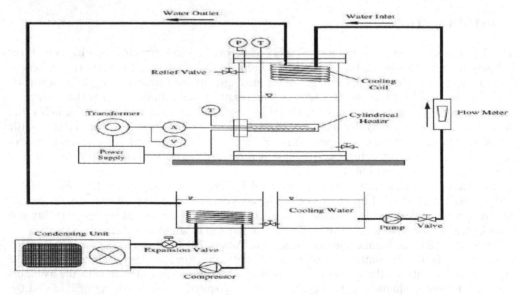

Figure 1.   A schematic of the test setup.

was applied on the heater surface. The test section assembly was held in the glass chamber, using a Teflon block.

2.2 Experimental procedure

The experimental procedure followed for the refrigerant was as follows:

1. The assembly of the Teflon block, copper tube, heater, and thermocouple was prepared. This assembly was then fitted in the boiling vessel.
2. A vacuum pump was turned on for few hours to evacuate the boiling vessel thoroughly and to ensure that there was no leak and the vacuum level of 600 mm of Hg was kept on hold for about 30 minutes.
3. The refrigerant was then charged in the boiling vessel and maintained up to 30 mm higher than the top of the heat transfer tube.
4. The liquid refrigerant was then heated gradually by increasing the heat input at an interval of 10 W using a dimmer. The copper tube wall temperature and liquid refrigerant temperatures are measured simultaneously and recorded using a data logger.
5. The copper tube was then changed and the above procedure (steps 1-4) were repeated after experimenting twice on each copper tube to ensure the repeatability of experimental results.
6. The pressure during the boiling process was maintained by circulating chilled water in the condenser coil using a chiller.

3 DATA REDUCTION

For each power input, the boiling heat transfer coefficient was calculated by the following equation (1)

$$h = \frac{Q}{A(T_W - T_{sat})} \tag{1}$$

where Q is the electrical heating power, T_{sat} is the refrigerant saturation temperature corresponding to the value of pressure in the boiling vessel, A is the surface area and T_w is the tube wall temperature which is obtained by taking a mean of the wall temperatures at the top, bottom and two sides of the copper tube as:

$$T_W = \frac{T_{top} + 2T_{side} + T_{bottom}}{4} \tag{2}$$

The power input Q, to the refrigerant from outside of the tube, is measured using a wattmeter.

A data acquisition system (Agilent make) was used for reading and recording the data. Four pre-calibrated T-type thermocouples circumferentially placed and were used to read the surface temperature of the test section and two RTD sensors were used to read the fluid and vapor temperatures during the experimentation process. The steady-state condition was considered for recording the readings when the temperature variation was less than 0.1 °C for a time duration of minimum 2 minutes. The power applied across the heater was indicated by a digital power indicator and a dimmer was used to vary the power while conducting the experiments.

An uncertainty analysis was carried out using a "propagation of errors" method. The uncertainties observed were: ±1 °C for temperature using T-type thermocouples and RTD sensors, ±5% for the applied wattage and 0.68% for the heat transfer coefficient.

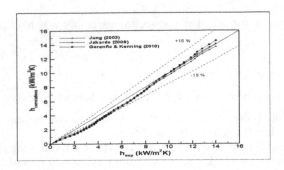

Figure 2. A comparison of experimental results for a plain tube with correlations.

4 EXPERIMENTAL RESULTS

4.1 Comparison with correlations

Figure 2 shows the validation of the experimental results using a plain surface with widely used empirical correlations in pool boiling heat transfer. The experimental heat transfer coefficient was compared with the heat transfer coefficient predicted by Jung(Jung, Song, Ahn, & Kim, 2003), Jabardo(Ribatski & Jabardo, 2003) and Gorenflo & Kenning(Gorenflo & Kenning, 2010) correlations. All the results obtained for the plain surface fall within the deviation of ±15%.

4.2 Effect of circumferential micro-grooves

The experiments were conducted on a micro-grooved surface (as per geometry given in Table 1) with R141b as a working fluid at atmospheric pressure. The photographs of smooth and micro-grooved surfaces are shown in Figure 3. To analyze the heat transfer performance of the micro-grooved surface, the experiment results were compared with that of the smooth surface. Figure 4 (a) and (b) show the comparison of the micro-grooved surface with that of a smooth surface in terms of heat flux and heat transfer coefficient respectively.

4.3 Effect of pressure

The experiments were also conducted on the micro-grooved surface using R141b at different pressures. At a given heat input of 350 W, the maximum heat transfer coefficient was observed to be 7.73 kW/m^{2}C at a wall superheat of 11°C and at atmospheric pressure. While at the pressure of 0.7 kg/cm^2 gauge, the maximum heat transfer coefficient was 10 kW/m^{2}C at a wall superheat of 10.2°C for the same surface as shown in Figure 5. It can be attributed to

Table 1. Geometric parameters of the micro-grooved cylindrical surface.

Depth (mm)	Fin spacing (mm)	Fin thickness (mm)	Pitch	Area Enhancement Factor
0.17	0.46	0.55	1.01	1.10

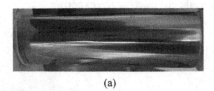

(a)

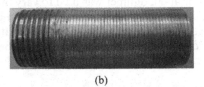

(b)

Figure 3. Photographs of (a) smooth surface (b) micro-grooved surface.

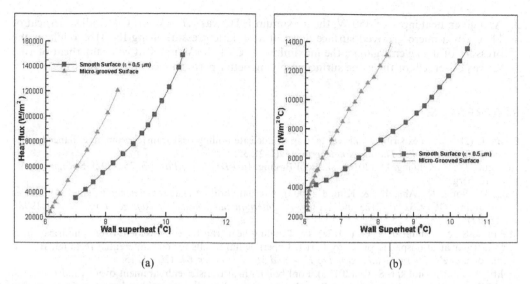

Figure 4. A comparison of heat transfer performance for various surfaces (a) Heat flux vs Wall super-heat (b) Heat transfer coefficient vs Wall superheat.

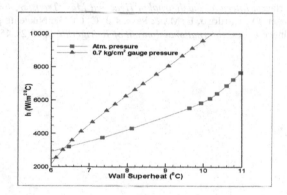

Figure 5. A comparison of the heat transfer coefficient at different pressures for the micro-grooved surface.

the decrease in surface tension with an increase in pressure. Therefore, the energy required to initiate the nucleation of bubbles on the heater surface decreases, thus giving an enhanced performance at 0.7 kg/cm^2 gauge pressure as compared to atmospheric pressure.

5 CONCLUSIONS

The following conclusions can be drawn from the experimental investigation of nucleate pool boiling of R141b on smooth and micro-grooved copper tubes:

1. The validation of the experimental results was done for the smooth surface of average roughness, ε= 0.5μm and the results obtained were in good agreement with Jung, Jabardo, and, Gorenflo & Kenning correlations within ±15%.
2. The heat transfer performance of the micro-grooved surface enhanced by 63.4% - 82.3% as compared to that for a smooth surface.

3. At a given heat input of 350 W, the maximum HTC was 7.73 kW/m^2°C at wall superheat of 11 °C for a micro-grooved surface at an atmospheric pressure using R-141b. While at the pressure of 0.7 kg/cm^2 gauge, the maximum HTC of 10 kW/m^2 K at wall superheat of 10.2 °C was observed for the same surface implying better performance at lower pressure.

REFERENCES

Chen, T. (2013). An experimental investigation of nucleate boiling heat transfer from an enhanced cylindrical surface. *Applied Thermal Engineering, 59*(1–2), 355–361.

Gorenflo, D., & Kenning, D. (2010). H2 Pool Boiling. In *VDI Heat Atlas* (pp. 757–792). Springer Berlin Heidelberg.

Jung, D., Song, K., Ahn, K., & Kim, J. (2003). Nucleate boiling heat transfer coefficients of mixtures containing HCFC32,HFC125,and HFC134a. *International Journal of Refrigeration, 26* (July1998), 764–771.

Mehta, J. S., & Kandlikar, S. G. (2013a). Pool boiling heat transfer enhancement over cylindrical tubes with water at atmospheric pressure, Part I: Experimental results for circumferential rectangular open microchannels. *International Journal of Heat and Mass Transfer, 64*, 1205–1215.

Mehta, J. S., & Kandlikar, S. G. (2013b). Pool boiling heat transfer enhancement over cylindrical tubes with water at atmospheric pressure, Part II: Experimental results and bubble dynamics for circumferential V-groove and axial rectangular open microchannels. *International Journal of Heat and Mass Transfer, 64*, 1216–1225.

Ribatski, G., & Jabardo, J. M. S. (2003). Experimental study of nucleate boiling of halocarbon refrigerants on cylindrical surfaces. *International Journal of Heat and Mass Transfer, 46*(23), 4439–4451.

Rocha, S. P., Kannengieser, O., Cardoso, E. M., & Passos, J. C. (2013). Nucleate pool boiling of R-134a on plain and micro-finned tubes. *International Journal of Refrigeration, 36*(2), 456–464.

Technologies for Sustainable Development – Mahajan, Patel & Sharma (eds)
© 2020 Taylor & Francis Group, London, ISBN 978-0-367-33737-7

Design of trolley attaching mechanism for unit load automated guided vehicle system

Jatin Panchal, Bimal Kumar Mawandiya, Dhaval Shah & Shashikant Joshi
Mechanical Engineering Department, Institute of Technology, Nirma University, Ahmedabad, India

ABSTRACT: In smart industries, automated guided vehicle (AGV) has become a top-notch automated material handling technology of a flexible manufacturing system which is the more adaptable, accurate, capable, efficient, versatile and secured medium of node interconnection. AGV's are the unmanned vehicle which maneuvers on pre-defined guide path in a warehouse to transport goods among different workstation. This vehicle automatically performs loading, route selection, vehicle speed, and unloading. AGV system is integrated mainly by a central controller, paths an electronic communication mechanism and routing strategies. In the present paper, the design for the attaching mechanism of trolleys for transportation of unit load has been proposed. The gripping mechanism drags the trolleys from their pick-up center to respective delivery centers. The manual experimental model of the gripping mechanism has been fabricated for validation purposes.

Keywords: AGV, trolley, attaching mechanism, gripper

1 INTRODUCTION

The process layout of an Automatic Guided Vehicle (AGV) has already been created for the continuous flow process type cement industry. Applications where manpower is needed, the Automatic Material Handling System (AMHS) is utilized in certain stages to carry out the handling of material for transportation of load. It is not a viable option to facilitate manpower with a forklift for stacking/unstacking of loads at initial stages of material handling. Therefore, AGV with the attachment of trolley; though the vehicle just needs to have a correct solution regarding the navigation of material. There are many ready-made trolleys attaching mechanisms are available in the market such as platform lifting, scissor lift, etc. These kind of engaging mechanisms completely lift the platform of the trolley from the ground, but in proposed concept caster wheels are attached permanently to the legs of trolleys to facilitate maneuverability through manpower in certain initial stages of material handling, instead of lifting the entire platform. The simple attaching gripper mechanism can drag the trolleys as a labourer.

Vosniakos and Mamalis (1990) discussed an issue related to designing and installing a system of automated guided vehicles in the vehicle manufacturing system. Two major unit load vehicles are one where all electronic components, devices, sensors, etc. are located and a work platform where palatalized work-piece is carried. They also suggested vehicle maneuverability configuration of the driving wheel located centrally with a differential steering wheel mechanism with two casters supported at front and rear. They described the flow of AGV either to have unidirectional or bi-directional flow pattern. Gerarod et al. (2015) developed an intelligent heavy autonomous vehicle system called RobuTAINeR used for carrying 20 to 40 feet containers inside the confided space of maritime terminals. They designed the maneuverability of the robot by placing four omni-drives at front left, front right, rear left, and rear right respectively. Wang et al. (2007) has shown a new way of vehicle maneuverability by developing omnidirectional wheel. This type of drive focussed on the cooperative interface between human-robot for material handling and transportation. The author also focused on safety issues with the automated guided vehicle where sensors are not accurate to avoid a collision in complicated paths. Banerjee and Bhattacharya (2005) discussed a versatile approach to optimize a manufacturing system using the Taguchi method. Taguchi proposed a three-step method for framing product and process design which are system design, parameter design, and tolerance design. The author employed the Taguchi method to uncover the optimal combination of factor level that maximizes the average utilization of the automated guided vehicle system (AGVs). Ueno et al. (2009) focused on harmonic omnidirectional mobile robot whereas Sanhoury et al. (2011) studied trajectory tracking of non-holonomic differential wheel mobile robot steering system.

2 CONSTRUCTION OF GRIPPER

The gripping mechanism for engaging/disengaging of AGV with the trolley has been designed by keeping few important characteristics in mind such as mechanism which should provide alleviation, self-engagement, mechanical advantage, be simple in construction and actuation and more economical to build.

After setting some benchmarks, a special shape has been selected to facilitate the fair amount of mechanical advantage to the gripper. For that, the gripping arm of gripper provided a certain slope (amount of inclination), which after attachment the dragging force of the vehicle imparts a self-engagement characteristic between trolley and vehicle. The degree of the slope has been adopted from the "concept of the wedge". Wedging action amplifies the dragging force of vehicle (x) on the engaging surface of the gripping arm (y). Mathematically,

$$y = \frac{x}{2 * (\sin \theta + \mu \cos \theta)} \tag{1}$$

According to the calculations, it can be observed that as on the decreasing degree of the slope for the manufacturing of gripper, more and more mechanical advantages can be achieved. For example, suppose 100 N dragging force is applied through the vehicle on the gripping mechanism (gripping mechanism has a slope of 2°), approximately 2.3 times amplified force will act on the surface of the gripping arm.

The minimum angle of 5° has been chosen for the manufacturing of the gripping arm made up of teak wood. As per the manufacturing point of view, less than 5° angle is not a practical option. The gripper mechanism has been designed by coercing a single side acting linear actuator to actuate both sides. To make this concept fruitful following design factor has to be taken into consideration:

Table 1. Slope selection for maximum wedging action.

Sr. No.	$\theta(°)$	x(in N)	y(in N)
1.	2	100	230.05
2.	3	100	221.33
3.	4	100	213.30
4.	**5**	**100**	**205.90**
5.	6	100	199.05
6.	7	100	192.70

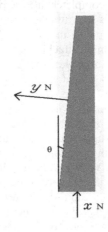

Figure 1. Arm of gripper.

1) Guiding pin should have a degree of freedom restricted to one. Only x-direction motion should be allowed.
2) The linear actuator should not be grounded.

The Figure 1, unveiled the concept behind the gripping mechanism of the gripper. The gripper is the assembly of the following listed items:-

a) Two gripping arms (given special shape to have a mechanical advantage).
b) Two back supporting plate (on which gripper arms are mounted).
c) Four guiding pins (welded on a rear supported plate of gripping arm) which facilitate to have a single degree of motion.
d) Linear actuator bolted on both sides of the rear support plate.
e) A base plate on which the entire structure is assembled.

3 WORKING OF GRIPPER

To make the system feasible and workable an additional load transfer mechanism will be needed to transfer the actuating force from the actual actuating side of the linear actuator to the non-actuating side of the linear actuator. The load transfer can be easily down with the help of tension spring. The tension spring always tends to bring the linear actuator to its middle position whenever expansion or retraction occurs.

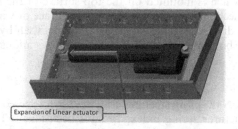

Figure 2.   Expansion by means of linear actuator.

Figure 3.   Opening of gripping mechanism.   Figure 4.   Closing of gripper mechanism.

The contraction of spring always tries to bring both the end points to the equilibrium position as per the construction of the gripper and contributes to the system to actuate from both the sides whenever expansion and retraction occurs. The mechanism can be done with the help of single spring mounted on the center just above or below the linear actuator mounted, but it may be possible that a guiding pin may experience more resistance on one side of the spring and another experiences less resistance. This will be due to the pin joint between the rear support plate and a linear actuator. To counteract this problem, two springs were chosen whose equivalent stiffness will be equal to the single spring.

Figure 5. Unbalanced forces counter acted by springs.

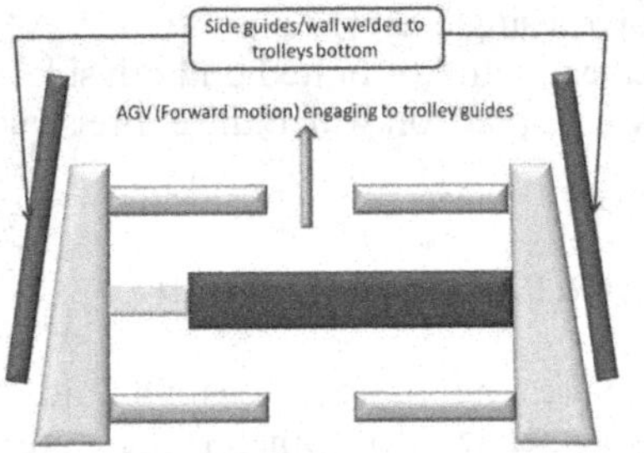

Figure 6. Engaging and disengaging of gripper mechanism with trolley.

The attachment of trolley is only possible when the trolley has a similar shape on its bottom surface. So, the gripper which is mounted on the top surface of the AGV can be open. These openings of the gripper mechanism will engage with the surface or wall of the extruded part on the trolley. At the time of disengagement, the gripper contract itself by retracting linear actuator and passes away from the bottom of the trolley leaving trolley stationary to its destination.

3.1 Spring calculations

To manufacture spring, EN42 spring steel material has been selected and according to the recommendation of manufacturer spring index, 10 have been taken. The modulus of elasticity, modulus of rigidity, Poisson's ratio, ultimate tensile strength and shear stress have been taken as 203 MPa, 79.84 GPa, 0.272, 927 MPa and 535 MPa respectively.

1) Spring index C = 10,
2) Shear stress factor $K_s = 1 + \frac{1}{2C} = 1.005$
3) Shear stress: $\tau = \frac{8WD}{\pi d^3} * K_s \therefore d = 1.5$ mm ≈ 2 mm
   Therefore, $D = 15$ mm ≈ 20 mm
4) Deflection: $\delta = \frac{8WD^3 n}{Gd^4} \therefore n = 35.64 ≈ 36$
5) Free length of spring: It is the length of the spring in unloaded condition

$$L_f = L_s + \delta_{max} + 0.15\delta_{max} = 72\,mm$$

4 EXPERIMENTAL MODEL

In the previous section, the gripper mechanism and working of a gripper have been thoroughly discussed. It seems simple in design but is unmanageable after manufacturing and also, it is more costly if tried to manufacture the same prototype as per construction of the gripper (which has flat rectangular cross-section guiding pins) mentioned above. The hike in manufacturing cost is due to the accuracy of machining required to maintain throughout the guideways that include tighter tolerances, clearance, and utilization of shim spacer to improve the surface condition of the base plate.

To avoid such an erratic task, an optimized designed has been remade. That makes the gripping mechanism more reliable, functional and more economical. Such change can be done by altering the rectangular cross-section to the conventional cylindrical cross-section of guiding pins. Further, the advantage of the cylindrical cross-section is improved mechanical properties in all direction as compared to the rectangular cross-section guiding pins. The opening and closing experimental model illustrate as follows.

Figure 7.   Gripping mechanism in closed position.   Figure 8.   Gripping mechanism in opened position.

5   CONCLUSIONS

The trolleys have been adopted to complete the transaction of material through the automated guided vehicle (AGV) after taking trial and evaluating several mechanisms. The proposed concept of a trolley will eliminate the requirement of complicated systems (such as AS/RS system, separate pickup and drop station, etc.) of the traditional automatic transportation system which not only acquire the huge space but also limits the capacity. To have movement of trolleys an attaching mechanism has been proposed which can be mounted on the AGV's top portion to get engaged and disengaged by a linear actuator. When the gripping mechanism would be an open phase the AGV will go beneath the trolley and get engaged by extruded surfaces of the trolley and at the time of retraction of gripper during delivery of trolley AGV will get disengaged and pass beneath the wheeled trolleys.

REFERENCES

Banerjee, D. and Bhattacharya, R. 2005. Robust Design of an FMS and Performance Evalution of AGVS. *International Conference on Mechanical Engineering*, Dhaka, Bangladesh.
Gerarod, A., Rui, L. and Rochdi, M. 2015. Multi-domain Model of Steering System for An Omnidirectional Mobile Robot. *Conference on Robotics and Biomimetics*, Zhuhai, China.
Sanhoury, I.M.H., Amin, S.M.H. and Hussain, A.R. 2011. Trajectory tracking of steering system Mobile Robot. *International conference on Mechatronics*. Kaula Lampur, Malasiya.
Ueno, Y., Ohno, T. and Terashima, K. 2009. The development of driving system with differential drive steering systemfor omni-directional mobile robot. *International conference on Mechatronics and Automation*. Changchm, China.
Vosniakos, G.C. and Mamalis, A.G. 1990. Automated Guided Vehicle System Design for FMS Application. *International Journal of Machine Tools Manufacture*. vol. 30, no. 1, pp. 85–97.
Wang, Z.D., Fukaya, K., Hirata, Y. and Kosuge, K. 2007. Control Passive Mobile Robots for Object Transportation. *International Conference on Robotics and Automation*, Roma, Italy.

Technologies for Sustainable Development – Mahajan, Patel & Sharma (eds)
© 2020 Taylor & Francis Group, London, ISBN 978-0-367-33737-7

Process capability analysis of worm gearbox manufacturing

Umang Prajapati, Bimal Kumar Mawandiya, Kaushik Patel & Dhaval Shah
Mechanical Engineering Department, Institute of Technology, Nirma University, Ahmedabad, Gujarat, India

ABSTRACT: This research work is carried out in an industrial gearbox manufacturing plant. It is observed that the variations in the critical dimensions of the output in the worm gearbox manufacturing process unit are very high. The worm shaft and worm wheel are the two main components of the worm gearbox. It is observed that the variations were not natural variations and hence, these variations were due to the assignable causes. This leads to the process out of statistical control and as a result produces non-conforming products, which in turn, increases the cost of the worm gearbox. In the present paper, detailed process analysis is carried out for the worm gearbox manufacturing unit to bring the manufacturing process of worm shaft and worm wheel under statistical control by eliminating assignable causes, rejection, rework, cycle time and scrap.

Keywords: Process capability analysis, statistical quality control, C_p, C_{pk}

1 INTRODUCTION

Process capability is the long-term performance level of the process after it has been brought under statistical control. Process capability study is a scientific and systematic procedure that uses control charts to detect and eliminate the unnatural causes of variation until a state of statistical control is reached. When the study will be completed, it will identify the natural variability of the process. The worm shaft and worm wheel are two main components of the worm gearbox shown in Figure 1.

Wu et al. (2009) described the normal distribution curve for the process spreads relative to specification interval for the normal distribution. A mean (average) that is always in the center between the upper specification limit and lower specification limit of the process. Sibalija and Vidosav (2009) implemented a case study on statistical process control and process capability analysis. The six sigma study for an observed manufacturing system was conducted according to DMAIC methodology. To analyze defects in the manufacturing process, Pareto analysis was performed. Kumar et al. (2012) performed machine parameter optimization of the grinding machine, which gives an idea of methodology for increasing the quality of a product manufactured in a machine, using various quality control tools such as cause and effect diagram, why-why analysis and statistical control charts. The authors also discussed different stages of action of the corresponding process capability index (C_{pk}) value. Favi et al. (2017) used a 4M approach: man, machine, method and material to optimize the manual assembly process. This work confirmed the usefulness of the approach to optimizing manual assembly. Rasoul et al. (2018) analyzed tolerances in manufacturing by process capability analysis. This research work provides an idea of process capability analysis calculations that required important assumptions such as statistically controlled processes, independent parameters, etc. The author also explained the performance of the process expressed in nonconforming parts per million (PPM).

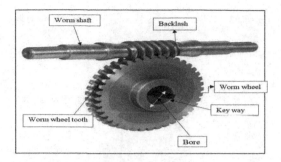

Figure 1.   Worm shaft and worm wheel.

2   STATISTICAL AND ROOT CAUSE ANALYSIS

2.1   *Pareto analysis*

Pareto analysis is a strategy for recording and breaking down data identified with an issue and with related causes, which clearly shows the most significant perspectives to be recognized.

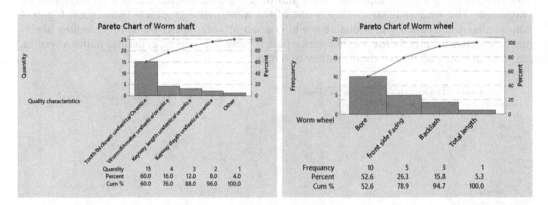

Figure 2.   Pareto analysis of (a) Worm shaft and (b) Worm wheel.

2.2   *Fishbone diagram*

The cause and effect diagram is a graphical tool used to identify possible causes of defect. The cause and effect diagram is also known as the fishbone diagram. Using fishbone diagram major causes were found in the worm shaft and wheel manufacturing process.

2.3   *Process performance*

Here, final tooth operation is considered as shaft tooth profile grinding. To identify variation in the tooth profile grinding process of the worm shaft, measured a long-term process capability with continuous observation of the worm shaft R 20:1 which have 3.4798 mm tooth thickness size and 50-micron tolerance. The process performance test is prepared by analyzing the several collected observations using Minitab software. The process performance of the worm shaft and worm wheel is shown in Figure 3.

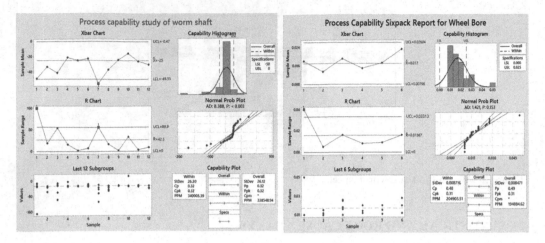

Figure 3. Process performance for (a) Worm shaft and (b) Wheel bore.

2.4 Normality test

No two product units produced by a process are identical, some variation is unavoidable. Statistics is the science of analyzing data and drawing conclusions, taking into account data variation. The stem-and-leaf display is one of the most useful graphical techniques. Figure 4 shows that the frequency of value at 6 mm is a major variation close to the grinding allowance. Whereas Figure 5 shows a time series plot of observation which, defines major variation to shaft grinding allowance that is nearer to 6 mm.

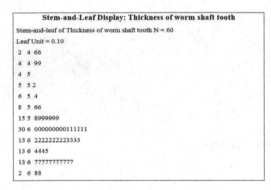

Figure 4. Stem-and-leaf display.

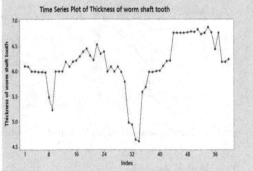

Figure 5. Time series plot.

From the process performance test of the worm shaft and worm wheel, it is shown that the process does not follow the distribution curve. Both processes, namely the profile grinding process of the worm shaft tooth and the turning process of worm wheel bore, are having variation. These two processes work in the presence of assignable causes, which is the reason for defects.

3 PROCESS IMPROVEMENT

From the Pareto charts and fishbone diagrams, the causes for the defect are discovered. For quality improvement, under the quality maintenance pillar of Total Productive Maintenance (TPM) various tools like 4M condition, why-why analysis, 6W2H action plan are used in this research work. This helps to improve the quality of the system. After revisited 4M audit, assignable causes were reduced and the process was in statistical control.

4 PROCESS CAPABILITY ANALYSIS AND RESULTS

The process capability test shows how well a process output when a variation occurred in the process. Several measurements were taken over one month and results generated from the process capability test.

4.1 Process capability of worm shaft manufacturing process

Figure 6 Shows the process capability of the ultimate worm shaft process that is worm shaft tooth profile grinding wherever major variation is founded. Conjointly results show that the C_p value is higher than 1.33 that shows that process is at an adequate level than the previously checked result. The normal distribution curve follows the specification limits throughout the process which indicates the process is under statistical control.

4.2 Process capability of worm wheel manufacturing process

Figure 7 shows that process capability (C_p) and process capability index (C_{pk}) of turning process of 'SNU worm wheel 2.25" R 30:1' that indicates (C_p) value 1.36. The results also show that process capability is improved as compared to the recent long-term turning process performance test.

4.3 Process capability index and defect rate

The process capability index (C_{pk}) for different processes were determined using nonconforming parts per million (ppm) for a process corresponding to values if the process mean were at the target. The value for the process capability index of the profile grinding operation is found as 1.71. This means that worm shaft profile grinding produces 0.3402 PPM shaft. So, productivity of worm shaft tooth grinding process is improved. Similarly, the worm wheel process, process capability (C_{pk}) value is 1.25 which indicates worm wheel production from

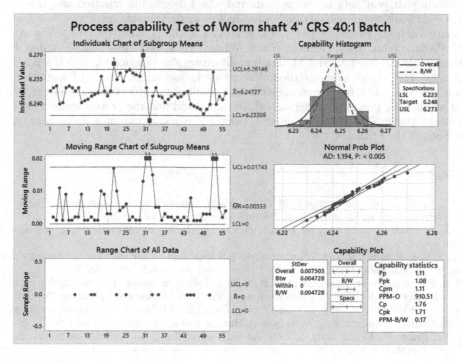

Figure 6. Distribution of measurements.

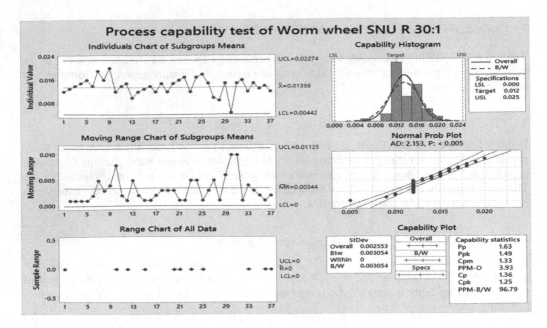

Figure 7. Distribution of observation.

turning operation that its process defect is 318.2914 PPM. It shows the improvement in the productivity of worm wheel turning process.

5 CONCLUSIONS

Interpreting past year data of worm shaft and wheel defect, the rejection and rework were higher which leads to unnecessary higher production cost. Applying quality control tools such as Pareto Analysis, Fishbone Diagrams, Process Capability Analysis, Why-Why Analysis, 6W2H (What, Who, Why, Where, When, Whom, How and How much) Action Plan, 4 M (Man, Machine, Method and Material) audit on regular practice leads to less possibilities of rejection, rework, process variation. Variations in various processes of worm shaft and worm wheel production have been reduced. As a result of the process capability test, it can be concluded that the process is under statistical control and that the process follows the normal distribution curve. Therefore, assignable causes, rework, cycle time have been reduced throughout all the processes.

REFERENCES

Favi, C., Germani, M. and Marconi, M. 2017. A 4M approach for a comprehensive analysis and improvement of manual assembly lines. *Procedia Manufacturing*. Vol. 11. pp. 1510–1518.

Kapadia, M. 2007. Measuring your process capability. *Quality & Productivity Journal*. Symphony Technologies.

Kumar, J. P. Indhirajith, B. and Thiruppathi, K. 2012. Quality improvement of a grinding operation using process capability studies. *Int. J. Mech. Eng. Robot. Res*. vol. 1. No. 1. pp. 76–90.

Mahshid, R., Mansourvar, Z. and Hansen, H.N. 2018. Tolerance analysis in manufacturing using process capability ratio with measurement uncertainty. *Precision Engineering*. Vol. 52. pp. 201–210.

Sibalija, T. and Vidosav, M. 2009. SPC and Process Capability Analysis–Case Study. Proceedings of the 5th International Conference on Total Quality Management & Excellence. 37. pp. 1–6.

Wu, C.W., Pearn, W. L. and Kotz, S. 2009. An overview of theory and practice on process capability indices for quality assurance. *International journal of production economics*. Vol. 117.2. pp. 338–359.

Technologies for Sustainable Development – Mahajan, Patel & Sharma (eds)
© *2020 Taylor & Francis Group, London, ISBN 978-0-367-33737-7*

Design and development of quality system for gear case casting

Vaibhav Suthar, Bimal Kumar Mawandiya, Kaushik Patel & Dhaval Shah
Mechanical Engineering Department, Institute of Technology, Nirma University, Ahmedabad, Gujarat, India

ABSTRACT: In today's competitive world, it is mandatory to sustain the quality of the product that the organization delivers to the customer. The present paper addresses the issue of finding the root cause of the casting defect with the help of a cause and effect diagram, quality assurance matrix and other quality control tools. The objective of the paper is to identify the defects using cause and effect diagram and other quality control tools especially casting defects. The preventive measures should be effective enough to achieve the desired quality of the product. The quality system is developed and rejection is reduced after implementing various quality control tools. This process helps to reduce the rejection and increase the sigma level of the gear case casting process.

Keywords: Fish bone diagram, cause and effect diagram, quality system, casting process

1 INTRODUCTION

Casting is one of the oldest manufacturing processes in which, the metal is melted and poured into a mould. The mould is left for the solidification, and after solidification, the casting is pulled out from the mould. The excess amount of casting is removed by machining. To ensure defect-free casting, a quality system should be implemented. The quality system incorporates the aim of operation, specification, allocating responsibility to the respective and controlling the cause of the defect. Various quality control tools like the fishbone diagram can be used for examining the root cause of the defects. Various castings defects like blow hole, porosity, inclusion, cold shut, mismatch, shrinkage, etc. can be occurred due to the variation in process parameters. Hence, one has to control the process parameters for the prevention of the defects. In this research paper, the main focus will be on to prevent the cause of defect occurrence. Subsequently, through a case study, a quality system has been discussed that helps to prevent cause of defect and maintain the quality.

Borowiecki et al. (2011) used the Pareto chart to find out which defect drives the most percentage as compared to others. It was found that the majority of the defect that occurs due to the faulty gating system design. In the process of the gating system, improvement enables proper directional solidification, the velocity of the molten metal, prevents misrun, etc. Shivappa et al. (2012) focused on the analysis of the casting defect and its remedial measure. In this paper, the casting process was observed and found that the four defects i.e. blowhole, mismatch, sand drop and oversized were repetitive at a particular location. To nullify the casting defect, the mould was cleaned properly, six locators were provided to avoid mismatch and modification in the gating system. The given remedial measure was adopted and production was carried out and the reduction in rejection of casting was observed. Jadhav and Jadhav (2013) observed that in each case, the gating system design was not responsible for the casting defect. The casting defect can be reduced by controlling alloy composition and pouring temperature. The seven quality control tools were used to reduce defects which include check sheet, Pareto chart, scatter diagram, flow chart, etc. Siekanski et al. (2003) performed the analysis of foundry defect and remedial measure for the same defect was suggested. Ishikawa diagram was used to find the risk of failure. The cause and effect diagram represent the factors responsible for the defect. The Pareto chart shows some defect which drives the majority of the percentage of the defects like displacement, slag, misrun, shrinkage, etc. Pandey and Jain (2016) used the six sigma methodologies for reducing the defect of the ingot mould in foundry. The quality tools like flow chart, why-why analysis, cause and effect diagram, etc. were used. Using the Taguchi method, the optimum level of parameters that affect the casting process was identified.

2 METHODOLOGY

A pilot study has been conducted to eliminate casting defects in the incoming casting material for gear case. Figure 1 shows the process flow in a foundry for the gear case manufacturing process. The wooden pattern has been developed as the first step in the casting process. The same pattern has been inspected to check all the required parameters. In the next subsequent stage, mould preparation and mould inspection have been performed to check the condition of the mould and gating system design. After mould inspection, the mould is closed properly and the molten metal is poured. After pouring the mould is left for solidification. As the casting solidifies, it will be pulled out from the mould and the excess material from the casting is removed by machining. The parameters due to which the defect occurrs have been investigated.

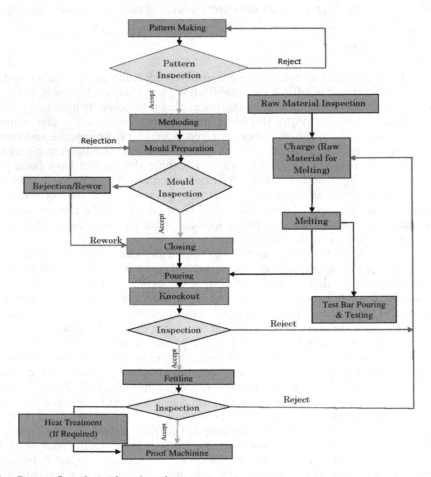

Figure 1.   Process flow chart of sand casting process.

340

3 CAUSE AND EFFECT DIAGRAM

The cause and effect diagram is a tool to display graphically the causes of any problem which is also called a fishbone diagram or Ishikawa diagram. The cause and effect diagram shows all the possible modes of causes for the defect and the factors which affect the quality.

Table 1. Cause of defects and preventive action taken for the defect.

Component	Defect	Cause	Preventive Action
Bearing Housing	Shrinkage	Earlier solidification of molten metal	Placing thermostatic sleeve so that to maintain temperature and feed the metal
Gear Case	Mismatch	Improper setting of cores	Providing locators and using proper closing pins
Gear Case	Cold Shut	Less fluidity of molten metal	Maintain proper pouring temperature
Gear Case	Sand Inclusion	Less core strength, Erosion of mould wall	Increase strength of core and high content of binder.
Gear Case	Flash	Excessive clearance between top and bottom of mould, cavity present in pattern	Ensure cavity to be filled, sealing the mould at parting line
Gear Case	Blow Hole	Low permeability of sand, High turbulence during filling	Adequate venting

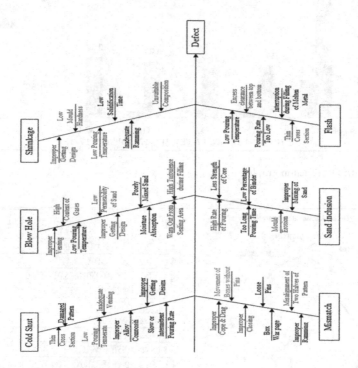

Figure 2. Fishbone diagram for gear case casting process.

The fishbone diagram has been prepared as shown in Figure 2. The observations have been made on the casting process and regarding the particular casting defect, preventive action has been taken. After finding the root cause of the defect with the help of the fishbone diagram

341

and applying corrective actions accordingly helps to prevent the cause of defect occurrence. Thus the number of rejections of gear cases have been reduced.

4 RESULTS

After implementing the corrective actions, the reduction in the defects has been observed in the casting of the gear case. It can be seen that the implementation of the quality system has improved the sigma level compared to the previous two financial years. Also, Figure 3 shows the reduction in the rejections of castings. The sigma level has been increased from 2.45 sigma to 3.06 sigma. Figure 4 shows a radar chart and the improvement in the ratings given to the vendor. This rating was given after the improvement steps have been taken. This improvement leads to produce a less defective gear case.

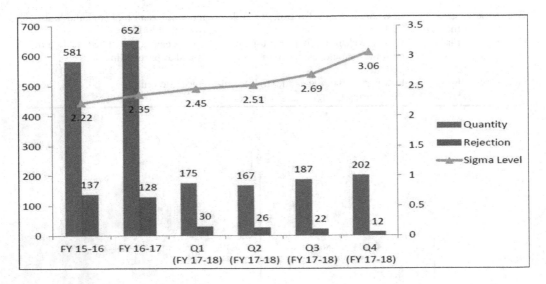

Figure 3.   Quantity vs rejection graph.

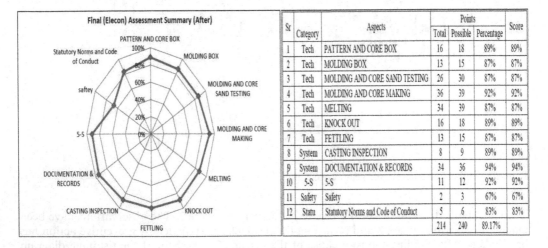

Sr	Category	Aspects	Total	Possible	Percentage	Score
1	Tech	PATTERN AND CORE BOX	16	18	89%	89%
2	Tech	MOLDING BOX	13	15	87%	87%
3	Tech	MOLDING AND CORE SAND TESTING	26	30	87%	87%
4	Tech	MOLDING AND CORE MAKING	36	39	92%	92%
5	Tech	MELTING	34	39	87%	87%
6	Tech	KNOCK OUT	16	18	89%	89%
7	Tech	FETTLING	13	15	87%	87%
8	System	CASTING INSPECTION	8	9	89%	89%
9	System	DOCUMENTATION & RECORDS	34	36	94%	94%
10	5-S	5-S	11	12	92%	92%
11	Safety	Safety	2	3	67%	67%
12	Statu	Statutory Norms and Code of Conduct	5	6	83%	83%
			214	240	89.17%	

Figure 4.   Radar chart and ratings given to vendor.

5 CONCLUSION

In this case study, the fishbone diagram has been used to reduce the number of rejection in gear case casting. A detailed fishbone diagram has been prepared and after identifying the root cause preventive action has been taken. Thus the root causes of the defects have been eliminated and hence the castings are defect-free. The implementation of the quality system shows good results. Therefore, similar type of quality system can be implemented for other range of products. The same type of quality system can also be prepared for the in-house material. This can help to improve the quality of the product.

REFERENCES

Borowiecki, B., Borowiecka, O. and Szkodzinka, E. 2011. Casting Defect Analysis by Pareto Method. *Archives of Foundry Engineering*. Vol. 11. Issue 3. pp 33–36.

Jadhav, B.R. and Jadhav, S. 2013. Investigation and Analysis of Cold Shut Casting Defect and Defect Reduction by Using 7 Quality Control Tools. *International Journal of Advance Engineering Research and Studies*. Vol. II. pp: 28–30.

Pandey, A. and Jain, K.K. 2016. Six Sigma in Manufacturing of Ingot Moulds in Foundry and Pattern Shop by Improving Sand Quality. *American Journal of Engineering Research*. Vol. 5, No. 4. pp: 209–217.

Shivappa, D.N., Rohit, and Bhattacharya, A. 2012. Analysis of Casting Defects and Identification of Remedial Measure: A Diagnostic Study. *International Journal of Engineering Inventions*. Vol. 1, No. 6. pp: 01–05.

Siekanski, K. and Borkowski, S. 2003. Analysis of Foundry Defects and Preventive Activities for Quality Improvement of Castings, *Metabk*. Vol. 42. No. 1. pp: 57–59.

Technologies for Sustainable Development – Mahajan, Patel & Sharma (eds)
© 2020 Taylor & Francis Group, London, ISBN 978-0-367-33737-7

Numerical study of resin flow in VARTM composite manufacturing process

Pratik Patel, Tushar Gajjar, Dhaval Shah, Shashikant Joshi & Bimal Kumar Mawandiya
Mechanical Engineering Department, Institute of Technology, Nirma University, Ahmedabad, India

ABSTRACT: Vacuum Assisted Resin Transfer Moulding (VARTM) process is widely used for decades because of its cost-effectiveness and easy availability in composite manufacturing. The flow of the resin plays a vital role in the VARTM process. In this research paper, the flow of resin has been simulated by taking two dimensional and three-dimensional geometry with different shapes using ANSYS Fluent software. The resin inlet at the surface and edge point has been considered as an input parameter. The void content and mould filling time have been measured as an output parameter from simulation results. The effect of resin flow inlet locations and shape of the components have been considered as a process parameters. The complete mould filling pattern and velocity of resin have been observed during simulation in each case. The flat, as well as a curved shape, have been considered for the study. The simulation results have also been verified with experimental results for each case.

Keywords: VARTM process, resin flow, mould filling time, FLUENT

1 INTRODUCTION

The composite materials are widely used in the last few decades because of their advance material properties like specific weight and density ratio. Carbon fibre reinforced polymer (CFRP) is one of the human made composite materials which is primarily used in space applications. There are many manufacturing techniques like hand layup process, autoclave process, resin transfer moulding (RTM) process, etc. used for manufacturing of composite material. Since the last few decades, the VARTM process is extensively used for manufacturing because of its cost-effectiveness with good dimensional accuracy. The selection of resin inlet depends on component size, length and thickness of the laminates which can be considered as process parameters. Since the last few decades, the large sized components made of composite material have been manufactured using the VARTM process and the majority chances of failure are because of void portion and unfilled area that inspired to study on flow measurement (Yang et al., 2014).

Many researchers performed experimental studies as well as applied the simulation approach for the VARTM. The simulation plays a vital role in manufacturing because it gives an idea about the behaviour of the material. Yenilmez et al. (2009) studied on thickness variation and concluded that thickness increases with distance because of the pressure gradient. Correia et al. (2005) developed an analytical formulation of governing equations for the flow of incompressible fluids through compacting porous media and their application to vacuum infusion of composite materials. The effect of process parameters such as inlet and outlet pressures, fibre architecture and lay-up were quantified. The analytical study performed in Liquid Injection Molding Simulation (LIMS) developed by the University of Delaware but it is not useful for the large component (Goren and Atlas, 2008; Lee et al. 2011; Correia et al. 2002). The PAM-RTM plugin with CATIA was used for resin flow and pressure gradient measurement for a large component (Andersson et al. 2003). The ANSYS fluent is one of the alternative ways for simulating a resin flow and easy to operate and modelling. The percentage error

344

with experimental and simulation results was observed within a 10 % range. In this research paper, the resin flow in the VARTM process has been simulated in ANSYS Fluent software. The VOF model has been used for tracking the flow. The location of the resin has been selected as a process parameter. The simulation has been performed in two-dimensional and three-dimensional rectangular geometry. The three-dimensional curved shaped geometry has also studied for observing a flow. The surface inlet and edge point inlet has taken in two-dimensional geometry. The remaining void area has been observed in both cases. Similarly, the three-dimensional study on the rectangular shape has been studied for observing the complete filling of the area. Moreover, the flow of resin in a curved shape has been studied for comparison with rectangular geometry.

2 MATHEMATICAL FORMULATIONS

The pressure gradient has been determined as per Darcy's law as Equation 1 in the VARTM process. The negative sign represents decreasing the pressure with distance.

$$\bar{u} = \frac{-K.\nabla p}{\mu} \tag{1}$$

Where, $\bar{u}$ is Darcy's velocity, K is compaction state permeability, μ is resin viscosity and p is fluid pressure. The thickness varies during the VARTM process with respect to changing pressure gradient. The relation has been shown in Equation 2. Where, h is the thickness of the mould cavity and t is the time.

$$\nabla.\bar{u} = -\frac{1}{h}\frac{\partial h}{\partial t} \tag{2}$$

The VARTM process works on vacuum pressure. The difference between resin pressure and compaction pressure is equal to the atmospheric pressure, so the atmospheric pressure has been determined as per Equation 3.

$$P_{atm} = P_r - P_{comp} \tag{3}$$

The volume of fluid (VOF) method is mostly used Eulerian technique for interface tracking. In this method, resin location tracked by the volume fraction F_r in a computational cell. If $F_r = 1$ means the cell is full of resin and if, $F_r = 0$ means the cell is full of the other phase is like air, etc. and if $0 < F_r < 1$ means the cell contains the interface between the two phase. In the VOF method, the velocity field solved by the single set momentum Equation 4. Where, ρ is density, g is the gravitational acceleration for the incompressible fluid the continuity can be express by using Equation 5. The interface between the two phases is tracked by solving a continuity equation for the volume fraction according to Equation 6.

$$\frac{\partial(\delta_{fluid}u)}{\partial t} + \nabla.\left(\rho_{fluid}u \times u\right) = -\nabla p + \nabla.\left[\mu_{fluid}\left(\nabla u + \nabla u^T\right)\right] + \rho_{fluid}g \tag{4}$$

$$\nabla.u = 0 \tag{5}$$

$$\frac{\partial F}{\partial t} + \nabla.(uF) = 0 \tag{6}$$

For the porous medium, the velocity has been replaced by Darcy's velocity. Neglecting the gravity term in the VOF momentum equation for the resin flow in the infusion process and the substitute Darcy's velocity. The permeability K has been determined by Equation 8,

$$i\frac{\partial(\delta_{fluid}\bar{u})}{\partial t} + \nabla\cdot\left(\rho_{fluid}\bar{u}\times\bar{u}\right) = -\nabla p + \nabla\cdot\left[\mu_{fluid}\left(\nabla\bar{u}+\nabla\bar{u}^{-T}\right)\right] + \frac{\mu_{fluid}}{K}.\bar{u} \qquad (7)$$

$$K = k\frac{(1-V_f)^3}{V_f^2} \qquad (8)$$

Where k is a Kozeny – Carman constant and V_f is the volume fraction. So the continuity equation for the VOF model becomes

$$\frac{\partial F}{\partial t} + \nabla\cdot(\bar{u}F) = 0 \qquad (9)$$

3   SIMULATION PROCEDURES

The simulation has been carried out in ANSYS Fluent software. The flow diagram for the entire procedure has been described in Figure 1. The various case studies have been considered related to the variation of rein inlet locations. The 2D and 3D flow in flat plate geometry have been observed with respect to the pressure gradient for a different type of laminate thickness. Moreover, the resin flow has also been observed in curved shaped components. The VOF model has been used to perform the simulation in software. The rectangular geometry, size 150 mm X 50 mm has been used for making a flat plate model. The frozen area has been selected as material in the geometry section. Meshing has been generated with the minimum size of the mesh as 0.25 mm. After completing geometry and meshing, the VOF model has been implemented at the time of application of boundary condition as shown in Figure 2. The resin injection vent has been modelled as a left side surface of the edge. The pressure inlet boundary conditions have been defined at the inlet injection. Similarly, the right side edge or surface of the plate has been taken as a resin outlet and the pressure at outlet boundary conditions has been defined. The remaining face or edge has been selected as a wall. There is no slip condition defined for the wall. The pressure at the inlet port has been taken as 1.01325 bar which is atmospheric pressure and the outlet pressure port has been set as 15 Pa which observed from the experimental studies.

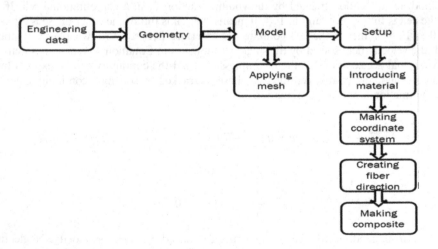

Figure 1.   Flow diagram of ANSYS ACP composite lay-up.

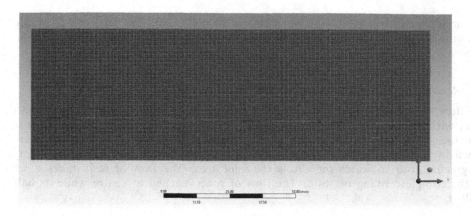

Figure 2.   The meshed model for flat plate.

4   RESULTS

4.1   *Resin flow in flat plate with two dimensional geometry*

The meshed geometry has been considered as input in Fluent. The properties of the material have been supplied according to requirements. The location of the resin inlet plays a vital role in pouring the mould cavity. So as a consequence, filling the mould is depends on the inlet location. Two different cases have been performed for observing the flow. The left side edge in the first case, while two points in the second case have been given as inlet. The flow of resin has been covered the entire portion within a short duration in the first case compared to other cases. There is no void portion that has been found in the first case while the corner portion remains empty in the second case. The inlet flow is linear with distance in case 1, while the flow is fluctuating with increasing distance in case 2. The flow of resin has been compared at distance of a 5 mm, 10 mm and 15 mm as shown in Figure 3. According to this observation,

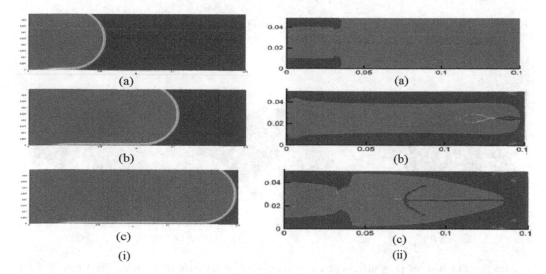

Figure 3.   The simulation flow in flat plate using Fluent at (a) 5 mm, (b) 10 mm (c) 15 mm distance for (i) edge inlet and (ii) two points inlet.

the entire edge inlet is more suitable compared to point inlet. The large boundary area has remained as the void in the second case while the entire area has been covered easily in other cases.

4.2 Resin flow in flat plate with three dimensional geometry

The flow in three-dimensional geometry gives complete information about mould filling along with the thickness of flat components. The resin must pass through each lamina in entire thick components. The complete three-dimension flow of resin has been shown in Figure 4. The red colour indicates a complete filling of the component while the green colour indicates remaining the void area. The flow of resin takes some time for the complete filling of resin. The flow of resin reached at centre while its half area is filled.

4.3 Resin flow in curved shape components

The resin flow is different in different geometry so it is very important to observe the flow for different geometry. The curved shape has been selected for the study of resin flow as shown in Figure 5. It is observed that the centre has been filled first completely, while boundary area stats filling later. The edge inlet has been given in this curved shaped component.

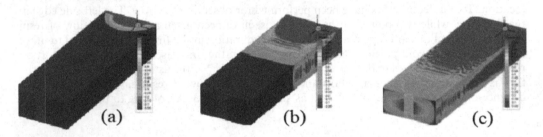

Figure 4.   Three dimensional flow of resin at distance of (a) 3 mm (b) 8 mm and (c) 15 mm.

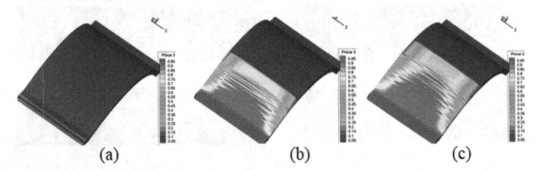

Figure 5.   The flow of resin in curved shape component at distance of (a) inlet, (b) 4 mm, and (c) 10 mm.

5 CONCLUSIONS

The flow of resin in the VARTM process has been simulated in ANSYS Fluent software. The rectangular geometry has been studied with the entire edge and two points as an inlet in two-dimensional geometry. The large void area has been observed in two point inlet case while edge inlet case covers each portion of the geometry. Moreover, the rectangular shape has been studied in three-dimensional geometry and it has been observed that the complete fill area starts from the centre and after that, it expands to the boundary. The entire edge inlet takes less time compared to the point inlet. Moreover, when the two inlets take less time compared to the single inlet. The flow of resin in curved shape has also been studied with three-dimensional geometry.

REFERENCES

Yang, B., Jin, T., Bi, F. and Li, J. 2014. Modeling the resin flow and numerical simulation of the filling stage for vacuum-assisted resin infusion process. *Journal of Reinforced Plastics and Composites.* Vol. 33. No. 21. pp. 1976–1992.

Yenilmez, B., Senan, M. and Sozer, E.M. 2009. Variation of part thickness and compaction pressure in vacuum infusion process. *Composites Science and Technology.* Vol. 69 No. 11-12. pp. 1710–1719.

Correia, N.C., Robitaille, F., Long, A.C., Rudd, C.D., Simacek, P. and Advani, S.G. 2005. Analysis of the vacuum infusion moulding process: I. Analytical formulation. *Composites Part A: Applied Science and Manufacturing.* Vol. 36. No. 12. pp. 1645–1656.

Goren, A. and Atlas, C. 2008. Manufacturing of polymer matrix composites using vacuum assisted resin infusion molding. *Archives of materials Science and Engineering.* Vol. 34. No. 2. pp. 117–120.

Lee, Y., Jhan, Y., Chung, C. and Hsu, Y. 2011. A prediction method for in-Plane permeability and manufacturing applications in the VARTM process. *Engineering.* Vol. 3. No. 7. pp. 691–699.

Correia, N.C., 2002. Use of resin transfer molding simulation to predict flow, saturation, and compaction in the VARTM process. *International Mechanical Engineering Congress and Exposition.* pp. 61–68.

Andersson, M.H., Lundström, T.S. and Gebart, B.R. 2003. Numerical model for vacuum infusion manufacturing of polymer composites. *International Journal of Numerical Methods for Heat & Fluid Flow,* Vol. 13. No. 3, pp. 383–394.

Technologies for Sustainable Development – Mahajan, Patel & Sharma (eds)
© 2020 Taylor & Francis Group, London, ISBN 978-0-367-33737-7

Enhancement in mechanical strength of knob of molded case circuit breaker

Chirag Kumar M. Vaghela & Reena R. Trivedi
Mechanical Engineering Department, Institute of Technology, Nirma University, India

ABSTRACT: The present work addresses improvement in mechanical strength of knob of molded case circuit breaker (MCCB). In MCCB, the operation like on, off, trip and reset are performed using mechanism. The purpose of deploying mechanism is to transfer the force to the next link in order to operate on, off/reset and trip. MCCB is operated on the basis of four-bar linkage mechanism. Motorized mechanism is an external accessory used to operate MCCB. It is directly mounted on MCCB. Comparing the standalone working of MCCB with the Combination (i.e. Motorized mechanism with MCCB), Total no. of operations in combination compared to stand alone MCCB is very less. The proposed work emphasizes on the improvement of the mechanical strength of the motorized mechanism and molded case circuit breaker.

1 INTRODUCTION

1.1 *Working of knob*

The knob is one of the important parts in MCCB to operate the ON, OFF and Reset position. Due to this, It is very necessary to have a higher strength of Knob [1]. As the mechanism and other parts like a fork, are live during the working of MCCB, it can contain more than 630A of current with 240 V. So, if the knob exerts more load on it. The knob can be broken. this breaking of a knob can create a condition of exposing live current. It can cause a high amount of hazardous effect.

Figure 1. Assembly of knob.

Figure 2. Exploded knob assembly.

1.2   Procedure for solution

- To change the geometry in order to increase its strength
- To carry out the boundary conditions
- Validating the results, by performing test on FEM software.

1.3   Cause and effect diagram

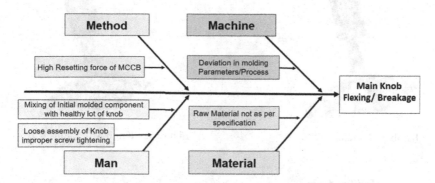

Figure 3.   Cause and effect diagram on knob breakage.

1.4   Comparison of proposed material

Table 1.   Comparison of proposed materials.

Parameters	101L NC010 Nylon Resin of DU PONT	33% GF Nylon 66 Zytel 70G33L	60% Glass filled BKV60EF
Contents	Lubricated polyamide 66% resin	33% GF reinforced PA66 resin for injection molding	60% GF reinforced PA6 resin for injection molding
Behavior	Ductile	Brittle	Brittle
Yield/Ultimate stress	82 MPa	200 MPa	230 MPa
Modulus of elasticity	3100 MPa	9300 MPa	20000 MPa
Poisson's ratio	0.41	0.39	0.39
Density	1140 kg/m^3	1390 kg/m^3	1700 kg/m^3

For this particular case, mainly two factors can be changed by considering all available resources for the enhancement of MET:

- By choosing higher strength of material than existing.
- By changing the existing design to design a knob more resistant towards the applied force.

2   PROPOSED DESIGN OF KNOB

2.1   Concept 1: Main knob with R3

As Shown in Figure (4), The radius of existing design with having R1 (1 mm radius) is increased up to R3 to strengthen the knob [2][3]. As it is observed that by increasing rib or fillet the material can be strengthened against the acting load.

2.2 Concept 2: Main knob with R5

As shown in Figure (5), The radius of existing design with having R1 is increased up to R5 to strengthen the knob [4][5]. As it is observed that by increasing rib or fillet the material can be strengthened against the acting load.

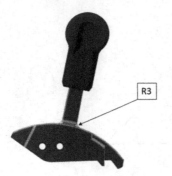

Figure 4.   Knob with radius R3.

Figure 5.   Knob with radius R5.

2.3 Concept 3: Modified fork

As shown in figure (6),(7),(8) and (9), the modified fork design has extra solid body. which will be assembled in to main Knob's hollowed part as shown in figure (9), so that total load applied on knob can be transferred on fork body [2][5].

Figure 6.   Existing fork.

Figure 7.   Modified fork.

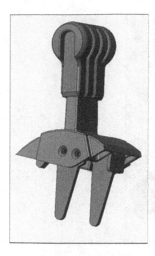

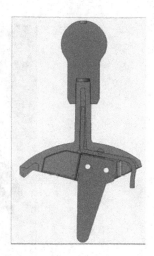

Figure 8.   Modified fork with knob.

Figure 9.   Cross-section view of knob assembly.

2.4   *Finite Element Analysis (FEA) of modified designs*

FEM analysis is a computational method used to analyse or study the engineering and physics problems in order to find the solution. In the present case, FEM can be used to find stress generation and deformation. To find stress value on particular parts, Simulation has been performed of variable cases of knob on computerized analysis tool ANSYS Mechanical.

3   RESULTS AND DISCUSSION

In this section results obtained by performing FEA simulation is discussed. Figure (10) shows Mechanical boundary conditions.

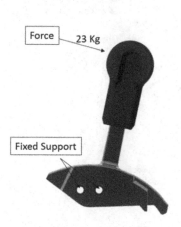

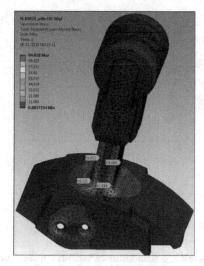

Figure 10.   Boundary condition for finite element analysis.

Figure 11.   Behaviour of knob in loading condition.

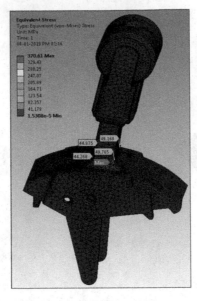

Figure 12. Behaviour of modified fork knob in loading condition.

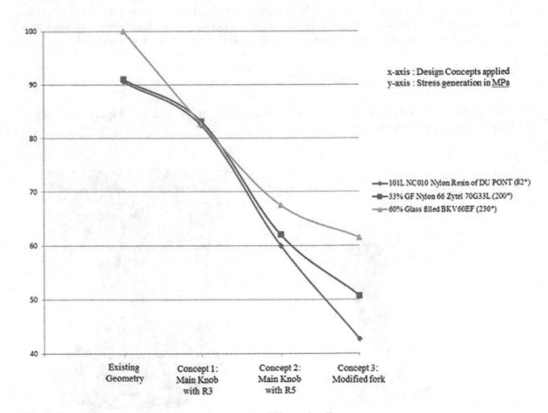

Figure 13. Behaviour of various concepts under different loading.

Table 2. Stress generation at the corner of knob.

Parameters	Extended Knob		
	101L NC010 Nylon Resin of DU PONT (82*)	33% GF Nylon 66 Zytel 70G33L (200*)	60% Glass filled BKV60EF (230*)
Existing Geometry	90.58 MPa	90.99 MPa	100 MPa
Concept 1: Main Knob with R3	82.40 MPa	83.16 MPa	82.76 MPa
Concept 2: Main Knob with R5	60 MPa	62 MPa	67.5 MPa
Concept 3: Modified fork	42.68 MPa	50.73 MPa	61.51 MPa

As Table 2 shows that, existing geometry and Concept 1: R3 are very near to that of yield stress value of 101L NC010, the yield stress limit for respective material is shown with suffix (*).

4 CONCLUSIONS OF BREAKAGE OF KNOB USING FEA

Performing finite element analysis it can be seen that the existing design is failing which is not safe practice when perform manually. Material 101L NC010 Nylon Resin having 82 MPa yield stress and after applying loading condition nominal stress as 90.58 MPa and 82.40 MPa. It is very near to the yield stress value.

But in case of the change in design as well as change in material shows significant change in the strength of the knob. From Table 2, The results of concept 3: Fork modified are very safe as it stands very far from the yield stress values with having factor of safety of 2.

REFERENCES

[1] Z. Chen, F. Che, M. Z. Ding, D. S. W. Ho, T. C. Chai, and V. Srinivasa, Drop Impact Reliability Test and Failure Analysis for Large Size High Density FOWLP Package on Package," Proc. - Electron. Components Technol. Conf., 2017, pp. (1196–1203).
[2] M. P. Savruk and A. Kazberuk, Stress concentration near sharp and rounded V-notches in orthotropic and quasi-orthotropic bodies," Theor. Appl. Fract. Mech.,vol. 84, 2016, pp. (166–176).
[3] M. P. Savruk and A. Kazberuk, Near V-Shaped Notches with Sharp and Rounded Tip *," Sci. York, vol. 43, no. 2, 2007.
[4] M. P. Savruk and A. Kazberuk, Two-dimensional fracture mechanics problems for solids with sharp and rounded V-notches," Int. J. Fract., vol. 161, no. 1,2010, pp. (79–95).
[5] M. P. Savruk and A. Kazberuk, Relationship between the stress intensity and stress concentration factors for sharp and rounded notches," Mater. Sci., vol. 42, no. 6.

Technologies for Sustainable Development – Mahajan, Patel & Sharma (eds)
© *2020 Taylor & Francis Group, London, ISBN 978-0-367-33737-7*

Design and development of double stage cyclone dust collector for incinerator

Sneh Patel & Jatin Dave
School of Engineering, Institute of Technology, Nirma University, Ahmedabad, India

ABSTRACT: With the increase in applications of FRP, disposal of waste has become a severe problem. So, incineration of the Fiber Reinforced Plastic (FRP) material waste is the alternate option to reduce the volume of waste. The incineration process produces small toxic dust particles. Double stage cyclone dust collector helps to remove small dust particles from flue gases. In this paper, different cyclone models are considered as 1D2D, 2D2D and 1D3D. Cyclone models are designed according to the functional requirement. Flow analysis is carried out in CFD software with parameters as velocity and barrel diameter of cyclone to study the dust collection efficiency.

1 INTRODUCTION

Fiber Reinforced Plastic (FRP) is made of reinforced materials and binders (synthetic resin) and have been utilized in a wide variety. The aggregate combined volume of end-of-life and creation squander produced by the FRP showcase is very large in different countries (Tittarelli and Shah 2013, Osmani,and Pappu, 2010). As indicated by Osmani (2010), in the United Kingdom, about 90% of the GFRP waste is landfilled and the reuse use rate is just 13% (Osmani,and Pappu, 2010). Incineration of FRP is suitable option for better waste management with the process of burning done in proper way as FRP contains toxic elements which harm the environment if FRP is disposed in open environment (Olisa et al. 2016). Burning of FRP produce ash which is toxic and harmful for open environment. So, it has to be removed from the flue gases before releasing in to the environment. Cyclones are the most widely used air pollution abatement equipment in the agricultural processing industry for removal of particulate matter like ash.

Normally, Cyclones are centrifugal separators. Design of cyclone separator consists of an upper cylindrical half noted as barrel and a lower conic half noted as cone. The vortex generated within the cyclone body due to the centrifugal force of flue gases. As air reaches to the bottom of the cone, it begins to flow radially inwards. Particles in the flue gases drop inside the dust collection chamber which is connected at the bottom of the cyclone. Cyclone separators are extremely economical for larger particles diameter (more than 10μm) and low efficiency for smaller diameter particles (less than 10μm). By employing a single cyclone will decrease overall efficiency and reduce the standard of air taking off. A second cyclone needed to extend the efficiency for the smaller diameter (Chen et al 2018). A two-stage cyclone separator is designed. The efficiency of first cyclone is more for large dust particles. The second cyclone has high separation efficiency for small dust particles. The two-stage cyclone separator is shown in Figure 1 and the geometry of a cyclone is set by dimensions as shown in Figure 2. In this paper velocity of air is considered 13 m/s.

The first cyclone (Lower cyclone) separator is the main cyclone that reduces most of the large particles from the polluted air. The diameter and length of each barrel and cone decide the efficiency of the cyclone. There are different models outlined with the size and dimension of the various parts of the cyclone.

Here, D = The body diameter of cyclone, H_t = the total height of the cyclone, Dx = the vortex finder(outlet) diameter, S = the vortex outlet length, a = the inlet height, b = the

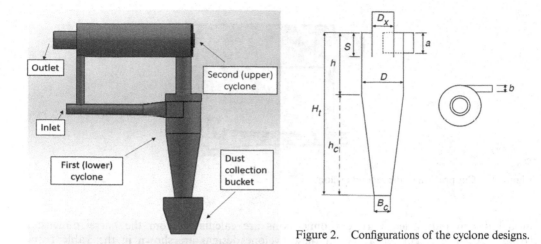

Figure 2.   Configurations of the cyclone designs.

Figure 1.   Cyclone separator.

inlet width, h_c = the height of the conical shape, h = the height of the cylindrical cylinder, B_c = bottom dust exit diameter

Figure 3 shows vanes inside of the cyclone on the outlet of the cyclone. Vanes are provided an outlet of the cyclone to prevent large particles going out of the first cyclone. It also prevents flue gases directly flow from inlet to outlet without creating a vortex inside of the cyclone. This increases the efficiency of the cyclone for large particles. The geometry of the second cyclone is a horizontal cylindrical shape with a diameter of 19 inches (Figure 4). The cylinder is Inlet of the second cyclone is placed upon the outlet of the first cyclone to avoid pressure drop. The inlet of the second cyclone is placed in the way that it creates cyclonic motion in the cylinder (Jadhav 2014). Due to centrifugal force, small particles are forced to go towards the wall of the cyclone. To promote circulation, motion the hollow cylindrical part is fixed in the centre of the cyclone at the inlet. A Venturi is placed before the exit of the second outlet of the second cyclone. Venturi stops the back pressure creating in the outlet of the second cyclone. Due to the venturi, a vacuum is created at the other side of venturi which creates suction in the outlet. This increases the efficiency of the second cyclone.

The common configurations of cyclones are 1D2D (Simpson and Parnell) cyclone, 2D2D (Shephrd and Laple) and 1D3D (Parnel) cyclone styles used for dust apparatus (Safkhani 2010). The D's in the name of 1D2D, 1D3D and 2D2D cyclones consult with the diameter of the barrel of the cyclone. The numbers of the D's are related to the length of the cylindrical portion and conical portion. Dimensions of the cyclone model are given per standard cyclone

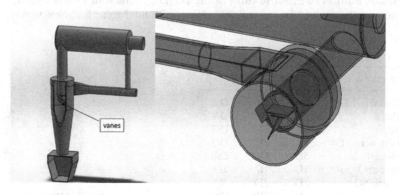

Figure 3.   Cut section of cyclone separator.

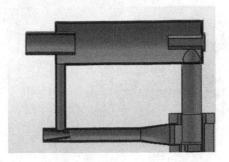

Figure 4.   Cut portion of the second cyclone.

model design parameter. Different dimensions are calculated from the barrel dimension (Ithape 2017). The configurations of these cyclone designs are shown in the Table 1 and Figure 5 shows the CAD models of the same.

In this paper, the inlet pipe diameter is 6 in. and velocity are around 13 m/s. The required diameter of the barrel is around two to three times than the inlet pipe diameter. Here the dimension of the inlet pipe is 6 in. So, the required barrel size is 12 inches or 18 inches. By considering 12 inches and 18 inches as barrel size, other dimensions can be obtained.

2   METHOD OF ANAYLSIS OF CYCLONES

The Reynolds stress model (RSM) and K-Epsilon are used in this work to know the flow separation characteristics. The different models are designed by using 3D modelling software. These models have been meshed in FEA software. Results are obtained for particles track file, trapped and escape particles, the pressure drop in the cyclone and velocity variation (Elsayed 2010, Funk,2010, Demir 2014).

There are 3 types of variation are considered in this analysis. 1) The diameter of the barrel – 12 inches and 18 inches 2) Model type – 1D2D model, 1D3D model, and 2D2D model 3) Velocity of flue gases – 8 m/s, 13 m/s and 20 m/s.

Input and two output sections are pre-named and the particles are calculated by considering number of particles cross a particular section. Solutions give the count of particles for different sections. This result shows escaped and trapped particles numbers which provides the efficiency of the cyclone model. This solution provides the path lines of the different particles which show behaviour inside the cyclone with different velocity.

Figure 6 shows a particle flow path of the 1D3D cyclone model of 12 inches barrel diameter the particles in the flue gases at the inlet and the path of particles. The velocity of the flue gases at the inlet is 13 m/s. A number of Effective turns of the 1D3D cyclone with 13 m/s velocity is around 6. The efficiency of the cyclone is around 86%. Figure 7 and Figure 8 shows vortex in a 1D3D

Table 1.   Configurations of cyclone models.

Parameters	1D2D	1D3D	2D2D
Cyclone diameter, D	D	D	D
Inlet height, a	D/4	D/4	D/4
Inlet width, b	D/2	D/2	D/2
Exit pipe diameter, Dx	D/1.6	D/2	D/2
Exit pipe length, S	5D/8	D/8	D/8
Cylindrical part height, h	1D	1D	2D
Cone part height, h_c	2D	3D	2D
Cone-tip diameter, B_c	D/2	D/4	D/4

Figure 5.   1D3D, 1D2D and 2D2D cyclone.

cyclone. Mainly two cyclones created inside the main cyclone. One at outside to the wall which collects the particles at the bottom. Second, generated in the centre which lets clean air to an outlet. Size and shape of inner cyclone provide information about efficiency and behaviour of air at a particular velocity. Here at 13 m/s, the vortex thickness is small which shows the good efficiency of a cyclone. Figure 9 and Figure 10 shows the CFD analysis of the second cyclone. The result is about particle path of respect to time. Particles enter from the inlet of the second cyclone travel in a cyclonic motion due to the centrifugal force. They collected on the second side and come out from outlet 2. Purified flue gas comes out from the outlet 1. The inner extended portion at outlet 1 prevents particles. The efficiency of the cyclone is 90%. Pressure drop inside of the horizontal cyclone is very less. This increases the efficiency of the cyclone.

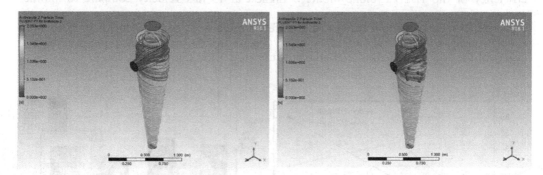

Figure 6.   Particles in a 1D3D cyclone of 12 inches barrel diameter and 13 m/s velocity of air.

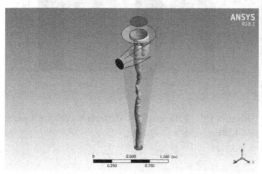

Figure 7.   Vortex in a 1D3D cyclone of 12 inches at 13m/s.

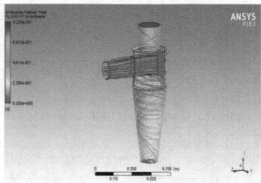

Figure 8.   1D2D cyclone of 12 inches and 13 m/s.

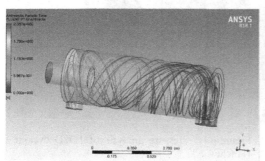

Figure 9.   CFD analysis of the second cyclone.

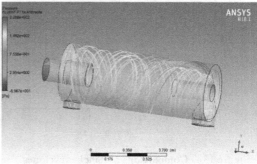

Figure 10.   Pressure analysis of the Second cyclone.

3 RESULTS AND CONCLUSION

The CFD analyses are performed for two cyclone types 12inches and 18inches with different geometries. Below some of the comparisons of a different cyclone are shown in graphs.

Figure 11 shows the graph of the comparison of the 1D3D cyclone model with two different diameter 12inches and 18inches. Figure 12 shows the efficiency comparison of the different cyclone of 18inches barrel diameter with respect to velocity.

Variation of efficiency is more in 18inches diameter cyclone compare to 12inches cyclone. Velocity is more important in larger diameter cyclone and it affects more to efficiency. 18inches cyclone achieves more efficiency compare to 12inches cyclone. The efficiency of the 1D2D cyclone model is highest at 13 m/s velocity compare to another cyclone.

Figure 13 shows a comparison graph of different cyclones at 13 m/s velocity. Pressure drop is also considered while designing the cyclone (Figure 14). As pressure drop increase the efficiency of the cyclone decreases. Here the pressure drop is less than 100 Pascal.

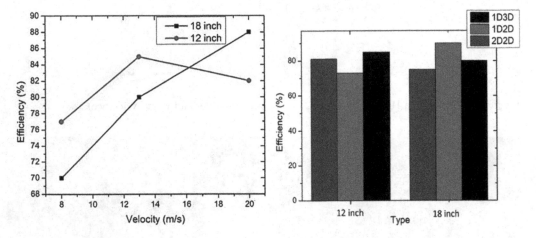

Figure 11. Comparison of 1D3D cyclones.

Figure 12. Efficiency of different cyclone model at 13m/s.

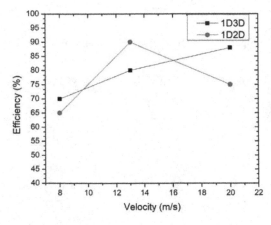

Figure 13. Comparisons of 18inches cyclones.

Figure 14. Pressure analysis of 1D2D cyclone of 18inches.

REFERENCES

Chen, J, Zhong, J. and Chen J. 2018. Effect of Inlet Air Volumetric Flow Rate on the Performance of a Two-Stage Cyclone Separator. *ACS Omega* 3.10: 13219–13226.

Demir, S. 2014. A practical model for estimating pressure drop in cyclone separators: An experimental study. Powder technology 268: 329–338.

Elsayed, K and Lacor C. 2010. Optimization of the cyclone separator geometry for minimum pressure drop using mathematical models and CFD simulations. *Chemical Engineering Science* 65.22: 6048–6058.

Funk, P. A. 2015. Reducing cyclone pressure drop with evases. *Powder technology* 272: 276–281.

Ithape, P. K., et al. 2017. Effect of Geometric Parameters on the Performance of Cyclone Separator using CFD. *IJCET*: 288–292.

Jadhav, M. R. 2014. Design of cyclone and study of its performance parameters. *International Journal of Mechanical Engineering and Robotics Research* 3.4: 247.

Osmani, M.and Pappu, A. 2010. An assessment of the compressive strength of glass reinforced plastic waste filled concrete for potential applications in construction. *Concrete Research Letters* 1.1: 1–5.

Olisa, Y. P., Amos, A. E., and Kotingo, K. 2016. The Design and Construction of a Step Grate Incinerator. *Global Journal of Human-Social Science Research*.

Safkhani, H., et al. 2010. Numerical simulation of ow field in three types of standard cyclone separators. Advanced *Powder Technology* 21.4: 435–442.

Tittarelli, F and Shah, S. P. 2013. Effect of low dosages of waste GRP dust on fresh and hardened properties of mortars: Part 1. *Construction and Building Materials* 47: 1532–1538.

Technologies for Sustainable Development – Mahajan, Patel & Sharma (eds)
© *2020 Taylor & Francis Group, London, ISBN 978-0-367-33737-7*

Dynamic modelling of a projectile launcher with controlled double inverted pendulum

P.S. Savnani, M.M. Chauhan*, A.I. Mecwan & R.N. Patel
Institute of Technology, Nirma University, Ahmedabad, Gujarat, India

ABSTRACT: The aim of this paper is to form a mathematical model to simulate the dynamics of a unique mechanism which can be used in a projectile launching machine. This mechanism is a bi-linkage mechanism with two sets of pivots having rigid and flexible links respectively and the torque to drive the mechanism is applied at the first pivot via a pneumatic cylinder. The mathematical model of this system is derived assuming the system to be a controlled double inverted pendulum and the trajectory of the ball, resulting from the combined motion of the links, is assumed to follow a circular path for a small segment. The launch parameters i.e. launch velocity and launch angle are derived from the numerical analysis of the mathematical model using simulation software and these parameters are experimentally verified to a great degree of accuracy.

Keywords: Circular path, controlled double inverted pendulum, flexible link, launch parameters, mathematical model, projectile launching machine

1 INTRODUCTION

Projectile launchers have been a topic of research since the dawn of mankind. From toys to warfare siege machines, these mechanisms have been of great importance to us. From lever operated catapults (Rihll, 2007) to gun powder cannons (Dupuy & Nevitt, 1990), people use these machines to launch different projectiles. Most of these machines utilize the principle of mechanical advantage in their operation. Here, a unique bi-linkage mechanism is used to launch projectiles. Compared to conventional projectile launchers, this mechanism provides certain extra parameters to control the projectile's trajectory, thus it is very versatile.

The objective of this report is to establish a correlation between the system and the projectile's desired outcomes which include its range and height. This report tries to derive a mathematical model to predict the outcomes and the potency of the simulated results of the model are tested experimentally. The simulation is carried out using the 'OCTAVE' software and the experimental values are analysed in a video analysis software 'TRACKER'.

The mechanism is assumed to be a controlled double inverted pendulum (Zhong & Rock, 2001) as shown in Figure 1 and the mathematical model is generated on this hypothesis. A double inverted pendulum is a system having two pendulums attached end to end with each other and it exhibits periodic, quasi-periodic and chaotic behaviour (Rafat, Wheatland, & Bedding, 2009). Here, the first pendulum is a continuous mass system while the second pendulum is a lumped mass system. Double pendulum's motion is highly dynamic and nonlinear in nature (Ohlhoff & Richter, 2000), therefore its trajectory is difficult to predict. Here, Euler-Lagrange Equations (Morin, 2007) (Widnall, 2009) are used to derive the mathematical model of the dynamic system.

After, the mathematical model is derived in terms of second order differential equations (Hand & Finch, 1998) (Polking, Boggess, & Arnold, 2005), the software Octave is used for

---

* Corresponding author: Mihir.chauhan@nirmauni.ac.in

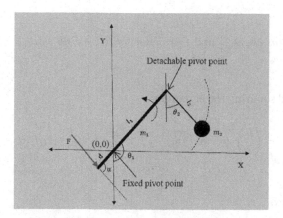

Figure 1.   Double inverted pendulum controlled using an external force.

numerical simulation and to plot the trajectories and velocities of both links. Then, coordinate transformation of the trajectory points is done to coincide the assumed circle's centre with the origin and the instantaneous velocity and the launch angle at the desired point are found and these resultant parameters are used to predict the outcomes. Finally the theoretical values are compared with experimental values to check the accuracy of the mathematical model. The experimental setup is an autonomous machine designed on the basis of this mechanism, which can launch the projectiles at different input parameters thus, facilitating in verification process.

2   MATHEMATICAL MODELLING

As shown in Figure 1, the first link is an aluminium extrusion, which is rotated about a fixed pivot point and it is extended to apply an external force (F) on the link i.e. torque to the fixed pivot. The external force is applied using a pressure filled pneumatic cylinder which is mounted on a rigid frame in the system. The second link is a string suspended ball which rotates about the detachable pivot point due to the torque applied at the fixed pivot point. The second link is detached once the launch angle is reached. The model is based on certain underlying assumptions. Here, the frictional forces are neglected in the model. Also, the suspended ball causes no air resistance. Apart from this, the first link is assumed to be rigid and is considered as a continuous mass system and the string in second link is non elastic and massless. The centre of mass of first link is midway between origin and the detachable pivot (i.e. distance d is very small compared to l_1). Also, it is assumed that the centre of mass of second link moves in a near circular trajectory for a small path.

The dynamic modelling of the system is derived using the Lagrange formulation,

$$\frac{d}{dt}\left(\frac{\partial L}{\partial \dot{\theta}_i}\right) - \frac{\partial L}{\partial \theta_i} = \tau_i \tag{1}$$

Where, L = Lagrange function, θ_i = Generalized coordinate, τ_i = External Torque, $i = 1, 2$

2.1   Position of masses

As shown in Figure 1, the centre of mass of first link lies at $\frac{l_1}{2}$ distance from origin and centre of mass of second link is at distance l_2 from detachable pivot, as string is considered as massless. The position vectors of centre of masses of first and second links are $\overrightarrow{r_1}$ and $\overrightarrow{r_2}$ respectively.

$$\overrightarrow{r_1} = \left[\frac{l_1}{2}sin(\theta_1) - \frac{l_1}{2}cos\ cos(\theta_1)\right] \text{ and } \overrightarrow{r_2} = [l_1sin\ sin(\theta_1) + l_2sin\ sin(\theta_2) - l_1cos\ cos(\theta_1) - l_2\theta_2)] \tag{2}$$

Differentiating and squaring the equations:

$$\left|\overrightarrow{r_1}\right|^2 = \left(\frac{l_1}{2}\right)^2 \dot{\theta_1}^2 \text{ and } \left|\overrightarrow{r_2}\right|^2 = l_1^2\dot{\theta_1}^2 + l_2^2\dot{\theta_2}^2 + 2l_1l_2(\theta_1 - \theta_2)) \tag{3}$$

2.2 Energy equations

The system has two different forms of energy: kinetic energy (the energy of motion) and potential energy. T_1 and T_2 are kinetic energies and V_1 and V_2 are potential energies for the first and second links respectively. Kinetic energy for each link is the summation of translational and rotational energies and potential energy for each link is the potential energy of their centre of masses with respect to the origin.

$$T_i = \frac{1}{2}m_i\left|\overrightarrow{r_i}\right|^2 + \frac{1}{2}I_i\dot{\theta_i}^2 \text{ and } V_i = m_igy_i \tag{4}$$

Where, m_i = Mass of link, $\overrightarrow{r_i}$ = Position vector, I_i = Moment of Inertia, θ_i = Generalized coordinate, g=Gravitational Constant, y_i = y coordinate of centre of mass.
Substituting (2) and (3) in equation (4):

$$T_1 = \frac{1}{6}m_1l_1^2\dot{\theta_1^2} \tag{5}$$

$$T_2 = \frac{1}{2}m_2\left(l_1^2\dot{\theta_1}^2 + l_2^2\dot{\theta_2}^2 + 2l_1l_2(\theta_1 - \theta_2)\right) \tag{6}$$

$$V_1 = -m_1g\left(\frac{l_1}{2}cos\ cos(\theta_1)\right) \tag{7}$$

$$V_2 = -m_2g(l_1cos\ cos(\theta_1) + l_2cos(\theta_2)) \tag{8}$$

2.3 Lagrange function

It is the difference of total kinetic energy and potential energy in the system. Following equation gives the Lagrangian:

$$L = T_1 + T_2 - (V_1 + V_2) \tag{9}$$

Substituting (5), (6), (7) and (8) in equation (9) and simplifying:

$$L = \left(\frac{1}{6}m_1 + \frac{1}{2}m_2\right)l_1^2\dot{\theta_1}^2 + \frac{1}{2}m_2l_2^2\dot{\theta_2}^2 + m_2l_1l_2(\theta_1 - \theta_2) + \left(\frac{1}{2}m_1 + m_2\right)gl_1cos\ cos(\theta_1)$$
$$+ m_2gl_2cos(\theta_2) \tag{10}$$

Differentiating equation (10) with respect to $\dot{\theta}_i$ and then differentiating the result with respect to time, also differentiating equation (10) with respect to θ_i and putting these values in equation (1) for i =1,2 respectively:

$$\left(\frac{1}{3}m_1 + m_2\right)l_1{}^2\ddot{\theta}_1 + m_2l_1l_2\cos\ddot{\theta}_2\cos(\theta_1 - \theta_2) - m_2l_1l_2(\theta_1 - \theta_2)(\dot{\theta}_1 - \dot{\theta}_2) + m_2l_1l_2\dot{\theta}_1\dot{\theta}_2$$
$$\sin\sin(\theta_1 - \theta_2) + \left(\frac{m_1}{2} + m_2\right)gl_1\sin\sin(\theta_1) = \tau_1 \tag{11}$$

$$m_2l_2{}^2\ddot{\theta}_2 + m_2l_1l_2\cos\ddot{\theta}_1\cos(\theta_1 - \theta_2) - m_2l_1l_2(\theta_1 - \theta_2)(\dot{\theta}_1 - \dot{\theta}_2) - m_2l_1l_2(\theta_1 - \theta_2)$$
$$+ m_2gl_2\sin(\theta_2) = \tau_2 \tag{12}$$

The torque $((\tau_1))$ in equation (11) can be determined using Figure 2, where F is the external force applied to the link at a distance 'd' from the origin. Here, 'φ' is the angle of piston with horizontal and 'α' is the angle of force with the link. And, 'b' is the distance of origin with X axis and cylinder's intersection.

$$\tau_1 = Fd\,\sin\sin\alpha \text{ and } F = \frac{P\pi D^2}{4} \tag{13}$$

Where, P = Pressure in pneumatic cylinder, D = Bore diameter of pneumatic cylinder
Using Figure 2,

$$b = d\,\cos\cos\left(\theta_1 - \frac{\pi}{2}\right) + a\,\cos\cos\varphi \text{ and } a = -\left(\frac{d}{\sin\sin\varphi}\right)\cos\cos\theta_1 \tag{14}$$

$$\sin\sin(\alpha) = \frac{b\sin\sin\varphi}{d} \tag{15}$$

Substituting equation (14) in equation (15) and simplifying the equation:

$$\sin\sin(\alpha) = \sin\sin\theta_1\,\sin\sin\varphi - \cos\cos\theta_1\,\cos\cos\varphi = (\theta_1 + \varphi) \tag{16}$$

Substituting equation 16 in (13),

$$\tau_1 = -Fd\,\cos\cos(\theta_1 + \varphi) \tag{17}$$

Comparing equation (11) and (17) and solving for $\ddot{\theta}_2$,

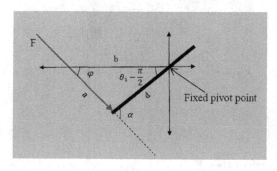

Figure 2. The force correlation triangle.

$$\ddot{\theta}_1 = \frac{-m_2 l_2 cos\ddot{\theta}_2 cos(\theta_1 - \theta_2) - m_2 l_2 sin\dot{\theta}_2{}^2 sin(\theta_1 - \theta_2) - \left(\frac{m_1}{2} + m_2\right) g sin(\theta_1) - \frac{F}{l_1} d cos cos(\theta_1 + \varphi)}{\left(\frac{1}{3}m_1 + m_2\right) l_1}$$

(18)

The torque (τ_2) in equation (12) is zero, as no external torque is applied at the detachable pivot point. Substituting $\tau_2 = 0$ in equation (12) and solving for $\ddot{\theta}_2$,

$$\ddot{\theta}_2 = \frac{-l_2 cos\ddot{\theta}_1 cos(\theta_1 - \theta_2) + l_1 \dot{\theta}_1{}^2 sinsin(\theta_1 - \theta_2) - gsin(\theta_2)}{l_2}$$

(19)

The equations (18) and (19) are coupled, therefore they are made independent in the following steps.

Using equations (18) and (19),

$$\ddot{\theta}_1 = -m_2 l_2 \dot{\theta}_1{}^2 sin\, sin\,(\theta_1 - \theta_2)\, cos\, cos\,(\theta_1 - \theta_2) + m_2 g sin(\theta_2) cos\, cos\,(\theta_1 - \theta_2)$$
$$\frac{-m_2 l_2 sin\dot{\theta}_2{}^2 sin\,(\theta_1 - \theta_2) - \left(\frac{m_1}{2} + m_2\right) g sin\, sin\,(\theta_1) - \frac{F}{l_1} d cos\, cos(\theta_1 + \varphi)}{\left(\frac{1}{3}m_1 + m_2\right) l_1 - m_2 l_1\,(\theta_1 - \theta_2)}$$

(20)

and,

$$\ddot{\theta}_2 = m_2 sinsin(\theta_1 - \theta_2) coscos(\theta_1 - \theta_2) + \left(\frac{m_1}{2} + m_2\right) gsinsin(\theta_1) coscos(\theta_1 - \theta_2) + \left(\theta^1 - \theta^2\right)$$
$$\frac{coscos\left(\theta^1 + \varphi\right) + \left(\frac{1}{3}m_1 + m_2\right) l_1 \dot{\theta}_1{}^2 sinsin\left(\theta^1 - \theta_2\right) - \frac{1}{3}m_1 + m_2 gsinsin(\theta)}{\left(\frac{1}{3}m1 + m_2\right) l_2 - m2l2cos^2(\theta_1 - \theta_2)}$$

(21)

Converting these second order differential equations into four first order differential equations using state space representation and solving using computer software, the solution of these equations at each time step within a defined time span is found. The observed time span decides the operating range of θ_1 and θ_2. The values of θ_1 and θ_2 are inserted in equation (2) to get the values of trajectory co-ordinates. Now the centre of mass of second link $((\vec{r_2}))$ is assumed to move in a near circular trajectory for a small path as shown in Figure 3.

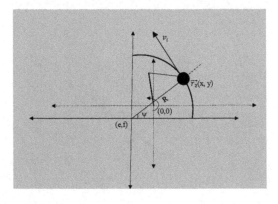

Figure 3.   Coordinate transformation.

Inputting the trajectory points of $(\vec{r_2})$ in the equation of a circle, the radius and the centre of the arc are found. Making the origin and centre of the circle coincidence and transforming the trajectory coordinates, the values of ψ are found for each trajectory point as:

$$\psi = \left(\frac{y}{x}\right) = \left(\frac{y-f}{x-e}\right) \qquad (22)$$

Here, x and y are transformed coordinate of arc, e and f are coordinate of centre point. Plotting ψ vs t, the instantaneous slope at ψ_{final} is found, which will give $\dot{\psi}$. Thus,

$$\text{Final launch angle} = \theta = \frac{\pi}{2} - \psi_{final} \text{ and Final launch velocity} = v_i = R\dot{\psi} \qquad (23)$$

3   SIMULATION RESULTS AND EXPERIMENTAL VALIDATION

Here, the double inverted pendulum's trajectory is simulated using a computer program and finally the angles and velocities of both the links are plotted with respect to time. The simulation is carried out using the ordinary differential equations obtained in the mathematical model.

Here, the governing input parameters are Force (F), Length of first link (l_1) and Angle of first link (θ_1). Furthermore, the Force is dependent on other primary parameters like Pressure in cylinder (P), Bore diameter (D). Every other parameter (e.g. m_1, m_2, θ_2, d etc.) is fixed for the analysis. Table 1 shows a set of values for which the experiment was carried out and finally it is compared with the simulation results.

3.1   Simulation results

To evaluate the performance of the system, the mathematical model is used to plot the positions of both links with respect to time. Initial angles of first link and second link are 125^0 and 0^0 respectively with vertical axis. The curves obtained are non-linear and without any period-

Table 1.   Input parameters.

m_1	0.4kg
m_2	0.1kg
l_1	0.57m
l_2	0.35m
g	$9.81\frac{m}{s^2}$
d	0.08m
P	$5.6*10^5$ pascal
φ	27^0
t	0.5s
D	0.025m

Table 2.   Initial boundary conditions.

θ_1	125°
θ_2	0°
$\dot{\theta_1}$	0
$\dot{\theta_2}$	0

Table 3.   Comparison of simulation and experimental results.

Output Parameters	Final Angle ($\theta = \frac{\pi}{2} - \psi$)	Final Velocity (v_i)
Simulation	44.88^0	$6.68\frac{m}{s}$
Experimentation	45.3^0	$6.59\frac{m}{s}$

367

icity. From Figure 4, it is inferred that both the curves have increasing trends but the rate of increase of second link's angle is greater than the rate of increase in first link's angle. Here, the first link's final angle is constrained and the time it took to reach that angle is found. This time decides the launch position, angle and velocity of the projectile.

In Figure 5 the profile of angular velocity vs time of both links have a high degree of non-linearity. Here, the first link shows non-periodic oscillatory motion, while the second link shows a general increasing trend with one sided oscillations. Figure 6 shows the angle of projectile (centre of mass of second link) with horizontal axis after the coordinate transformation. The curve shows rapid increasing trend with time. This curve is derived from the projectile's trajectory using the transformed coordinates.

The curve fitting tool is used to fit this curve into a four degree polynomial with R square value = 1 and RMSE = 0.0582. The analysed time span is less than 0.25 seconds because that is the experimental limit on the developed physical setup. Following equation is the result of curve fitting data:

$$\psi = p_1 t^4 + p_2 t^3 + p_3 t^2 + p_4 t^1 + p_5 t^0 \tag{24}$$

Here, $p_1 = 16620$, $p_2 = -5073$, $p_3 = 1530$, $p_4 = -14.34$, $p_5 = 1.824$

Substituting the value of time in this equation will give the angle (ψ) at required instant i.e. launch angle will be $\theta = \frac{\pi}{2} - \psi$.

As the limit for first link is 190^0, so from Figure 4 the time required to reach this point is 0.198 seconds. Inputting this value in above equation yields $\psi = 45.12^0$, thus $\theta = 44.88^0$. And substituting the value of time in its derivative will give the angular velocity at that instant i.e. launch velocity will be $v_i = R\dot{\psi}$. Inputting value of time yields, $\dot{\psi} = 510.935^0 s^{-1}$ or 8.91 $\frac{rad}{s}$ and R = 0.75m from circle's equation so $v_i = 6.68 \frac{m}{s}$.

3.2 Experimental validation

The results from the dynamic model have been tested on a physical system within tolerable limits. The projectile launcher is built using a ISO 6432/CETOP RP52P standard pneumatic cylinder having bore diameter 25mm and stroke length 125mm and a 5/2 solenoid to control its actuation .The first link is made of a hollow aluminium extrusion of cross section 22*22mm with thickness 1mm. The second link consists of a non-elastic thread and a rigid ball as the projectile at the end. The fixed pivot point has bearing support with two 8mm SKF bearings. The joint between the first link and the second link is a detachable pin joint which is detached from the system, when the launch angle is reached.

Here, a video analysis software 'TRACKER' is used to record the actual trajectory of the projectile. The trajectory is divided into two phases i.e. when the projectile is attached with the system, in Figure 7 from start of red marks to the intersection of magenta lines and after that.

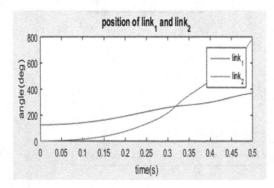

Figure 4. θ_1/θ_2 vs time. Figure 5. $\dot{\theta}_1/\dot{\theta}_2$ vs time.

368

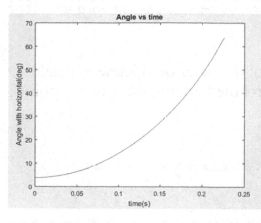

Figure 6. ψ vs time.

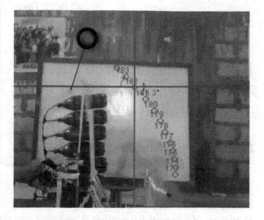

Figure 7.   Actual trajectory of the projectile.

After the intersection point the projectile leaves the system and follows the projectile motion with the output parameters of the system i.e. angle and velocity. The actual trajectory matches with the simulation results and gives same output parameters as predicted by the mathematical model within tolerable limits. Thus, it is observed that the experimental values match with the simulation results within tolerable limits. The small variations can be attributed to the assumptions made in the mathematical model.

4   CONCLUSION

The system is analysed using a mathematical model derived using Lagrangian mechanics approach to the double inverted pendulum incorporating the dynamics of the system. Furthermore, the simulation results match with the experimental results thus, proving the applicability of this model. The equation obtained from the curve fitting data for the input parameters used in the simulation will vary with another set of data. As, this model predicts the trajectory and the launch parameters accurately, it has widespread applications in industrial launchers, weapons technology, sports training machines etc. The present model can be further extended by considering the frictional forces and air resistance to predict more realistic behaviour of the launching mechanism.

REFERENCES

Dupuy, & Nevitt, T. 1990. *The Evolution of Weapons and Warfare*. New York: Da Capo Press.
Hand, L. N., & Finch, J. D. 1998. *Analytical Mechanics*. NewYork: Cambridge University Press.
Morin, D. 2007. The Lagrangian Method. In *Introduction to Classical Mechanics, With Problems And Solutions*, VI1 - VI55. Cambridge University Press.
Ohlhoff, A., & Richter, P. H. 2000. Forces in double pendulum. *Journal of Applied Mathematics and Mechanics, 80*(8).
Polking, J., Boggess, A., & Arnold, D. 2005. *Differential Equations with Boundary Value Problems*. New Jersey: Pearson Prentice Hall.
Rafat, M. Z., Wheatland, M. S., & Bedding, T. R. 2009;. Dynamics of a double pendulum with distributed mass. *American Journal of Physics*.
Rihll, T. E. 2007. *The Catapult: A History*. Yardley: Westholme Publishing.
Widnall, S. 2009. *LectureL20- EnergyMethods: Lagrange'sEquations*. MIT OpenCourseWare.
Zhong, W., & Rock, H. 2001. Energy and passivity based control of the double inverted pendulum on a cart. *Proceedings of the 2001 IEEE International Conference on Control Application*, 896–901. Mexico: IEEE.

Technologies for Sustainable Development – Mahajan, Patel & Sharma (eds)
© 2020 Taylor & Francis Group, London, ISBN 978-0-367-33737-7

Experimental study on effect of tool diameter on thickness, height and geometric accuracy of flange in hole flanging using single point incremental forming

Rudreshkumar Makwana
Assistant Professor, Mechanical Engineering Department, Institute of Technology, Nirma University, Ahmedabad, India

Bharat Modi* & Kaushik Patel
Professor, Mechanical Engineering Department, Institute of Technology, Nirma University, Ahmedabad, India

ABSTRACT: Single Point Incremental Forming is a forming technique in which sheet metal is formed by single point tool on vertical milling center. In this work, the principle of Single Point Incremental Forming is applied to perform hole flanging operation. Formability of Aluminium 1050 sheet in terms of Limit Forming Ratio is obtained. Experiments were performed to check effect of tool diameter on thickness distribution and average thickness of flange as well as on flange height.

1 INTRODUCTION

Hole flanging operation is performed to form flange from pre cut hole on sheet metal. Flange increases strength of edge and it also improves aesthetic appearance. In some parts of automobile, flange is required to join the other part. Hole flanging is conventionally performed by punch and die. To get various flanges with varied dimensions, it is required to use punch and die of appropriate size. The conventional method is costly when job production or prototype development is required. Single Point Incremental Forming (SPIF) can be performed without dedicated die and punch and thus it is feasible for such applications. The process has been explored by many researchers. In conventional forming, sheet metal undergoes necking before crack appears whereas in SPIF the necking is suppressed and hence formability improves [1]. Forming Limit Curve which is used to predict formability in case of conventional forming cannot be used for SPIF. A Fracture Forming Line (FFL) is used for formability prediction. FFL indicates that the failure in SPIF occurs at very large strains as compared to conventional forming. Researchers have explored use of SPIF for hole flanging operation, it offers high formability as compared to conventional hole flanging using punch an die [2]. Amount of deformation in hole flanging can be measured by the ratio of final flange diameter to the pre cut hole diameter which is known as Hole Expansion Ratio (HER), and the maximum HER is termed as formability in hole flanging which is also known as Limit Forming Ratio (LFR) [3]. Hole flanging using SPIF can be performed by multistage strategy [4] and single stage strategy [5]. In multistage strategy the hole flanging operation is performed in more than one stage. It involves movement of the tool in helical path and the diameter of helical path changes in every stage. In single stage strategy, tool is moved on a single helical path which directly gives final flange. In hole flanging using SPIF, cylindrical tool with ball nose is used to form the sheet. The tool is held in the spindle of Vertical Milling Center and then moved on

---

* Corresponding author:

required path with predefined step depth. Many researchers have worked on development of new tool or on analyzing effect of tool on formability [6, 7, 8]. M.Borrego et al [5] used three tools with different radius to investigate effect of tool radius on formability in single stage hole flanging using SPIF. As the tool radius increases, the formability increases.

However effect of tool diameter on thickness distribution and flange height need to be addressed. In this work, Aluminium 1050 sheet metal of 1.5 mm thickness were formed by single stage strategy after making precut hole on it. Three tools with diameter of 8 mm, 10 mm and 12 mm were used. Effect of tool diameter on geometric accuracy, thickness distribution, flange height and outer diameter of flange has been analyzed and presented in this paper. The parameter values were selected by the range of values used for such parameters in [1], [4], [5] and [6].

2 EXPERIMENTAL SET UP

The experiments were performed with Aluminium 1050 sheet of 100mm x 100mm and 1.5 mm thickness on Vertical Milling Centre. The experimental setup is shown in Figure 1. Three SPIF tools with diameter 8 mm, 10 mm and 12 mm were used. For cutting hole on the sheet before flanging operation, end mill cutter was used. CNC programs were prepared to move SPIF tool on helical path with step depth of 0.5 mm and feed of 300 mm/min.

3 RESULTS AND DISCUSSION

Experiments for hole flanging were performed on Aluminium sheets using SPIF process with the tool having diameter 8 mm by varying precut hole diameter from 28 mm to 37 mm to find LFR. It was observed that successful flange formation took place with precut hole diameter 37 mm and the sheet metal with precut hole diameter 36 mm failed during the process which are shown in Figure 2 (a).

So the LFR is worked out to be 1.56. Similarly, experiments were performed using SPIF process with tool having diameter of 10mm and 12mm by varying precut hole diameter from 28mm to 34mm. It was observed that in case of SPIF tool with diameter 10 mm and 12 mm, successful flange formation took place with precut hole diameter 34mm and 32mm respectively, the flanges are shown in Figure 2 (b) and Figure 2 (c).

So the LFR is worked out to be 1.70 and 1.81 respectively. The reason behind increase in LFR with increase in tool diameter can be attributed to the area of contact between the SPIF tool and the sheet metal. In case of larger tool, the contact area is more as compared to the smaller tool which reduces stress on the sheet metal. The thickness distribution on the flange

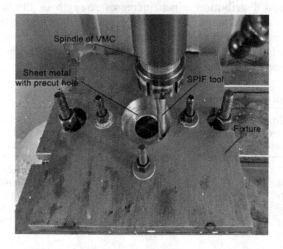

Figure 1.   Experimental set up for hole flanging using SPIF.

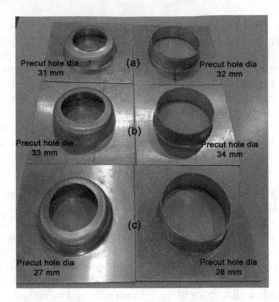

Figure 2.   Flanges formed using SPIF process with tool. (a) 12 mm diameter (b) 10 mm diameter (c) 8 mm diameter.

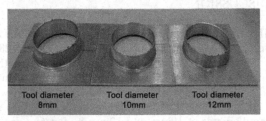

Figure.3.   Flanges obtained by the tools from precut hole diameter 37 mm.

obtained by three different tools from same precut hole diameter is indicative of formability. The flanges formed with precut hole diameter of 37 mm with three SPIF tools are shown in Figure 3. Thickness on the flanges were measured by pointed anvil micrometer (0.001 mm least count) at the interval of 2 mm from the flange root on four places (circumferentially 90° apart).

The results are tabulated in Table 1 and presented in Figure 4. It is observed that larger tool gives uniform thickness distribution which increases strength of the flange. The pattern of thickness distribution remains same in the three cases, thickness near the root remains high then it starts decreasing up to mid zone and then it starts increasing towards the edge.

It was observed that as the tool diameter increases the average thickness of flange increases. With larger tool diameter, the stresses induced are lower and the associated strains are also

Table 1.   Thickness distribution on flanges obtained by the three tool.

Distance from root of flange (mm)	Thickness on flange obtained by tool of 8 mm diameter	Thickness on flange obtained by tool of 10 mm diameter	Thickness on flange obtained by tool of 12 mm diameter
2	1.077	1.214	1.255
4	0.814	1.031	1.107
6	0.680	0.991	1.084
8	0.740	0.999	1.071
10	0.894	1.075	1.126
12	1.042	1.056	1.111
14	1.081	–	–

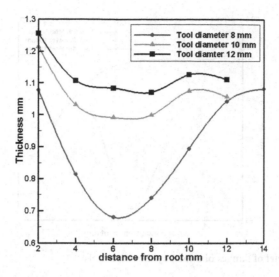

Figure 4. Thickness distribution on flanges obtained by the three tools.

lower. This results in lower flange height as well. Heights of all three flanges were measured by vernier height gauge (0.01mm least count) at four places (circumferentially 90° apart) and then average height was calculated. The results are shown in Table 2. The average height obtained is 14.78 mm, 13.17 mm and 12.28 mm for tools with 8 mm, 10 mm and 12 mm diameter respectively.

The variation of height and distribution of thickness affects the geometric accuracy of the flange. Apart from this, the outer diameter of flange along the flange height also varies which is also a parameter of deciding geometric accuracy. The outer diameter of flanges were measured by digital vernier caliper (0.01 mm least count) at interval of 2 mm from the root of the flanges. The results are shown in Table 3 and Figure 5.

It is observed from the results that most uniform geometry is obtained by the tool of diameter 12 mm and least uniform was obtained with 8 mm diameter tool. It is clear from the results that larger tool gives uniform geometry. In case of 8 mm diameter tool formation of bulge was observed in the lower zone (between 4 mm to 6 mm from root) and in the upper zone (between

Table 2. Flange height obtained by the three tool.

Tool diameter (mm)	8	10	12
flange height (mm)	14.78	13.17	12.28

Table 3. Outer diameters of flanges obtained by the three tools.

Distance from root of flange (mm)	Outer diameter on flange obtained by tool of 8 mm diameter	Outer diameter on flange obtained by tool of 10 mm diameter	Outer diameter on flange obtained by tool of 12 mm diameter
2	59.63	60.09	60.11
4	58.99	59.73	59.98
6	58.93	59.79	59.99
8	58.16	59.91	59.96
10	59.34	59.81	59.75
12	59.25	59.25	59.20
14	58.70	–	–

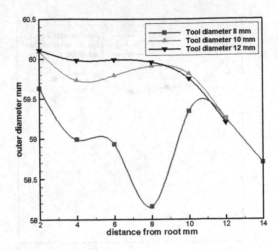

Figure 5.   Outer diameter of flanges obtained by the three tools.

8 mm to 14 mm from root). This can be attributed to excessive thinning of the flange which in turn led to bulge out. In case of 10 mm diameter tool bulge was also observed in the upper zone (between 6 mm to 10 mm from root) but in this case only one bulge was observed. Bulge formation was not observed in case of flange obtained by 12 mm diameter tool.

4   CONCLUSIONS

Experiments were performed to find effect of tool diameter on LFR, thickness and height of flange. Following conclusions can be drawn from the study.

- Higher LFR was obtained by increasing the diameter of tool.
- Increase in tool diameter results into thicker and shorter flange.
- Uniform thickness distribution was obtained when larger tool was used.
- In the flange obtained by hole flanging operation using SPIF, thickness near the root is more and it decreases up to the mid zone. After mid zone it increases towards the edge of flange.
- High geometric accuracy was obtained when larger tool was used.

ACKNOWLEDGEMENT

This work was supported by Nirma University, Ahmedabad, Gujarat, India. (vide office order no: NU/Ph.D/MRP/IT/2018-19/6439)

REFERENCES

[1] G.Centeno et al, Hole flanging by incremental sheet forming. International Journal of Machine Tools & Manufacture vol. 59:pp. 46–54, 2012.
[2] M.B.Silva et al, Incremental Forming of Hole-Flanges in Polymer Sheets, Materials and Manufacturing Process., vol. 28, no. 3, pp. 330–335, 2013.
[3] J.Mackerle, Finite element analyses and simulations of sheet metal forming processes, Engineering Computations., vol. 21, pp. 891–940, 2004.
[4] Z.Cui et al, Studies on hole-flanging process using multistage incremental forming. CIRP Journal of Manufacturing Science and Technology,vol.2, pp.124–128, 2010.

[5] M. Borrego et al, Experimental study of hole-flanging by single-stage incremental sheet forming, Journal of Materials Processing Technology, vol. 237, pp. 320–330, 2016.

[6] M. Bambach et al, A New Process Design for Performing Hole-flanging Operations by Incremental Sheet Forming, Procedia Engineering., Elseveir, vol. 81, pp. 2305–2310, 2014.

[7] Tingting Cao et al, Investigation on a new hole flanging approach by incremental sheet forming through a featured tool, International journal of machine tools and manufacture, vol.110,pp.1–17, 2016.

[8] T. Wen et al, Bi-directional dieless incremental flanging of sheet metals using a bar tool with tapered shoulders, Journal of Material Processing and Technology, vol. 229, pp. 795–803, 2016.

Technologies for Sustainable Development – Mahajan, Patel & Sharma (eds)
© 2020 Taylor & Francis Group, London, ISBN 978-0-367-33737-7

Numerical simulation to study fluid flow and heat transfer characteristics of conical spiral tube

Dipak Saksena & Vikas Lakhera
Institute of Technology, Nirma University, Ahmedabad, India

ABSTRACT: In heat exchangers, conical spiral tubes are used to enhance heat transfer. The variation of Nusselt number around the circumference of tube cross section is not widely reported in the literature related to conical spiral tube. The present study focuses on the velocity and temperature profiles at 45 and 225° angle from the start of the fluid flow using numerical simulation results. The coil with pitch circle diameter 100 mm, tube diameter 20 mm and cone taper angle 20° was modeled for simulation of boundary conditions of 0.5 m/s inlet velocity, 300 K inlet fluid temperature and 320 K constant wall temperature. The results show that the velocity profile is complex at 225° angle as compared to 45° angle. The highest value of Nusselt number is observed at the outermost side of the tube while the lowest is partially offset from the inner side of tube.

1 INTRODUCTION

Spiral and helical coils are widely used in heat exchangers to enhance heat transfer and the examples of use include power plants, petrochemicals, chemical and many other industrial applications. Various researchers have found that helical and spiral coils have better heat and mass transfer rates due to the curvature in geometry. The fluid flowing though the curved pipes face centrifugal force which tends to move the fluid particles away from the center of curvature thereby forming another flow across the main flow known as secondary flow. Dean [1] studied and mathematically described flow through curved pipes and found that the curvature ratio and Reynolds number govern the secondary flow. The Dean number is used to describe the flow in curved pipes and is given by the following equation:

$$De = Re\sqrt{\frac{r}{R_c}}$$

(1)

where *Re* is the Reynolds number, r is the tube radius and R_c the radius of coil curvature.

As compared to straight tubes, the flow inside the coiled tubes remains in the viscous regime up to a much higher Reynolds number. (Srinivasan et al. [2]) The curvature induces the helical vortices which tends to suppress the onset of turbulence and delays the transition. The critical Reynolds number, which describe the transition of laminar flow to turbulence flow is given by correlations available in the literatures. Jay Kumar et al. [3] have studied the variation of local Nusselt number around the circumference of tube for helical coil.

Based on the literature reviewed, it is observed that while several studies have been conducted on heat transfer and flow in helical coils, the spiral and conical spiral tube have received little attention despite having better heat transfer behaviour than helical coils. Because of the complexity of the structure only few researchers have studied the flow and heat transfer through a conical spiral tube. Yan Ke et al.[4] found numerically and experimentally that the cone angle has significant effect and pitch has nearly no effect on enhancing the heat transfer coefficient. The study also concluded that the temperature gradient of the outer side is larger than the inner side. This clearly indicates that there is a variation of Nusselt number around the circumference. The present study is based on the turbulent flow in the conical spiral tube and focusses on the nature of variation of Nusselt number around the circumference at a cross section in a conical spiral tube.

2 MODELING AND NUMERICAL SIMULATION

2.1 *Geometry of conical spiral tube*

Figure 1 shows the geometrical details of the modeled conical spiral tube prepared using Solidworks software. The pitch circle diameter of the bigger end (R_i) at inlet was considered as 100mm, pitch (H) as 30mm, tube diameter (d) as 20mm, and the cone taper angle as 20°s. The curvature varies along the tube. The tube cross section was considered to be circular and constant. For the analysis of results, the upper side is denated as outer and the bottom side as inner. The inner most point on a tube cross section is taken as zero° angle and then the anticlockwise angle is measured. The IGS file was imported in Fluent 17.1 for simulation.

Using numerical simulation, the turbulent flow behavior through conical spiral tube was studied using ANSYS FLUENT 17.1 software.

2.2 *The grid independency check*

A grid independency test for the conical spiral coil was carrie out in order to check the mesh quality. The number of elements were increased with an increase in partition of the circumference of the cross section. Table 1 shows the details of the grid independency test. The grids were generated by dividing the circumference of tube cross section into equal divisions. Simulation were performed for 20-80 equal division of circumference. For more than 80 equal partitions of the circumference, the results show less than 0.5% difference in the value of Nusselt Number. So the grid with 80 equal partitions of the periphery of the cross section was selected for simulation.

2.3 *The discretization scheme and conversion criteria*

The SIMPLEC scheme was used for pressure velocity coupling. For flows involving rotation, as the boundary layers are under strong adverse pressure gradients, separations and

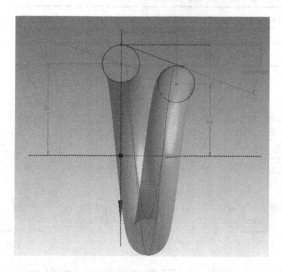

Figure 1.   Geometry of a conical spiral tube selected for simulation.

Table 1. The results of the grid independency check.

Partition of circumference	Number of Elements	Heat Transfer Coefficient	Nusselt number	% variation
20	28056	2696.43	89.88107	-
30	57967	2307.85	76.92847	14.41082
40	86190	2586.82	86.22757	12.08798
50	116620	2566.352	85.54507	0.79151
60	149454	2544.033	84.8011	0.869678
70	183654	2531.297	84.37657	0.500622
80	229104	2519.359	83.97863	0.471616

recirculation, the realizable k-ε, two equation turbulence model is used for computation (Shih Tsan-Hsing et. al) [5]. For turbulent kinetic energy and dissipation rate equations, the Power law scheme of discretization was used. The criterion for convergence was 1.0e-5 for velocities, continuity, k and ε. The energy equations was discretized with third order QUICK scheme. The convergence criterion of 1.0e-07 was set for energy balance.

2.4 Velocity, temperature and pressure at inlet

For inlet velocity, the value of 0.5 m/s was set for the simulation. The wall temperature 320K was set for constant temperature boundary condition. The temperature of fluid at inlet was set at 300K. The fluid pressure at inlet was set as atmospheric at 101325 Pascal [Table 2].

3 RESULT AND DISCUSSIONS

3.1 Velocity contours at 45° and 225° angle

Figure 2 (a & b) show the overview of the velocity contours at 45° and 225° from the start of the flow. Once the simulation is completed, the plane cutting the coil at 45° and 225° angle was generated in post CFD mode available in Fluent. At 45° angle, a symmetric velocity

Table 2. The boundary conditions for the simulation run.

Sr. No.	Inlet velocity (m/s)	Tube wall Temperature Constant (Steel) (K)	Inlet Temperature (K)	Pressure (Pascal)
1	0.5	320	300	101325

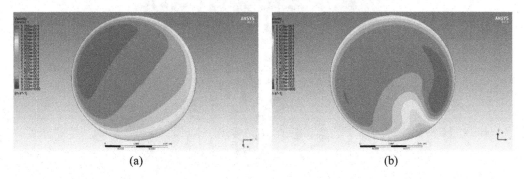

| (a) | (b) |

Figure 2. Velocity contours at 45° (a) and 225° (b) angle from start of flow.

profile is observed and the center of the contours are offset from the center of the cross section. The high velocity contours can be seen near the outer side on the top at 45° cross section. The high velocity region covers about half of the cross section area. The velocity contours at 225° angle in Figure 2(b) show the high velocity region shifted from left side to right side of the outer side. As the fluid particle moves in the upward direction, the gravitation force acting on the fluid particles keeps the high velocity cold particle at the bottom. This decreases the heat transfer at outer side near 270° angle from inlet. The torsion and buoyancy forces shift the high velocity region at 50° angle. High heat transfer takes place in this region. The velocity counters are not symmetric and also most of the cross section area is covered by the velocity from 4.84 to 5.18 m/s. Two centers of the contours with highest velocity about 5.53 m/s and one center of the contour with low velocity about 2.7 m/s can be seen. The velocity contours get changed from 45° to 225° of fluid flow from inlet which affect the heat transfer. The gravitation, centrifugal and torsion forces acting on the fluid particles play the major role in the changes of velocity profiles.

3.2   *Temperature contours at 45° and 225° angle*

Figure 3 (a & b) show the temperature contours at 45° and 225° angle. At the 45° angle, the temperature seems to be uniform and heating of the fluid start from the tube wall. Since the wall is set at 320K and fluid inlet is 300K, the fluid gets heated from the inlet. The high temperature region can be seen at inner side of the tube which is at bottom side of the cross section in Figure 3(a & b). Higher heat transfer takes place at the outer side of the tube than that at inner side of the tube as mentiond by different authors in available literature for curved tube. The hottest region is at 50° angle from the inner side which is different than helical coil.

3.3   *Local Nusselt number around the circumference of tube at 45° and 225° angle*

Figure 4 (a) show the nature of the local Nusselt number around the circumference of the cross section at 45° angle from the start of fluid flow. The inner side of the cross section is assigned 0° angle and further in anticlockwise direction, the angle is defined up to 360°. Hence the outer side of the tube is at 180° angle there by implying that 0° corresponds to inner side at bottom in cross section and 180° corresponds to outer most side at the top of the cross section. The nature of the curves appear to be sinusoidal and a peak can be seen at near 180° showing a sudden increase in Nu. This is because the centrifugal force becomes dominant as the cold fluid moves towards the outer side due to curvature of the coil. This enhances the heat transfer at the outer side of the tube and raises the value of Nu. Hence the percentage of circumference of cross section having lower Nusselt number is predominant.

The local Nusselt Number around the circumference of cross section at 225° angle is shown in Figure 4(b). As the flow develops, as compared to the cross section at 45°, the Nu at 225° angle is higher for most part of the circumference of cross section. This

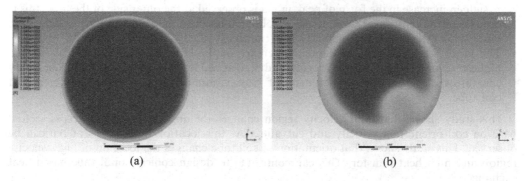

| (a) | (b) |

Figure 3.   Temperature Counters at 45(a) and 225(b) ° angle from start of flow.

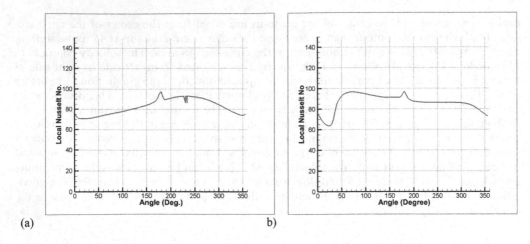

(a)   b)

Figure 4.   The local Nu around the circumference of tube cross section at (a) $\theta = 45°$ and (b) $\theta = 225°$.

shows that as the flow develops, the nature of variation of the Nusselt number around the circumference of the cross section also changes. Only at angle below 50° in both the cross section shows the lowest Nu. A sudden increase in Nu in the form of a peak can be seen at about 180° in both the cross sections. This is the outer most side of the tube. As the flow develops centrifugal force becomes dominant and cold fluid particles move toward outer side with high velocity. The higher temperature gradient results into high Nu at outer side. Also the high velocity region is settled at about 50° from inner side and shows higher values of Nu at 225° angle cross section.

4   CONCLUSIONS

The simulation for constant wall temperature condition was carried out for the conical spiral tube in the turbulent regime. The velocity, temperature profiles and variation of local Nu around the circumference was studied at 45° and 225° angle from the start of the fluid flow. The following are the major conclusions:

- The velocity profile is symmetric at 45° angle. It changes at 225° angle and the high velocity region shifts from the left to right of the outer side in the tube cross section.
- A uniform temperature of the fluid is observed at 45° and more variation in the fluid temperature is found at 225° angle. The hottest region can be seen little off set from the inner side.
- A sudden increase in the form of peak in Nu is observed at the outer side of the tube, this is because of the centrifugal force acting at cold fluid particles moving with high velocity near outer side of the tube which enhances heat transfer.
- The local Nu varies around the circumference of the tube cross section and it has the highest value at the outer most side and lowest value near inner side. Further study can throw more light on development of Nusselt number around the circumference of coil tube.

This study shows that high velocity region changes the angular position at cross section base on coil orientation velocity and curvature. At this location high values of Nu can be observed. This study shows that orientation of coil tube changes the location of high velocity region and high heat transfer. This can contribute to design conical spiral tube based heat exchangers.

REFERENCES

1. W. R. Dean, The stream line motion of fluid in a curved pipe, Philos. Mag.5 (1928) 673–695.
2. Srinivasan, P.S., Nandapurkar, S. S., & Holland, F. A., Friction factor for coils, Transactions of the Institution of Chemical Engineering, 48, (1970) T156–T161.
3. Jayakumar, J.S., Mahajani, S. M., Mandal, J. C., Vijayan, P. K. & Bhoi, R. Experimental and CFD estimation of heat transfer in helically coiled heat exchangers. Chemical Engineering Research and Design, 86(3), (2008) 221–232.
4. Yan Ke, Ge Pei-qi, Su Yan-cai, Meng Hai-tao, Numerical simulation on heat transfer characteristic of conical spiral tube bundle, Applied Thermal Engineering 31 (2011) 284–292.
5. Shih, Tsan-Hsing, Liou, W.W., Shabbir, A., Z., & Zhu, J. A new k-ε eddy viscosity model for high Reynolds number turbulent flows. Computers and Fluids, 24(3), (1995) 227–238.

Technologies for Sustainable Development – Mahajan, Patel & Sharma (eds)
© *2020 Taylor & Francis Group, London, ISBN 978-0-367-33737-7*

Electromagnetic field simulation of microwave heating in microwave oven

Yashashi Gupta & Niraj Chopde
Student, Mechanical Engineering Department, Nirma University, Ahmedabad, Gujarat, India

Shruti Bhatt*
Assistant Professor, Mechanical Engineering Department, Nirma University, Ahmedabad, Gujarat, India

Nilesh Ghetiya
Associate Professor, Mechanical Engineering Department, Nirma University, Ahmedabad, Gujarat, India

ABSTRACT: Microwave processing of materials attain constant due to its certain features like volumetric heating, less processing time and less power consumption as compared to conventional heating. Microwave heating of materials inside oven involves electromagnetic and thermal phenomena. These two phenomena should be solved simultaneously using a coupled approach so microwave heating needs to be simulated as electromagnetic and heat transfer equation in cyclic manner. Electromagnetic phenomenon causes power dissipation due to which heat generation takes place inside the material. Because of heat transfer, there is temperature change in the charge, which further results into dielectric material properties, which then again influences heat generation and thus completing the cycle. Thus to understand this concept of electromagnetic field interaction with nonferrous material aluminum has been simulated using comsol multiphysics software. Simulation results shows that the temperature in the aluminum material reaches up to 284 K in 15 seconds and surface electric field norm at 8×10^3 V/m in 15 seconds by exposure of microwave radiation at 1 KW.

1 INTRODUCTION

Microwaves are electromagnetic (EM) waves with frequencies between 300MHz and 300GHz and wavelength (Λ) in the range from 1mm to 1m. Microwave heating has many applications in the field of material processing like metals, composites and alloys due to its notable advantages compare to conventional heating. Compared with conventional heating, microwave heating has many advantages such as simultaneous heating of a material in its whole volume, higher temperature homogeneity, and shorter processing time [1]. Microwave interaction with material and heat generation rate depends upon the dielectric properties and electromagnetic field distribution inside the materials. By the exposure of microwave radiation, dipole rotation occurs in dielectric materials containing polar molecules having an electrical dipole moment. Interactions between the dipole and the EM field result in energy in the form of EM radiation being converted to heat energy in the materials. This is the principle of microwave heating for each material. Complex relative permittivity (ε^*) and loss tangent (δ) can be expressed as below equation 1 and 2 respectively.

$$\varepsilon^* = \varepsilon' - i\varepsilon'' \ \dots (1) \ \text{And} \ tan\delta = \varepsilon''/\varepsilon' \dots \quad (2),$$

---

* Corresponding author: shruti.mehta@nirmauni.ac.in

where the real part is a measure of the stored microwave energy within the material, and the imaginary part is a measure of the dissipation (or loss) of microwave energy within the material. The complex permittivity is usually a complicated function of microwave frequency and temperature [2]. To analyze multiphysics coupling of microwave heating process numerical simulation is an essential step. Mathematical model consists of EM field equation and heat transport equation. To simulate the heating process, we need to model the above two equations together with their coupling relation and then solve it using any numerical method. The most two common methods widely applied are: Finite-Difference Time-Domain (FDTD) and Finite Element Method (FEM).

2 LITERATURE REVIEW

J. Clemens et al [3] have developed numerical model to predict electromagnetic field inside waveguide and cavity by which they have approached an\bout power deposition and temperature distribution in process sample. They have represented coupled non-liner approach to correlate temperature and electromagnetic properties of the alumina material.

Bansal et al [4] have developed electromagnetic field simulation for the graphite and silicon carbide having two different dielectric properties. In their model, FEA based approach has been used to find relation between temperature and time with reference to losses in domestic microwave oven. They have reported that silicon carbide is more effective then graphite as a dielectric absorbing material.

Santos et al [5] mathematically modelled permittivity dependence on temperature, using the Maxwell equations and the boundary conditions, which can be solved using the Finite Element Method. To break down the historical of the heating procedure, the development of the electromagnetic field, the temperature and the skin depth, were reproduced dynamically in an ceramic sample. The assessment of thermal runaway was also made. This is the most critical thing seen in the sintering of ceramic materials since it causes deformations, or even melting in certain cases.

Warren et al [6], in this research, have used material characterization properties at high temperature for modelling of microwave oven. It was seen from experimental data that the materials degraded at high temperatures and not enough data was available for modelling. To gather data the newly designed probe with an air-filled coaxial line made out of invar and a thin silicon nitride support washer was used and modelling was performed. Thermal property characterization was conducted using an instantaneous triangular heat pulse from a Laser Heat Flash Line TM 5000 Diffusivity System. COMSOL was used to develop the multi-physics EM model. The experiments were carried out for casting of copper for different configurations and it was found that the experimental results were in close symmetry with the simulation.

3 SIMULATION OF MICROWAVE HEATING

Simulation of the aluminum material has been carried out in this present work. The mechanism of microwave interaction with these materials were simulated using COMSOL Multiphysics FEA based software and results are presented in this paper. Modelling and simulation study of microwave heating process was carried out to analyze the effect of microwave exposure on an aluminum to reach the melting temperature of the charge. Simplified model was developed to

Table 1. Parameters of the stimulation model.

Parameters	Value
Frequency	2.45 GHz
Power	1kW
Cavity Size - 2D	26cm × 26cm
Cavity Size - 3D	41cm × 38cm × 21cm

simulate the microwave heating process considering the following assumptions in order to reduce the complexity of the problem: 1) The nonferrous materials used in the analyses were homogeneous and isotropic. 2) The ambient temperature of the system was 300 K. 3) The mass transfer during the study was absent. 4) The chemical changes in the materials are assumed negligible. 5) The walls of wave guide and cavity were made of copper which works as a perfect conductor as it indicates the electric filed intensity in tangential direction is zero.

The models were built by integrating the physics of heat transfer in solids and electromagnetic wave frequency domains. Frequency domain and time dependent studies were used for simulation of the models. The Maxwell's equations were used to solve propagation of electromagnetic wave inside the cavity.

$$\nabla \times \mu_r^{-1}(\nabla \times E) - K_0^2 \left(r - \frac{j\sigma}{\omega} \right). E = 0 \tag{3}$$

and

$$\sigma C_P \, \partial T / \partial t - \nabla . K \nabla T = q \tag{4}$$

The electric field (E) inside the cavity is given in equation 3.where, the terms ω, $\varepsilon 0$, μr, εr, and σ are angular frequency, permittivity of vacuum, relative permeability, relative permittivity and electrical conductivity respectively. Heat transport equation 4 where ρ is the density Cp is the heat capacity, k is the heat transfer coefficient, T is temperature and t is the time. This equation equates the power generated by microwaves inside the material to the rise in the temperature of the material and various losses.

The 2D and 3D geometries as shown in Figure 1 were made in COMSOL. They were made using basic shapes as well as operations such as union and intersection, available in COMSOL. The parts in the above cavities are 1. Dielectric material 2. Microwave cavity which is filled with air 3. Waveguide 4. The border of the cavity. The following are the materials used for the simulation: Air fills the cavity of microwave, the inner layer of the microwave is made of copper and the dielectric material is made of aluminum. The boundary conditions applied to the model are as follows: The

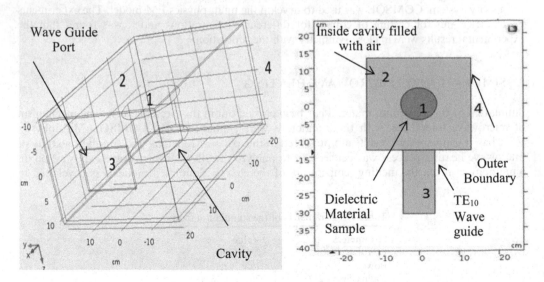

Figure 1.   3D and 2D microwave oven model developed in COMSOL multiphysics software.

walls in both 2D and 3D cavities are considered as perfect conductors. This boundary condition is given by the equation n × E =0 .This equation means that the tangential component of electric field is zero at the boundary of the cavity. Thus, no dissipation of energy of microwaves will occur at the boundaries and also, the microwaves will not be able to pass through the boundary. The Maxwell's equations are calculated throughout the microwave cavity. But the heat transport equation is only applied to the dielectric material because that is the main area where heat dissipation has to be studied. The port through with electromagnetic waves are excited is made a rectangular port with an input of 1kW and no phase change. Also the mode used for the port is TE_{10}.

There is no convective boundary condition applied at the boundary of dielectric material and air to simplify the problem. If applied, convective heat transfer would account for heat loss. The second term in the heat transport equation accounts for heat losses and that is the significance of negative term applied to it. This experiment is carried out from a time range of 0 to 15 seconds at an interval of 1 second and at a frequency of 2.45GHz. The meshing of 2D and 3D cavities was physics-controlled is shown in Figure 2

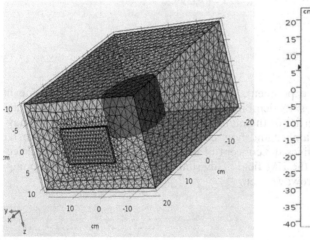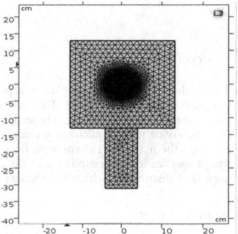

Figure 2.   Meshing of 3D and 2D cavity.

4   RESULTS

Microwave interaction of materials depends on the dielectric properties of the materials. Figure 3 represents the microwave heating, the energy which gets dissipated inside the core of the dielectric charge due to microwave interaction, forms the basis for heating of whole charge. The heat gets conducted from the core to the outer layers of the charge, resulting in temperature rise. The temperature of the charge results into change in the variation of complex permittivity.

The materials properties used in carrying out the simulation of the nonferrous material were illustrated in Table 1. The relative permittivity of the air was considered as 1, thus it act as a transparent to microwave radiation. The simulation results show the increase in temperature of 3 °C in a time duration of the 15 s at a power output of 1KW with resistive losses of the order of 1 x 108. W/m3. The temperature profile obtained in the is shown in Figure 3.

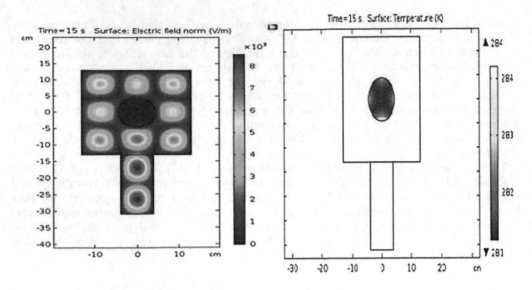

Figure 3.   Electromagnetic and heating progress after 15 sec interval.

5   CONCLUSION

To validate analysis for simulating experiments 2D and 3D modelling has been done in domestic microwave chamber. The results demonstrated that the significance of electromagnetic data and thermo-physical properties of the material. The study supports the electromagnetic behavior with in microwave chamber and how conduction and radiation heat transfer effect on the material. The study outlines the concept of resistive heating and how the dielectric properties lead to coupling with the EM field. This simulation proposes the electromagnetic field distribution within the microwave oven.

REFERENCES

1. Gupta, M., & Leong, E. W. W. (2008). Microwaves and metals. John Wiley & Sons.
2. Zhao X., Yan, L., & Huang, K. (2011). Review of numerical simulation of microwave heating process. Advances in Induction and Microwave Heating of Mineral and Organic Materials, 27–48.
3. Clemens, J., & Saltiel, C. (1996). Numerical modeling of materials processing in microwave furnaces. International Journal of Heat and Mass Transfer, 39(8), 1665–1675.
4. Bansal, A., & Sharma, A. K. (2014). 3D electromagnetic field simulation of silicon carbide and graphite plate in microwave oven. Int. J. Eng. Rob. Res., 1(1), 7–12.
5. Santos, T., Valente, M. A., Monteiro, J., Sousa, J., & Costa, L. C. (2011). Electromagnetic and thermal history during microwave heating. Applied Thermal Engineering, 31(16), 3255–3261.
6. Warren, B., Awida, M. H., & Fathy, A. E. (2012). Microwave heating of metals. IET microwaves, antennas & propagation, 6(2), 196–205.

Technologies for Sustainable Development – Mahajan, Patel & Sharma (eds)
© 2020 Taylor & Francis Group, London, ISBN 978-0-367-33737-7

Performance evaluation of downdraft gasifier for 'refuse derived fuel'

Mihir Lathigara & Darshit Upadhyay*
Department of Mechanical Engineering, Nirma University, Ahmedabad, Gujarat, India

Rohan Kadam & Rakesh Maheshwari
National Innovation Foundation, Grambharti, Amrapur, Gujarat, India

Rai Singh Dahiya
Enersol Biopower Pvt. Ltd., Jaipur, Rajasthan, India

Rajesh Patel
Department of Mechanical Engineering, Nirma University, Ahmedabad, Gujarat, India

ABSTRACT: A standalone gasifier coupled with I.C. Engine has proven its potential to fulfil the energy requirement of rural electrification. The purpose of this study was to estimate the potential of gasification process by combustion of bio-waste and municipal solid waste (MSW) also called as Refused Derived Fuel (RDF). Experiments were conducted on a 20 kWe downdraft gasifier at National Innovation Foundation-India (NIF-India), Gandhinagar. A screw conveyor was attached in the oxidation zone of a gasifier reactor to prevent it from chocking. After such modifications in the gasification system, experiments were conducted using different types of bio-mass including the bark of Babul, Teak and Neem tree. Performance parameters viz., the surface temperature of gasifier reactor, air flow rate, gas flow rate, engine output voltage, Engine oil pressure were recorded. The gas flow rate was observed over 5% higher during feeding of a combination of RDF and biomass as compared to biomass alone. The temperature of oxidation zone and flame was found over 6% higher of producer gas, resulted from the combination than the biomass alone. Ash content has increased marginally (2%) while char content was reduced drastically (27%). However, the tar content had increased significantly (64%) when the combination of RDF and biomass were used as feed in comparison to biomass alone. There is a need for developing a better filtration system to reduce the tar content in the producer gas.

Keywords: Biomass downdraft gasifier, refused derived fuel (RDF), rural electrification

1 INTRODUCTION

Being one of the fastest growing economies in the world and a developing county, India has huge electricity requirements (Thambi, Bhatacharya, & Fricko, 2018). Rural electrification has been a serious concern due to non-feasible and expensive process to establish grids in interiors areas (Patel, Upadhyay, & Patel, 2014). The rural population is majorly dependent on animal husbandry and farming. Energy from biomass is one of the methods to meet rural electricity requirement. Power from biomass from thermochemical conversion route such as gasification, combustion, pyrolysis, etc. reached above 9 GW of installed renewable energy systems (*Energy Statistics*, 2019). More than 50% of the land area is covered in forests and farms, making India rich in biomass

---

* Corresponding author

("https://knoema.com/atlas/India/topics/Land-Use/Area/Agricultural-land-as-a-share-of-land-area," n.d.) (Upadhyay, Makwana, & Patel, 2019). In interior area of India, a stand-alone gasifier coupled with a reciprocating engine may resolve rural electrification problems (Abe et al., 2007). The gasification process efficiently converts the majority of carbonaceous feedstock into combustible producer gas ($CO+H_2$) for applications such as furnaces, powering vehicles or engines ("Biomass Gasifier Definition," n.d.). Generally, gasifiers are available in a wide working range from as low as 51 kW up to 2 MW ("Current Status of Biomass Gasification in India," n.d.). Municipal Solid Waste (MSW) disposal is a major problem with the government in recent time. It includes various waste such as plastics, rubbers, metallic components, e-waste, dust etc (Materazzi, 2017). Refuse-Derived Fuel (RDF) comprises such waste with usable heating value. Utilizing these waste to generate electricity is beneficial for the social and environmental aspects (Ryu, 2010). As the production of MSW is increasing day by day, preparation of RDF from the MSW and its utilization for electricity generation is an important research area nowadays.

In the present study, experimentation and performance evaluation was done on a modified downdraft gasifier developed at National Innovation Foundation (NIF), India with three different feedstock. The influence of various feedstock on temperature profile, air flow rate, gas flow rate, output voltage, engine oil pressure, biomass fuel consumption was investigated.

2 EXPERIMENTAL SETUP AND TESTING

2.1 Feedstock selection

Chemical/Physical properties of the feedstock (babul wood and RDF) are characterized by proximate analysis, heating value, bulk density and particle size, as shown in Table 1. Proximate analysis and heating value were measured using IS 1350 (Part I)-1984 and IS 1250 (Part II)-1970 standards, respectively (*Indian Standard - Methods of Test for Coal and Coke, PartI: Proximate Analysis*, 1984; *Methods of Test for Coal and Coke, PartII: Determination of Calorific Value*, 1971).

2.2 Gasifier set up

A laboratory scale, atmospheric pressure downdraft gasifier (20 kWe) developed by Rai Singh Dahiya (Innovator, supported and recognized by NIF), Rajasthan was used for the experiments. The gasifier has a combination of different filters such as cyclone filter, charcoal filter, sawdust filers and cotton filter for filtering the producer gas and providing tar-free gas. The cooling system comprises of chilling tower and water tank with water scrubber mechanism including fan plates. Two cylinder petrol engine was used for generating electricity. The biomass was fed through the feed door and was kept in the reactor. The sub-stoichiometric air was allowed to enter in the oxidation zone through an air flow regulator. Bio-char was kept in the reduction zone which helps in the gasification of the combustion products. It also allowed the ash to fall in the ash chamber. The gasifier reactor was connected with the different cleaning systems such as

Table 1. Characterization of fuels.

Analysis Proximate	Wood	RDF
Volatile matter	65	8.78
Ash	6.95	9.15
Moisture	6.38	6.34
Fixed carbon[a]	21.67	75.73
Heating Value (MJ kg^{-1})	16.45	25.35
Bulk density (kg m^{-3})	549.67	348.16
Particle size (mm)	89 * 63.5 *63.5	NA

[a] by difference

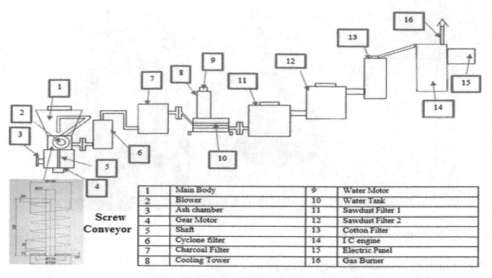

1	Main Body	9	Water Motor
2	Blower	10	Water Tank
3	Ash chamber	11	Sawdust Filter 1
4	Gear Motor	12	Sawdust Filter 2
5	Shaft	13	Cotton Filter
6	Cyclone filter	14	I C engine
7	Charcoal Filter	15	Electric Panel
8	Cooling Tower	16	Gas Burner

Figure 1. Schematic diagram of modified 20 kWe downdraft gasifier set up coupled with an engine.

cyclone filter, cotton filter, water scrubber, sawdust filters, cotton filter, and their accessories. Gas burner and I.C. engine with generator were kept after the cleaning system of the producer gas.

Hotwire anemometer with a data logger (Fluke make Amprobe TMA-21HW, ± 3%) was used to measure the air flow rate, gas flow rate, dry-wet bulb temperature, and relative humidity. Infrared thermal gun (Fluke 62 Mini Infrared Thermometer Gun, ± 1°C) was used for measuring the surface temperature of a gasifier reactor. Measuring cylinders were used to measure the tar liquid.

2.3   Gasification system procedure

Experimental system was cleaned properly before starting the experiments each time. Bio-char and fuel were added in the gasifier reactor. Ignition was carried out by using fire torch. The dust/particulate contents in the producer gas were captured in the cyclone separator. After that gas was washed and cooled in a venturi scrubber (with re-circulating cooling water in a cooling tower). This wet gas was supplied in sawdust filters, and a cotton filter to remove moisture content. Cooled, clean gas and air was sucked into the reciprocating engine generator set through butterfly valves arrangement. The quantity of the producer gas and air was controlled by a governor as per electrical load requirement. Finally, the output voltage from the electric panel of the engine system was observed.

2.4   Modifications in existing biomass gasifier

A few modifications were done in the existing gasifier to improve the output. One charcoal filter was added in the filtering system and the cooling tower was modified for improving the performance of the gasification system. A screw type conveyor shaft was also included for the proposed MSW gasifier for avoiding choking and blockage problems which used to occur in the oxidation zone. The Screw conveyor shaft was attached inside of the reaction chamber (oxidation zone) for rotation of Biomass material inside of gasifier during the working condition. The design of screw conveyor shaft was done considering capacity, fuel characteristics and load on oxidation zone of the reactor of biomass gasifier. Screw conveyor shaft and plate of reaction chamber both were rotated by a geared motor which was attached at the bottom of ash chamber.

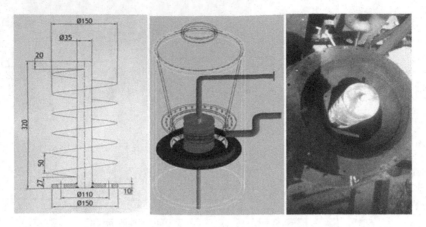

Figure 2.   Screw conveyor drawing.

2.5   *Performance evaluation of downdraft gasifier existing at NIF*

The performance of biomass gasifier was evaluated with and without the screw conveyor shaft attachment. Further, the performance with screw conveyor was evaluated using biomass/ wood logs and a mixture of RDF with biomass/wood as fuel.

Figure 3.   Biomass downdraft gasifier existing at NIF.

3   RESULTS AND DISCUSSION

3.1   *Thermal performance of biomass gasification system*

Performance of the system was evaluated by loading the reactor with material i.e. different types of wood pieces and RDF material in loose form. Gasification system was tested initially for Biomass/wood. Multiple replications were carried out to assess the performance of the Modified Gasification system. The results of the three replications are presented. Initially, the gasifier reactor was preheated in the range of 400°C, 600°C and 800°C temperature for three replications. About 35-40 kg of wood was fed to the reactor for gasification. Airflow diameter and gas output diameter were 1.9 cm and 4.1 cm respectively. The results obtained while feeding different types of wood pieces and RDF material in the modified Biomass Downdraft Gasification system are represented in Table 2.

Performance parameters of the gasification system were compared with and without using a screw conveyor mechanism as shown in the table. Different parameters such as air-gas flow rate, different zone temperatures, engine voltage, oil pressure, etc. were compared for wood

Table 2. Performance parameters of a gasification system.

Thermal performance of the system	Output		
	Without Screw attachment	With screw attachment	
Performance Parameters	Wood	Wood	Wood and RDF (50:50, wt%)
Initial Biomass mass fed in reactor, kg	36	37.85	19+19
Air Flow Rate, m^3 min^{-1}	0.320	0.355	0.370
Gas flow rate, m^3 min^{-1}	0.453	0.489	0.515
Initial temperature of Producer Gas, °C	204	217	288
Final temperature of producer gas, °C	40	42	47
Zone Temperatures, °C			
Drying Zone, °C	135	130	142
Pyrolysis Zone, °C	455	488	513
Oxidation Zone, °C	888	929	991
Reduction Zone, °C	567	585	617
Gas flame temperature, °C	182	221	235
Engine running time, min	—	53	43
Gasifier working time, h	3	3	3
Output Voltage, V	390	400	450
Engine oil Pressure, kg cm^{-2}	3.9	3	3.5
Ash, kg	2.502	2.630	2.683
Char, kg	1.948	1.63	1.177
Tar Liquid, Lit.	6.8	7	11.50

and wood with RDF material. It is observed that wood-RDF material offers the best results in terms of gasification temperature and engine voltage output.

3.2 Effects of air flow rate, gas flow rate, biomass fuel consumption on performance of a gasifier

Effects of different performance parameters with two different feedstock along with screw conveyor shaft attachment in the oxidation zone are described in Figure 3. Below graph indicates the different parameters such as initial temperature of producer gas, Biomass fuel capacity, Biomass fuel consumption, Air flow rate, gas flow rate, gasifier output, Tar, Ash and engine oil pressure on different feedstock by percentage wise and Temperature of different zones (drying zone, pyrolysis zone, combustion zone and reduction zone).

Figure 4 indicates that when 100% wood was used as feed, ash content was recorded as 4.26 kg while the same was reduced to 3.86 kg when combination (50% wood and 50% RDF) was used as fuel. Tar content was recorded 7 liters and 11.5 liters simultaneously while the

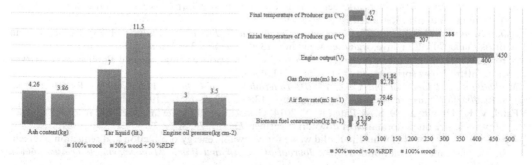

Figure 4. Performance parameters of gasifier with different feeds.

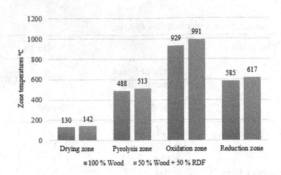

Figure 5. Variation in temperature of different zones of gasifier with different feeds.

engine oil pressure was recorded 3 kg cm^{-2} and 3.5 kg cm^{-2} in both the conditions. The gas flow rate was observed 91.86 m^3 h^{-1} with combination feed 82.78 m^3 h^{-1} with wood was used alone as feed.

3.3 *Variation of temperature in different zones*

Figure 5 indicates the temperature of the oxidation zone while feeding 100% wood is 929°C which was recorded 991°C while feeding 50% wood and 50% RDF, while the temperature of the reduction zone was 585°C, and 617°C, with these feedstock respectively.

4 CONCLUSIONS

The experiments were conducted on the 20 kWe modified downdraft gasifier with wood and wood – RDF (50:50, wt%) material. Inclusion of screw conveyer is helpful to improve the movement of feedstock inside the gasifier reactor helps to increase temperature and air-gas flow rate. RDF material helpful to increase the zone temperature, air-gas flow rate and engine voltage. The major conclusion from this study is that the RDF has the potential to generate electricity.

REFERENCES

Abe, H., Katayama, A., Sah, B. P., Toriu, T., Samy, S., Pheach, P., Grierson, P. F. 2007. Potential for rural electrification based on biomass gasification in Cambodia. *Biomass and Bioenergy*, *31*(9), 656–664.
Biomass Gasifier Definition. Retrieved May 7, 2019, from https://homeguides.sfgate.com/biomass-gasifier-definition-78891.html
Current Status of Biomass Gasification in India. Retrieved March12, 2019, from http://www.eai.in/ref/ae/bio/csbg/biomass_gasification_india_current_status.html
Energy Statistics. (2019), https://knoema.com/atlas/India/topics/Land-Use/Area/Agricultural-land-as-a-share-of-land-area. Retrieved July 17, 2019
Indian Standard - Methods of Test for Coal and Coke, PartI: Proximate Analysis. (1984). Retrieved from https://ia600905.us.archive.org/6/items/gov.in.is.1350.1.1984/is.1350.1.1984.pdf
Materazzi, M. 2017. *Clean Energy from Waste* (1st ed.). Springer International Publishing. Retrieved from https://www.springer.com/gp/book/9783319468693
Methods of Test for Coal and Coke, PartII: Determination of Calorific Value. (1971). Retrieved from https://ia800406.us.archive.org/19/items/gov.in.is.1350.2.1975/is.1350.2.1975.pdf
Patel, V. R., Upadhyay, D. S., & Patel, R. N. 2014. Gasification of lignite in a fixed bed reactor: Influence of particle size on performance of downdraft gasifier. *Energy*, *78*, 323–332.
Ryu, C. 2010. Potential of municipal solid waste for renewable energy production and reduction of greenhouse gas emissions in South Korea. *Journal of the Air and Waste Management Association*, *60*(2), 176–183.

Thambi, S., Bhatacharya, A., & Fricko, O. 2018. *India's Energy and Emissions Outlook: Results from India Energy Model*. Retrieved from https://niti.gov.in/sites/default/files/2019-07/India's-Energy-and-Emissions-Outlook.pdf

Upadhyay, D. S., Makwana, H. V., Patel, R. N. 2019. Performance evaluation of 10 kWe pilot scale downdraft gasifier with different feedstock. *Journal of the Energy Institute, 92*, 913–922. https://doi.org/10.1016/j.joei.2018.07.013

Author Index

Printed in the United States
by Baker & Taylor Publisher Services